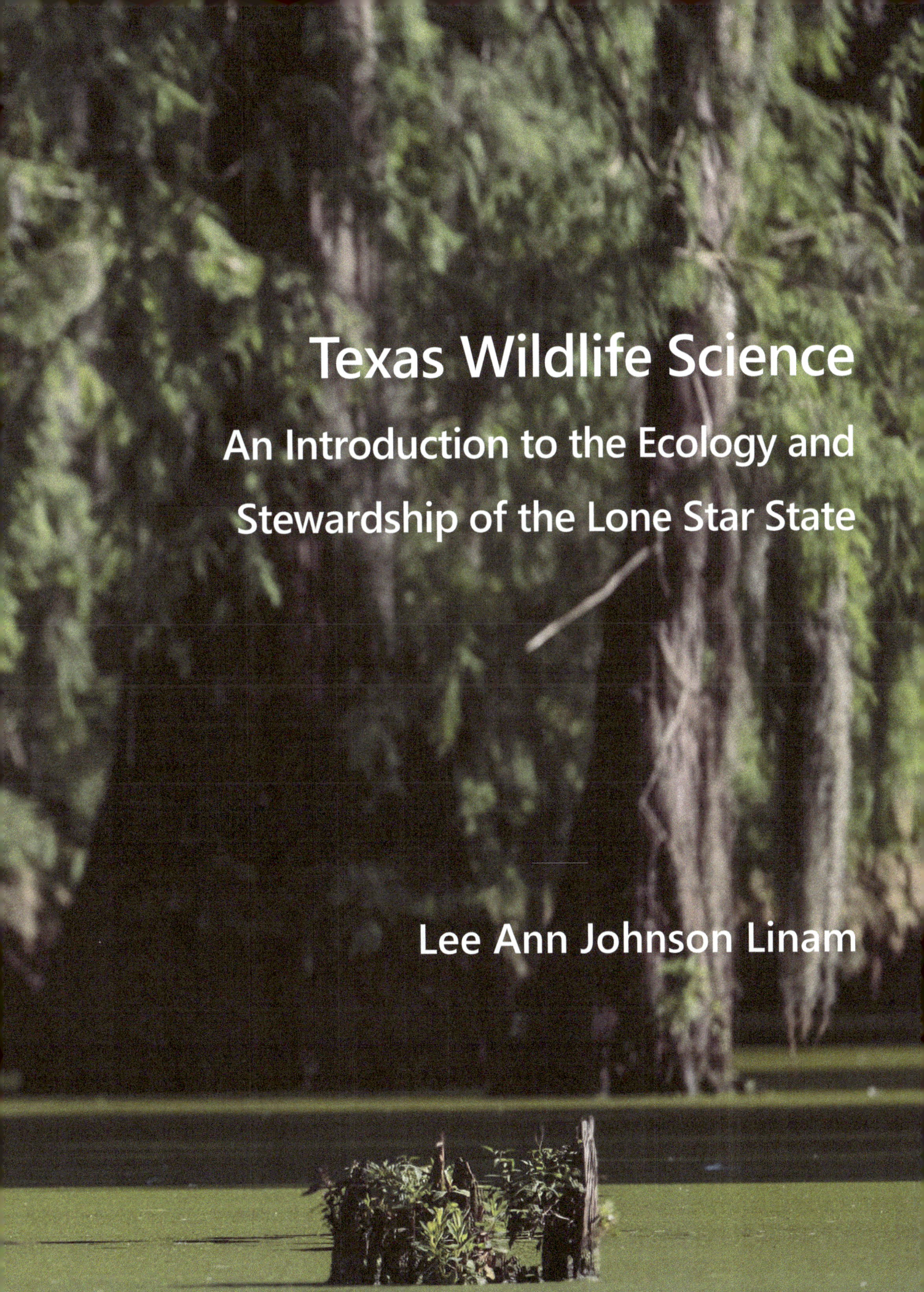

Texas Wildlife Science
An Introduction to the Ecology and Stewardship of the Lone Star State

Lee Ann Johnson Linam

Copyright © 2025 by

Lee Ann Johnson Linam.

All rights reserved.

First edition

Library of Congress Cataloging-in-Publication Data
Names: Linam, Lee Ann Johnson; author. Storey, Mark Alan; photographer.
Title: Texas Wildlife Science
Subtitle: An Introduction to the Ecology and Stewardship of the Lone Star State
Identifiers:
ISBN 979-8-9938696-0-5 (hardback)
ISBN 979-8-9938696-1-2 (paperback)

Related content: texaswildlifescience.com

Identifiers:

Front cover photograph (barred owl chicks) and title page photo (roseate spoonbill in cypress swamp) by Mark Alan Storey.

Back cover photo (Texas horned lizard) by Abigail Linam Bradbury.

DEDICATION

For Dr. Ray C. Telfair, II, who taught me to love wildlife ecology.

For Charles D. Stutzenbaker, who taught me to love wetlands more deeply.

And, most of all, for my father, Earl "Frank" Johnson, Jr., who taught me to love people by loving wildlife and wetlands.

LIST OF ABBREVIATIONS

Agrilife	Texas A&M Agrilife Extension Service
CC	Creative Commons license
WMA	Wildlife Management Area
NWR	National Wildlife Refuge
SP	State Park
TFS	Texas A&M Forest Service
TPWD	Texas Parks and Wildlife Department
USDA	United States Department of Agriculture
USFS	United States Forest Service (part of USDA)
USFWS	U.S. Fish and Wildlife Service
WC	Wikimedia Commons

CONTENTS

Introduction viii
Acknowledgments x
Chapter 1 – Understanding Wildlife Science 1
Chapter 2 – A History of Wildlife Science in Texas 9
Chapter 3 – Basics of Ecology 24
Chapter 4 – The Physical Environment of Texas 40
Chapter 5 – Ecoregions of Texas 51
Chapter 6 – Habitat Management 71
Chapter 7 - Texas Woodlands 88
Chapter 8 - Texas Grasslands 110
Chapter 9 – Texas Deserts and Shrublands 132
Chapter 10 - Texas Wetlands 147
Chapter 11 – Wildlife Populations 168
Chapter 12 – Texas Birds 188
Chapter 13 – Texas Mammals 213
Chapter 14 – Texas Amphibians 238
Chapter 15 – Texas Reptiles 256
Chapter 16 – Texas Invertebrates 277
Chapter 17 – A Role for Everyone 299
Appendix A – Reflection Questions 311
Appendix B – Vegetation Sampling Methods 317
Appendix C – Key References 319
Appendix D - Glossary 325
Appendix E – List of Scientific Names 332
Index 340

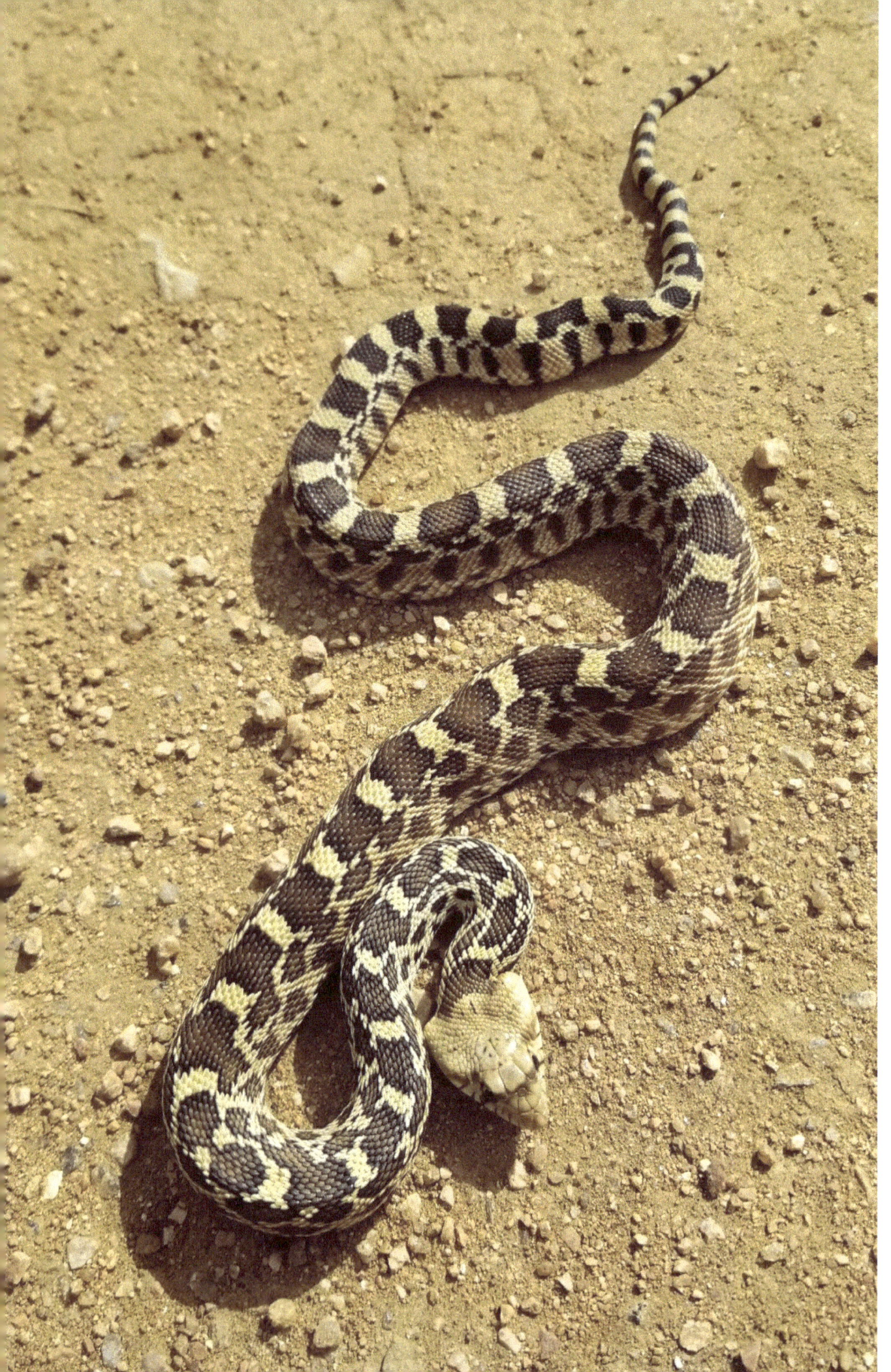

INTRODUCTION

Texas, a leader among U.S. states in biodiversity, hosts an amazing abundance of wildlife species, but this rich assortment of wild fauna does not exist on its own.

The physiography, geology, soils, and climate of the Lone Star State form the foundation for a variety of plant communities, from the starkly beautiful deserts of West Texas to the cathedral longleaf pine forests of East Texas and from the windswept shortgrass prairies of the Texas Panhandle to the lush subtropical palm forests of the Lower Rio Grande Valley. In these diverse and ever-changing habitats, birds, mammals, reptiles, and amphibians occupy their distinctive niches—predators follow prey, symbionts aid each other, keystone species create habitat for other members of the ecosystem, and populations rise and fall—all providing a calling for wildlife scientists.

Wildlife science incorporates the disciplines of biology, ecology, land management, economics, and even sociology to conserve or enhance wildlife species in their natural environment. Guided by a legacy of exploration, conservation ethics, and environmental law, wildlife scientists in Texas apply a variety of tools in managing wildlife habitats and wildlife populations in the state, working on public lands and with private landowners in the state to enhance valuable game populations, provide unparalleled wildlife viewing opportunities, and recover imperiled species. With few untouched landscapes in the state, wildlife scientists follow the adage of Aldo Leopold, one of the earliest leaders in the field of wildlife science, that habitat can be restored by some of the same tools—fire, axe, cow, plow, and gun—that first destroyed it.

A tour through the diverse vegetation communities of Texas offers a glimpse of how the conservation of species is interwoven with the conservation of habitats, application of science, and collaboration with Texas citizens. Quail management in the Rolling Plains of Texas produces healthy Texas horned lizard populations and supports reintroduction opportunities for this beloved state reptile. Application of fire in red-cockaded woodpecker habitats in East Texas creates a mature pine canopy and herbaceous understory wonderland in campgrounds of national forests, along with open space for restoration of Louisiana pinesnakes. Sound deer management in the forests and savannas of the Texas Hill Country protects golden-cheeked warblers, black-capped vireos, and Edwards Aquifer recharge. Springflow from the aquifer in turn supports aquatic endemics such as San Marcos salamanders and Cagle's map turtles and ultimately benefits coastal wetlands, whooping cranes, and commercial and recreational fishing and birding on the Texas coast. And in the Texas Panhandle, outreach and education lead to the coexistence of cows, prairie-chickens, and prairie dogs.

Although a diversity of careers abound in wildlife science, from law enforcement to land management to conservation genetics, wildlife conservation in Texas is not the purview of professionals alone. Opportunities exist for private landowners to collaborate on wildlife management plans, for citizens to volunteer as scientists, and for wildscaped homes to provide habitat for backyard wildlife. This book will explore the historical, physical, and biological basis for natural diversity in Texas, along with the roles that humans can play in restoring, conserving, and utilizing the terrestrial resources of the state. For the wildlife scientist, the ultimate goal is the adoption of Leopold's land ethic—helping Texas citizens see themselves as a part of Texas ecosystems and thus sustain wildlife populations for future generations.

(opposite) Bullsnakes are one of the state's largest snake species.
Photo: Mark Mitchell, TPWD.

ACKNOWLEDGMENTS

Preparation of this book was an opportunity to reflect upon the many individuals that shaped my career in wildlife and those who nurtured an enthusiasm for sharing my ecological fascinations with others. Many of them helped to improve the content and quality of this book.

My sincere thanks is offered to the individuals who reviewed chapters, including Patricia Morton, David Diamond, John Karges, Terry Blankenship, Linda Campbell, Ricky Maxey, Jim Neal, Jaime González, Marsha May, Robert Bradley, Lisa Bradley, Michael Smith, Mark Alan Storey, Brent Ortego, and David Bowles. A special thanks is offered in memorium to Dr. Ray C. Telfair, II, who taught me my first wildlife course and offered careful editing of the entire manuscript.

Many friends and former colleagues shared excellent photographs or provided permission to use existing graphics. I was especially blessed by the numerous photographic contributions from Aggie friend and science colleague, Mark Alan Storey. Attributions are found in the figure captions. Photos designated CC (Creative Commons) or WC (WC) Wikimedia Commons were marked as being available for commercial use. Photos or illustrations lacking an attribution are the work of the author.

Final mention of thanks goes to my husband, Gordon Linam, who offered encouragement and patiently supported long hours of immersion in this project that represented a life-long ambition, but did not look much like "retirement."

CHAPTER 1 – UNDERSTANDING WILDLIFE SCIENCE

The diversity of wildlife **species**[1] in Texas is awe-inspiring. Majestic white-tailed deer, intimidating alligators, and adorable baby quail capture the imagination. The field of wildlife science may seem romantic and rewarding, conjuring *National Geographic* images of biologists in stunning wild places sometimes facing down danger to study and protect these fascinating creatures. However, wildlife science encompasses much more than chasing down charismatic wild animals. It involves understanding complex, sometimes messy, **ecosystems** and working to conserve and manage species and their **habitats** in complex, sometimes messy, social settings. Effective wildlife scientists must have a grasp of **biology**, policy, education, marketing, and a host of other tools. And in Texas, effective wildlife scientists must be able to build bridges of partnership with private landowners who provide much of the habitat that wildlife needs.

Figure 1.2. Wetland habitats are often high in species diversity, especially for avian (bird) species. Photo: Brent Ortego, TPWD.

DEFINING WILDLIFE

The first step in understanding the field of wildlife science is examining what "wildlife" means. **Wildlife** can be defined as *terrestrial or semi-terrestrial non-domesticated vertebrates existing in a natural or semi-natural wild environment.* Wildlife includes amphibians, reptiles, birds, and mammals that can be found free-ranging on land or a combination of land and water. (Usually, the study of fish and other fully-**aquatic** organisms falls into the parallel disciplines of fisheries, aquatic, or marine science.) Some species may start as domesticated animals but may become wild if they escape captivity. Examples include pigs, cats, dogs, and burros. A species of wildlife

HIGHLIGHT 1.1 - WHAT IS BIODIVERSITY?

Biodiversity is a term often used to describe the richness and diversity of biological life. It can be examined in several ways:

- Species richness is the number of different species found within a given area.
- Species diversity is a calculation of the number of different species *and* how evenly distributed those species are.
- **Genetic diversity** is the variety of genotypes (or genetic combinations) found in a population of a species. Genetic diversity is crucial because it allows species to adapt to changing environments.
- Ecosystem diversity is the variety of ecosystems within a larger region.

(overleaf) Figure 1.1. A mountain lion cub is fitted with a radio transmitter collar as part of a wildlife science research project. Photo: Joe Herrera, TPWD.

---

[1] New terms in **bold** are defined in the glossary.

that transitions from domesticated to wild is called **feral**. Although wildlife scientists often focus on vertebrate animals (animals with backbones), biologists in these fields must have an understanding of a vast array of organisms that are part of the ecosystem; therefore, in recent years, the field of **wildlife management** has expanded to address overall **biodiversity** (see Highlight 1.1), including wild plants and invertebrates.

Texas is rich in biodiversity. According to a report prepared by NatureServe, a database compiled by The Nature Conservancy, Texas ranks second among U.S. states in the number of species (California is first). It ranks third in **endemics** –species found nowhere but Texas. Unfortunately, it also ranks fourth in the number of **extinctions** that have taken place within the state.

Species are identified by both their genus and species name, or their "scientific name," in a system called **binomial nomenclature** (Highlight 1.2). Thus, the scientific name for the bobcat is *Lynx rufus*. The name of the genus is capitalized, but the species portion is not. Both words are italicized. Scientific names are usually based on Latin words and often provide clues about the organism's characteristics. For example, the scientific name for the northern mockingbird, the official Texas State Bird, is *Mimus polyglottus*, which translates as "a mimic with many tongues." That's an accurate representation of the many sounds that mockingbirds can make as they imitate other birds, animals, and even man-made devices. Species are sometimes known by a variety of common names, such as redbird vs. cardinal or javelina vs. collared peccary. (Some taxonomists capitalize common names, whereas others do not.) Variable common names can lead to confusion, but standardized scientific names provide a consistent system of identifying species across different disciplines and languages. Occasionally, scientists revise common or scientific names. Appendix E provides a cross-reference between common names and scientific names as of the time of this printing.

HIGHLIGHT 1.2 - WHAT'S IN A NAME?

Taxonomy is the science of naming and classifying organisms. Taxonomists name, define, and classify groups of biological organisms based on shared characteristics. Originally proposed by Carl Linnaeus in the 1700s, the classification system starts with large, broad groups and works down to individual species. The example below shows the taxonomy of the bobcat.

Category:	Bobcat Example:
Domain	Eukaryota – all species with complex cells
Kingdom	Animalia – all animals
Phylum (plural = phyla)	Chordata – all animals with a dorsal nerve cord
Class	Mammalia – all animals with hair & milk (mammals)
Order	Carnivora – animals with prominent canine teeth
Family	Felidae – all cats
Genus (plural = genera)	Lynx – includes several medium-sized wild cats
Species	rufus – a Latin term meaning reddish

Figure 1.3. The bobcat, found across much of the state, fits into a family of animals that includes all wild cats. Photo: Kelly Munro Photography.

DEFINING SCIENCE

Wildlife conservation is also closely aligned with the pursuit of science. Science can be defined as *a system for discovering general truths or laws about the natural world as obtained and tested through the scientific method.* The **scientific method** involves several steps, including.

1. *Observation* – The scientist notes patterns in the natural world that seem to be repeated or require investigation.
2. *Question* – The scientist reads through other research and poses a question that would be valuable to explore.
3. *Hypothesis* – The scientist suggests an idea or statement of assumption that can be investigated.
4. *Experimentation* – The scientist designs and conducts an experiment that allows them to measure the conditions described in the hypothesis. The research must understand the variables, or the things that can change, in the experiment. The independent variable is the one factor that the scientist adjusts (the treatment). The dependent variable is the response that the scientist measures. The experiment should be repeated several times and include a control where no treatment is applied, and all other variables are the same.
5. *Analysis* – The scientist analyzes the data to draw conclusions. They use statistical tests to make sure that any differences in results reflect real differences, not just normal variability. The scientist may reject or fail to reject the hypothesis; however, they never state that they have "proven" the hypothesis, as new research in the future can shed new light on the results.
6. *Sharing Results* – The final step in the scientific method is to share results with others. Usually, this step involves peer review and publication. Peer review is when other scientists with similar expertise are invited to read and critically review the research report. Their input can help assess whether the experiment and analysis are rigorous enough to be shared more widely. If the paper is accepted and published, it can offer ideas for wildlife conservation. It can also inspire other scientists to investigate further and try to replicate the results.

HIGHLIGHT 1.3 - APPLICATION OF THE SCIENTIFIC METHOD TO THE CONSERVATION OF WHOOPING CRANES ON THE TEXAS COAST

The whooping crane is an endangered species that spends the winter in wetlands on the Texas coast. Whooping cranes form long-lasting pair bonds between males and females. Pairs, sometimes with their current year's chicks, usually defend a **territory** of **marsh** habitat on the Texas coast and spend most of their time there feeding on Atlantic blue crabs, clams, other small animals, and wetland plants. They also roost or sleep in the wetlands.

These same marshes (marshes are wetlands dominated by reeds and other grass-like plants) are popular with outdoor recreationists, including boaters, fishermen, and duck hunters. Whooping cranes (or "whoopers") are very alert to dangers in their surroundings. They may interrupt their feeding to move away from danger or defend their territory from other whoopers. **Ethologists** (biologists who study the behavior of animals) suggest that animals that must spend much time in alert behavior may be harmed because they have less time available for feeding, courtship, preening, resting, and other vital behaviors. Therefore, biologists involved in the conservation of the whooping crane conducted the study described below.

"Responses of Wintering Whooping Cranes to Airboat and Hunting Activities on the Texas Coast"

1. *Observation* – Wildlife biologists in Texas noticed the frequent presence of outdoor recreationists in habitats for endangered whooping on the Texas coast.

2. *Question* – Biologists wondered if the frequent presence of boats and hunters could be causing disturbance to whoopers so that they did not feed as frequently.

3. *Hypothesis* – Since whooping cranes returned to these areas year after year, biologists did not think a problem existed, so they hypothesized that whooping cranes in areas with boating recreation behave the same as whoopers in areas not frequented by boats.

4. *Experimentation* – Biologists decided to watch the behavior of whooping crane families during times when boats and hunters were absent and during times when they were present. All the observations focused on whooping crane pairs in the marsh with one chick. The researchers gathered time budget data for two hours, recording the behavior of the individual whoopers every 30 seconds in a way that could be calculated as percentages. Three different treatments were staged: a hunter hunting from an airboat, a hunter in an outboard, or an airboat just driving toward the cranes. After one hour of activity, the boat departed while the biologists watched for one more hour. Biologists also gathered data on families with no staged disturbance. Data were collected on seven days, totaling more than 266 hours of observation.

5. *Analysis* – The researchers chose to focus on the time spent in alert behavior. At first glance, the data seemed to indicate that all the boats caused increased alert behavior, but biologists noticed that other natural factors, such as the presence of other birds, could also cause the whoopers to be alert. When statistics were used, it was found that only the direct airboat disturbance caused the cranes to increase their alert behavior. Therefore, the biologists only rejected the hypothesis that airboats do not affect whooping cranes. They recommended that land managers monitor the number of airboats near whooping crane territories in case disturbances become too frequent.

6. *Sharing Results* – A scientific paper was written, submitted to the *Wildlife Society Bulletin*, and reviewed by other **ornithologists** (biologists who study birds). The reviewers suggested some changes in the analysis of the data. After the biologists made the recommended edits, the *Wildlife Society Bulletin* published the paper. At least 18 other researchers have mentioned this study in their own research.

Figure 1.4. Whooping crane family feeding in coastal marsh in Texas. Photo: Klaus Nigge, USFWS, CC 2.0.

Sometimes, it is challenging for scientists to use the scientific method in wildlife conservation, especially in wild environments. For instance, if biologists want to study how drought affects quail reproduction, they cannot change the rainfall variable. In these cases, wildlife biologists can gather observations and data from many different sources and circumstances over a long time to draw conclusions. This approach is known as **induction**. They can then use inductive reasoning to conduct experiments or make land management decisions based on these observations and adjust their strategies based on the results. This process is called **adaptive management.**

DEFINING WILDLIFE SCIENCE

What, then, is "wildlife science?" **Wildlife science** is the study of animals and their environments, with a focus on how to apply ecological knowledge to balance the needs of wildlife with the interests of humans. Individuals working in the field of wildlife science may be called wildlife scientists, wildlife biologists, wildlife ecologists, or wildlife managers. As an applied science (the use of scientific knowledge to solve practical problems), the field of wildlife science draws from many different disciplines (Table 1.2). Data, experience, and intuition are all valuable in the field of wildlife science. One of the first professional wildlife scientists, Aldo

Leopold, defined **game management** not as science but as art, "the art of making land produce sustained annual crops of wild game for recreational use." Unlike some science disciplines, wildlife science acknowledges that people's goals have a role in applying science in this field. As such, wildlife science is not synonymous with the term "preservation," which often implies that an area is set aside and not used by people; instead, wildlife science often focuses on **conservation** or the "wise use" of natural resources.

WHY STUDY WILDLIFE SCIENCE?

Why is this field important? Throughout history, wild things have been a source of value to humans. These benefits range from use values—situations in which people derive benefit directly from using wildlife species—to existence values, which assign a value to wildlife simply because it exists. Several types of value have been defined, with overlap in some of these categories.

Consumptive Use Value (e.g. hunting) – Consumptive use involves the killing or harvest of a resource. Historically, much of the value of wildlife was rooted in its utilization for food, clothing, and other purposes. Early humans hunted and consumed wildlife, using their meat, skins, and bones. This subsistence use evolved into commercial industries focused on harvesting species for various needs, such as food, hides, feathers, and oils. Today, wildlife managers prioritize **sustainable harvest**, and hunting remains a popular sport with economic benefits. In Texas, over a million hunting licenses are sold yearly, contributing significant economic value to the state. A 2023 report by the Natural Resources Institute at Texas A&M University estimated that each year white-tailed deer hunting alone contributes about $10 billion to the state's economy, raises more than $500 million in taxes, and supports more than 60,000 jobs. Additionally, some hunters donate game meat to

Table 1.2. Disciplines that contribute to the field of wildlife science.

Discipline	Definition	Example in Wildlife Science
Biology	The study of living things	Investigating how chytrid fungus spores infect frogs
Ecology	The study of the relationships living organisms have with each other and their environment	Monitoring the effect that population cycles of muskrats have on marsh vegetation
Conservation	The wise use of natural resources	Identifying appropriate harvest rates for American alligators based on nesting density
Management	Implementing changes to populations, habitat, and human use to produce some desired result	Applying prescribed fire at the correct intervals to maintain red-cockaded woodpecker habitat
Economics	The study of how wealth is produced and distributed	Helping communities and landowners collect income from birdwatching
Sociology	The study of human society	Assessing trends in the public's support for wildlife conservation

combat hunger, bringing more benefits to people of the area.

- *Non-consumptive Use Value (e.g. wildlife watching)* – People enjoy wildlife watching activities like birdwatching, bird feeding, wildlife tours, and photography. In Texas, one of the top birding destinations in the country, wildlife watching is extremely popular, with 4.4 million participants in 2011 (Fig. 1.5). Many surveys show that wildlife watchers spend even more than hunters. Thus, non-consumptive use, including **nature-based tourism**, also contributes several billion dollars to the state's economy annually. Even those who currently do not engage in wildlife watching still value the opportunity to do so in the future. They want to know that they will have that option.

- *Scientific Value* – Wildlife species play significant roles in scientific research both in the laboratory and their natural habitats. Amphibians, for example, act as **sentinel species** due to their sensitivity to pollution. The antlers of white-tailed deer can be analyzed to help monitor environmental radiation levels. Toxins from certain toad species show promise in cancer treatment. Armadillos have been studied to better understand leprosy.

- *Ecological Value* – Different wildlife species play essential roles in maintaining healthy ecosystems. Examples include pollinators like insects and the Mexican long-nosed bat. Grazers like bison shape grassland communities. Burrowing animals such as gophers and prairie dogs aerate the soil and provide habitat for other species. Beavers and alligators create wetland habitats, and rodents help spread beneficial soil fungi. **Frugivores** (fruit-eaters), like raccoons, assist in seed dispersal. **Predators** like mountain lions help keep **herbivore** populations in balance. These diverse species contribute to the overall balance and well-being of ecosystems.

- *Existence Value* – Wildlife is valued by some people just because it exists, even if they don't use or interact with it. People may appreciate wildlife for its beauty, cultural significance, or the belief that all life is sacred. Many indigenous communities embrace a concept of **kincentricity**, which sees both humans and nature as part of an ecological family. Economic studies show that people are willing to pay for the existence of wildlife, even if they are not wildlife watchers or hunters. These people often donate to organizations that work to protect wildlife species.

Figure 1.5. Birdwatchers looking for whooping cranes on a guided tour of Texas coastal wetlands. Photo: James Alan Storey.

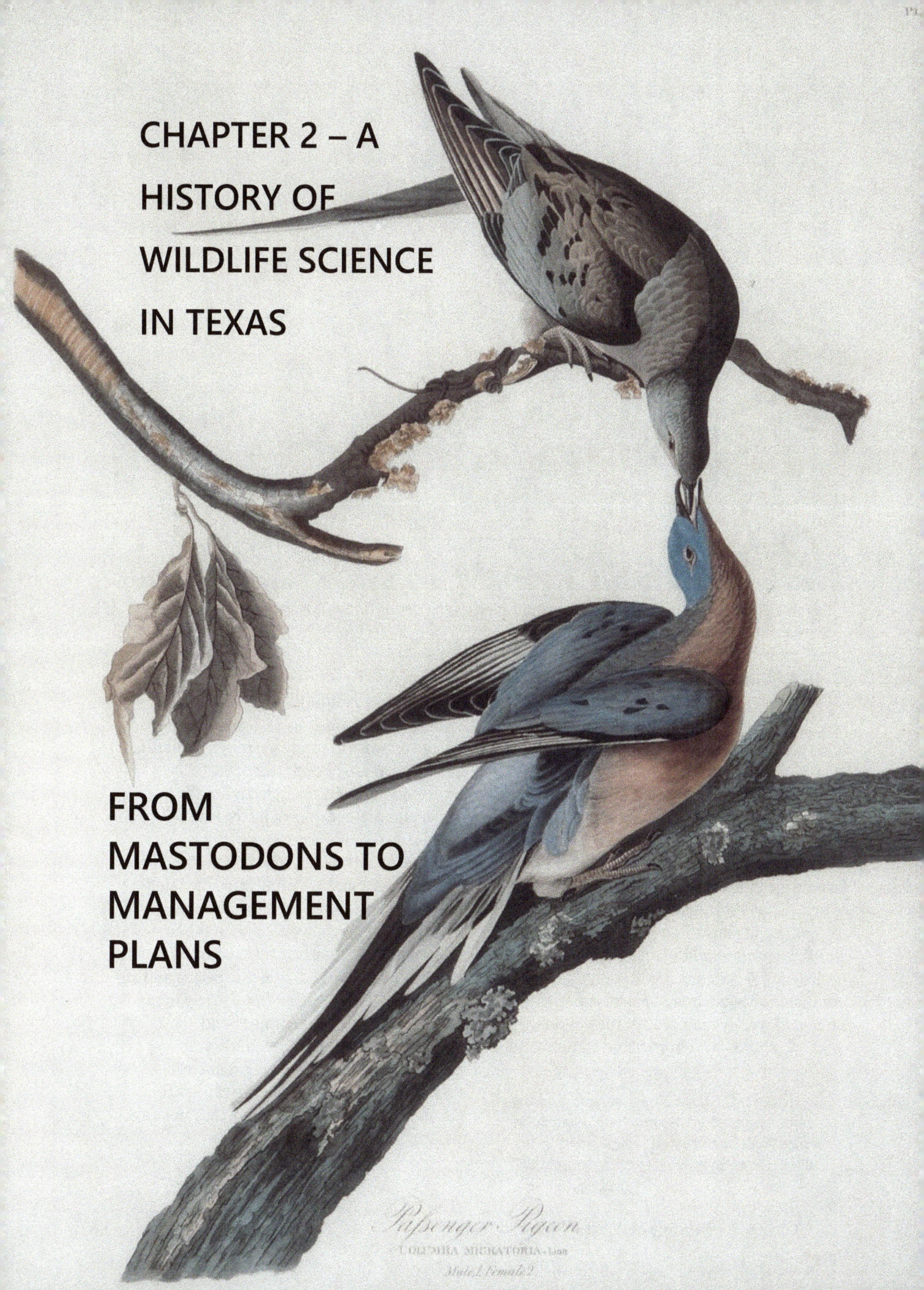

CHAPTER 2 – A HISTORY OF WILDLIFE SCIENCE IN TEXAS

FROM MASTODONS TO MANAGEMENT PLANS

Figure 2.2. The arrival of Spanish horses improved the ability of Native Americans to hunt bison. Source: Karl Bodmer, WC (public domain).

THE BEGINNINGS

For eons, humans have observed and interacted with wildlife, including the very first people to inhabit this continent. Native American attitudes toward wildlife were deeply rooted in their spiritual beliefs and practical needs. They relied on hunting and gathering for sustenance and often held a deep respect for nature, viewing animals as sacred and interconnected with the natural world. The first indigenous people survived by gathering plants and hunting both modern wildlife species and the megafauna of early millenia. (Megafauna are literally "large animals.") Megafauna in North America included woolly mammoths, mastodons, gigantic sloths, wild horses and camels, and enormous armadillo-like glyptodonts. Evidence of wildlife use by Native Americans has been found in many locations, including spear tips among bone fragments of harvested animals, duck decoys made of reeds, and pictographs of birds, reptiles, wolves, deer, and bison.

These early North Americans could be called Texas's first **naturalists** since they had to observe the species they hunted closely. They may have also been the first wildlife managers, using fire and other tools to shape landscapes to benefit species and aid their hunts. On the other hand, early nomadic hunters likely did not comprehend sustainable harvest at first, and some researchers believe hunting pressure combined with **climate** shifts caused the extinction of almost half of North America's megafauna.

(overleaf) Figure 2.1. Passenger pigeons once occurred across much of Texas. John James Audubon created this painting in the 1830s before they went extinct. Source: WC (public domain).

THE EXPLORATION ERA

European settlers in North America did not encounter megafauna, but they did find abundant wildlife during the 16th and 17th centuries. English settlers and French explorers noted plentiful fish, grizzly bears, stags (deer and elk), fox, beavers, **waterfowl**, and other birds. The most remarkable accounts were of the passenger pigeon. These birds numbered in the billions and may have been the most abundant wildlife species in the world at the time. Their flock movements were reported to darken the sky for hours or even days and fill the air with the sound of thunder. Massive flocks of curlews and plovers were also reported.

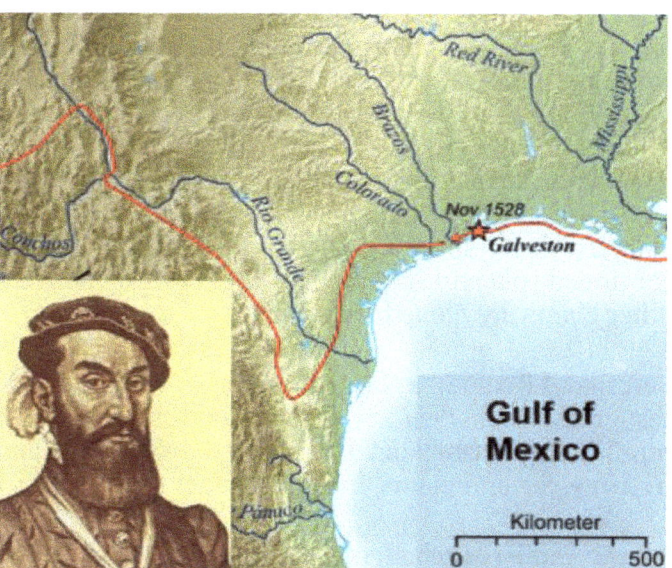

Figure 2.3. Cabeza de Vaca's route through southern Texas after shipwrecking on Galveston Island encountered significant hardships, including food shortages. Source: Lencer, CC 3.0.

Spanish explorers provided observations of wildlife in the southern half of the continent, including the Texas area. They mentioned animals such as bison, prairie dogs, plains grizzlies, turkeys, goats (possibly referring to mountain goats), and gray lions and leopards (referring to mountain lions and jaguars). In Texas, after Álvar Núñez Cabeza de Vaca shipwrecked on Galveston Island, he made his way across southern Texas (Fig. 2.3), documenting the use of local foods by Native American communities. Cabeza de Vaca listed oysters, deer, fish, spiders, snakes, blackberries, roots, prickly pear, pecans, bison, rabbit, mesquite beans, and a tart fruit (possibly wild persimmon) among the foods that the native tribes used; however, he also noted that Native Americans faced periods of severe hunger, indicating that natural resources may have been scarcer in certain seasons and some locations.

The unique wildlife in the New World attracted naturalists who documented new species through collections and artwork. Their observations, including major expeditions like Lewis and Clark's, contributed to the understanding of North America's vast biodiversity. Notably, ornithologist Andrew Wilson and artist John James Audubon played significant roles in documenting the continent's birds and other wildlife species in the early 1800s. Audubon visited Texas in 1837, and his son collected specimens here in the 1940s. Audubon's paintings of the Texas turtle dove (white-winged dove), orange-bellied squirrel (eastern fox squirrel), hispid cotton rat, collared peccary, and black-tailed hare (black-tailed jackrabbit) are based on specimens collected in Texas.

Other naturalists also flocked to Texas as a new frontier for ecological collections. Often working as collectors for renowned botanists or as part of boundary survey teams, explorers spent extensive time in Texas, creating species lists for the new territory, then republic, and then state. Notable naturalists in Texas during the 1800s and 1900s included:

- Jean Louis Berlandier came to Texas in 1827 from Switzerland. He traveled across many parts of Texas collecting specimens of **flora** and **fauna**. He settled in Matamoros and continued collecting information about the plants, animals, and Native Americans of the region until 1851. The Texas tortoise (*Gopherus berlandieri*) is named for him.

- Ferdinand Jacob Lindheimer, also known as the "Father of Texas Botany," came to Texas from Germany in 1836. He collected plant specimens for botanists from Saint Louis and Harvard University. He settled in New Braunfels, where he started a newspaper and continued collecting plants. The Texas prickly pear cactus (*Opuntia lindheimeri*) and many other plants are named for him.

- Ferdinand Roemer, a German geologist, came to Texas in 1845 to survey mineral resources. He made many observations on the Texas landscape, Native Americans, and reptiles and amphibians. He returned to Germany and wrote a book on Texas **natural history**. Several plants, including cedar sage (*Salvia roemeriana*), bear his name.

- Joseph Daniel Mitchell, the first Texas-born naturalist, was born in the Coastal Bend of Texas in 1848. He studied insects and collaborated with the U.S. Department of Agriculture. He also collected birds' eggs and nests, reptiles, mollusks, minerals, and Native American relics. The Mitchell's wentletrap (*Amaea mitchelli*), a seashell found only on the coast of Texas and Mexico, is named for him.

- Thomas Volney Munson, a horticulturalist and botanist, came to Texas in 1876. He was keenly interested in native grape species. He found Texas grapes that were resistant to insect infections, a discovery that helped save the French wine industry from collapse. Those Texas rootstocks are still in use in Europe and California today.

- John Kern Strecker moved to Waco, Texas, in 1887 when he was 13. At the age of 18, he started collecting specimens for the Baylor University Museum. He became the curator of the museum and published papers on Texas wildlife. The Strecker's chorus frog (*Pseudacris streckeri*) is named for him.

- Vernon Orlando Bailey and Florence Merriam Bailey were a ground-breaking husband and wife team of naturalists. They conducted collecting and exploratory trips throughout Texas between 1889 and 1905, focusing on mammals and birds, especially in West Texas. Florence was one of the first ornithologists to collect data on birds using binoculars instead of collecting with a gun. The Bailey's pocket mouse (*Chaetodipus baileyi*) is named for Vernon; Bailey's mountain chickadee (*Parus gambeli baileyae*) is named for Florence.

Figure 2.4. John Kern Strecker began his naturalist career in Texas at a young age. Source: University of North Texas Libraries, The Portal to Texas History, Private Collection of T. B. Willis.

Figure 2.5. As this 19th-century painting by George Caitlin depicts, elk and bison were once common in Texas. Source: WC (public domain).

THE EXPLOITATION ERA

Impressed by the abundance of wildlife, European settlers in North America mistakenly believed that the resource was unlimited, leading to a heavy harvest of animals. Even when Audubon witnessed the mass harvest of passenger pigeons, he thought that their vast numbers meant they would never disappear. Over time it became clear that overharvest and the destruction of habitat were causing the decline of wildlife populations. Unfortunately, it took many years for people to realize that wildlife was not inexhaustible, and, as a result, some species would go extinct.

The first species to go extinct in North America following European settlement was the great auk. These large, flightless birds lived in the northern seas and were similar to penguins. They used to gather in large numbers on islands near Nova Scotia and Iceland. Sailors mentioned that the birds were so abundant that they couldn't even walk between them. However, because they were easy to catch and harvest for their meat, eggs, oil, and feathers, their populations rapidly declined. Efforts to protect them came too late, and the last known great auk was collected from the North Atlantic in 1840.

During early colonization, many animals were also harvested for their fur. This led to extensive exploration and trade, with beaver skins being the most sought-after. Tens of millions of beaver skins were shipped to England for making felt hats. By the 1830s, beaver populations had significantly declined. Deer were also heavily hunted, both for their meat and hides. Hunting regulations were established, but by the late 1800s, white-tailed deer were nearly **extirpated** in some areas. The Merriam's elk **subspecies** in Texas went completely extinct, as did the elk subspecies found in the eastern U.S.

Waterfowl were also once abundant, and colonists harvested large numbers of geese, ducks, and teal for commercial sale (Fig. 2.6). The unsustainable hunting strategies (sometimes using massive punt guns that could kill dozens in one shot) led to the decline of waterfowl populations. Feathers from birds were also in high demand for fashion and cosmetics, leading to the overharvest of egrets, herons, spoonbills, and swans. Approximately 5 million birds were killed annually during the peak of the feather trade. Alarm at declining bird numbers resulted in the formation of the Audubon Society in 1896.

Figure 2.6. Heavy harvest of waterfowl was common in the United States in the 1800s and early 1900s. Source: USFWS, WC (public domain).

Some of the species noted by Spanish explorers as being most abundant are among those that declined most precipitously in the state due to high levels of exploitation. Eventually, bison, passenger pigeons, wolves, jaguars, and bears disappeared from the state, and the number of deer, prairie dogs, mountain lions, pelicans, and prairie-chickens in Texas dropped dramatically.

As the country and Texas prepared to enter the 20th century, a very different story of wildlife abundance was evident. Species like passenger pigeons, Carolina parakeets, bison, beaver, white-tailed deer, prairie-chickens, and wild turkey, once considered innumerable, were nearly gone. For some it would be too late. Passenger pigeons and Carolina parakeets went extinct in the early 1900s. For others there was still time. It seemed the country was finally ready to respond. The time was right for the profession of wildlife science to emerge.

Texas, too, once hosted an abundance of wildlife that must have seemed inexhaustible, but unregulated harvest during the early settlement of the state caused many species here to decline or disappear. For example, during a seven-year period in the 1850s, one trading post in Waco shipped 75,000 deer hides to New York in addition to many bear and buffalo pelts. In 1877, a San Antonio newspaper reported a "successful" two-week hunt by two men who managed to harvest "three bears, sixty-seven deer, two hundred and nineteen turkeys, … four geese, forty-six ducks, thirty quails…" In 1879, a Corpus Christi newspaper reported the harvest of 3,000 brown pelicans from a nesting island in the Laguna Madre.

Figure 2.7. As this postcard mailed in 1907 in Uvalde demonstrates, deer harvest levels were high in Texas in the early 1900s. Source: DeGolyer Library, Southern Methodist University, WC.

THE CONSERVATION ERA

By the end of the 19th century, naturalists and policy-makers began to call for an effort to save species and habitats. Early voices included authors George Perkins Marsh, Henry David Thoreau, and Ralph Waldo Emerson. John Muir, an ardent and eloquent naturalist, advocated for wilderness and helped form the Sierra Club. William T Hornaday, a zoologist at the United States National Museum, raised alarm at the possible extinction of the buffalo. Florence Merriam Bailey encouraged observation of birds with binoculars instead of guns. Theodore Roosevelt helped form the Boone and Crockett Club to promote wildlife conservation and ethical hunting. In response to these calls for action, three primary strategies to conserve wildlife emerged—land conservation, wildlife laws, and wildlife science.

LAND CONSERVATION

Wildlife scientists understand that habitat is crucial for wildlife conservation. Fortunately, by the late 1800s, policy-makers also began to recognize the value of setting aside wild lands. In 1872, President Ulysses S. Grant established Yellowstone National Park, the first national park. Yosemite and Sequoia National Parks were formed shortly after with the help of John Muir. The National Park Service was established in 1916 to better manage the parks. It now oversees more than 400 park units. These units range from small historical sites to large protected areas like Yellowstone National Park. Important wildlife habitats in Texas are found in Big Bend National Park, Guadalupe Mountains National Park, Padre Island National Seashore, and Big Thicket National Preserve, all managed by the National Park Service.

In 1891, Congress created **forest** reserves through the Forest Reserve Act. These lands were initially used for timber production, but their purpose shifted towards conservation under President Theodore Roosevelt and Gifford Pinchot. Pinchot, often known as the "Father of American Forestry," became the first chief of the United States Department of Agriculture Forest Service (USFS). Today, USFS manages over 190 million acres across 154 national forests and 20 national grasslands. There are four national forests in East Texas—Davy Crockett, Sabine, Angelina, and Sam Houston—and one national grassland, Caddo-Lyndon Baines Johnson National Grassland, found in Northeast Texas.

As President, Teddy Roosevelt (Fig. 2.8) was a strong advocate for conservation in many ways. In 1903, he created Pelican Island Federal Bird Reservation in Florida to protect water birds from the feather trade. This later became the first national wildlife refuge, part of a system managed by the U.S. Fish and Wildlife Service (USFWS). Today, the National Wildlife Refuge (NWR) System covers over 150 million acres of wildlife habitat across nearly 600 refuges. These refuges aim to conserve and manage fish, wildlife, and plant resources for the benefit of present and future generations of Americans. Texas has about 20 national wildlife refuges. Most focus on providing habitat for migratory waterfowl. However, Balcones Canyonlands NWR near Austin is dedicated to protecting endangered **songbirds,** and refuges in South Texas focus on rare subtropical species.

Figure 2.8. President Theodore Roosevelt and John Muir, pictured here at Yosemite National Park. Both were leaders in the early wildlife conservation movement. Source: WC (public domain).

The Bureau of Land Management (BLM) is the largest manager of public land in the United States. It was established in 1812 to encourage settlement in the American West. Today, the BLM oversees 245 million acres of land, allowing for various activities such as energy development, livestock grazing, mining, timber harvest, wildlife conservation, and outdoor recreation. The BLM only manages one tract in Texas, as Texas retained the public lands within the boundaries that were set for it as a state in 1845. The BLM's 12,000-acre Cross Bar Management Area near Amarillo was once part of the country's helium reserves.

WILDLIFE LAWS

With overharvest as a major problem, laws were also needed to protect wildlife. The Lacey Act, passed in 1900, was the first national law in the United States to protect wildlife. It prohibited the interstate and international trade of illegally-killed animals, helping the federal government enforce state regulations. Subsequent federal laws addressed interstate wildlife issues, funded state wildlife programs, and addressed cross-border environmental concerns. Some of these laws included:

- 1918 - Migratory Bird Treaty Act - This important act involves international treaties between the U.S. and Canada (signed in 1916), Mexico (1936), Japan (1972), and Russia (1976). It serves to protect nearly all native bird species, allowing USFWS to set boundaries on the harvest of waterfowl and other migratory species such as doves. State laws must conform to that regulatory framework.

- 1934 – Migratory Bird Hunting Stamp Act – This Act requires migratory bird hunters everywhere in the country to buy a federal "duck stamp." Funds from duck stamps are then used to fund refuges and other activities that benefit migratory birds. Beginning in 1981, Texas hunters were also required to purchase a Texas Migratory Game Bird Endorsement, essentially a state stamp for waterfowl and dove hunting.

- 1937 – Federal Aid in Wildlife Restoration Act (Pittman-Robertson Act) – This Act placed a federal excise (sales) tax on the purchase of firearms and ammunition. The funds collected federally are then distributed back to the states, where the state wildlife agencies contribute matching funds. The Pittman-Robertson Act is the main source of funding for wildlife conservation around the country. Texas receives nearly $50 million per year in Pittman-Robertson funding and provides its match primarily from the sale of hunting licenses.

- 1964 – Land and Water Conservation Act – This Act collects funding from offshore drilling leases to provide monies to purchase land for recreation and endangered species conservation. Proceeds are shared with states. Over the last 40 years, about $600 million was directed to projects in Texas. The state then offers some of these funds as grants to local governments for parks and recreation.

- 1966 – Endangered Species Act – Modified in 1969 and 1973, this Act recognizes a list of endangered and threatened species that are at risk of extinction. Animal species on the list cannot be killed or sold and their habitat may be protected. The Act also provides grants to states for research and conservation of endangered and threatened species. Texas has about 120 federally-listed endangered or threatened species.

- 1972 – Clean Water Act – This Act regulates discharge and water quality standards in "waters of the United States." Implemented by the U.S. Environmental Protection Agency, it has played a major role in cleaning up U.S. rivers. Because it prohibits discharge into regulated waterways, it has also helped to protect wildlife habitat in wetlands from being filled.

- 1973 – National Environmental Policy Act – This Act requires federal agencies and agencies receiving federal funding to consider the impact of major actions on environmental quality. It requires documents such as Environmental Assessments and Environmental Impact Statements that can help to lessen the impact of projects on wildlife habitats.

- 1985 – Farm Bill – Conservation Reserve Program – Beginning in 1985, this agriculture law adopted annually by Congress began to include funding to help farmers prevent erosion, protect water quality, and conserve wildlife habitat on former farmlands. It provides an economic benefit to farmers for protecting wildlife habitat. Texas currently has about 3 million acres enrolled in the Conservation Reserve Program.

- 2000 – State and Tribal Wildlife Grants Program – Beginning in 2005, Congress allocated grants to be distributed to states to "help keep common species common." Each state is required to complete a comprehensive Wildlife Action Plan to guide priorities in funding and is required to provide matching funding under the plan. Efforts have been made to enhance this program in recent years. As part of the planning, states identify "Species of Greatest Conservation Need."

WILDLIFE SCIENCE

It is often said that the profession of wildlife science began in 1933 when a forester in Wisconsin named Aldo Leopold published a book entitled *Game Management*. Leopold became the first professor of game management at the University of Wisconsin and has come to be known as the "Father of Wildlife Management." He believed in using scientific principles to manage land and wildlife to ensure long-term productivity and biodiversity. His ideas gained traction, and by 1937 the *Journal of Wildlife Management* began publishing peer-reviewed articles in the field of wildlife science. Cooperative Wildlife Research Units were established at universities across the country, including

Figure 2.9. Aldo Leopold, shown here on a trip to Mexico, is often called the "Father of Wildlife Management" for his efforts to incorporate science into wildlife conservation. Source: USFS, CC 2.0.

Texas A&M University. (The Texas Cooperative Wildlife Research Unit is now housed at Texas Tech University.) In 1938, Texas A&M awarded its first graduate degree in wildlife research and game management to Dan Lay, who later became one of the first wildlife biologists employed by the state.

HISTORY OF WILDLIFE CONSERVATION IN TEXAS

Federal laws and agencies provide a framework for conservation of migratory species and endangered species, but the responsibility of implementing wildlife conservation mostly lies with state governments. In Texas, the exploitation of wildlife was unregulated until 1895 when the Fish and Oyster Commission was created to regulate fishing. In 1907, the Game Department was added, leading to the agency being called the Texas Game, Fish, and Oyster Commission. Hunting licenses were introduced in 1909, and the first game wardens were hired in 1919 to enforce game laws. Wildlife management initially focused on predator control, leading to the elimination of wolves and jaguars and a decline in bears and mountain lions. However, by the 1930s the Commission began embracing the more holistic science of wildlife management, funding its activities with revenue from the Pittman-Robertson Act, license sales, and permits for natural resource extraction

Phil Goodrum was the first head of the Commission's Wildlife Restoration Program. He hired Dan Lay, Val Lehmann, and Rollin Baker as some of the first field biologists in the state. These men surveyed wildlife populations to help the Texas Legislature set appropriate hunting rules. They worked with County Wildlife Planning Boards for data and recommendations. Thanks to their efforts, some vulnerable species, such as the Attwater's prairie-chicken, were protected. They also reintroduced overharvested species like beaver, deer, turkey, and quail. Although they did some research on non-hunted species, such as the nine-banded armadillo that had just arrived in Texas, their main focus was on game species.

The Texas Commission biologists worked with hunting clubs, planning boards, and private landowners to share information. They also started translocating wildlife to restore populations, collaborating with the famous King Ranch, which had already begun to effectively manage its wildlife.

In 1945, the state biologists established the first wildlife management area (WMA) called Sierra Diablo WMA near Van Horn. Nowadays, there are 47 WMAs in Texas, covering every ecological region in the state. These state-owned properties are used for research, education, and outdoor recreation activities like hunting, hiking, and birdwatching. They provide valuable insights into managing wildlife in real-life conditions, such as hunting and grazing on private land.

In 1963, the modern Texas Parks and Wildlife Department (TPWD) was formed by merging the State Parks Board and the Game and Fish Commission; however, there was a lack of proper authority for managing wildlife. To address this, the legislature passed the Wildlife Conservation Act in 1983. This act gave TPWD the power to manage fish and wildlife resources across all counties in Texas. It allowed for better consistency in setting hunting seasons and limits. TPWD could now use scientific data to set hunting zones, establish harvest limits, and protect

Figure 2.10. Black Gap Wildlife Management Area, located on the Texas-Mexico border, was the second WMA established in Texas. Photo courtesy TPWD © 2025, (Earl Notthingham, TPWD).

endangered species by closing certain areas during specific seasons.

The focus of wildlife management in Texas has expanded over the years. Initially, it was mostly about game and fur-bearing animals that could be harvested. Still, in the 1960s Dan Lay initiated work on the red-cockaded woodpecker, a non-hunted species that he recognized was vulnerable because it relied on mature fire-maintained pine forests. Eventually, in 1973, TPWD hired its first biologists to focus on **nongame** species. The department also established regulations to protect threatened and endangered species. Since then, the responsibilities of biologists across the state have expanded to include both game and nongame animals. In 1983, the department started selling a nongame stamp to raise funds and more recently began offering a Texas horned lizard license plate. Funds from these sources support a group of nongame specialists in the Wildlife Conservation program and research designed to manage habitats to support both game and nongame species.

Although refuges and WMAs exist, Texas is predominantly a private lands state (about 95% of Texas is in private ownership). These private landowners are key partners in conserving wildlife in the state. In 1972, the Private Lands Assistance and Technical Guidance Program was established to collaborate with landowners in conserving and managing wildlife habitat. The program offers various resources such as technical guidance biologists that work with landowners, Wildlife Management Associations, habitat improvement grants for private lands, Managed Lands Deer Permits, tax incentives, conservation easements, and recognition through the Lone Star Land Stewards award program. These initiatives aim to acknowledge and support private landowners in their efforts to protect and manage wildlife on their properties. The USFWS, U.S. Department of Agriculture (USDA) Natural Resources Conservation Service, Texas A&M Agrilife Extension Service (Agrilife), Texas A&M Forest Service (TFS), and other agencies and nonprofits also have programs to collaborate with private landowners in wildlife conservation.

In Texas and most other states, hunting plays a crucial role in funding wildlife conservation. TPWD uses funds from hunting licenses, the Pittman-Robertson Act, and the sale of Migratory Bird and Upland Bird stamps to support wildlife projects. TPWD offers hunting opportunities on WMAs, some state parks, and areas leased under its Public Hunting Permit program. Private landowners also offer hunting leases that help generate revenue for wildlife habitat improvement on their own lands. To ensure safety and conservation, new hunters in Texas are required to take a hunter education course. TPWD and its partners also organize mentored Texas Youth Hunts to introduce potential new hunters to the sport.

In addition to TPWD and federal agencies, many other organizations contribute to wildlife conservation in Texas. Universities and entities like the Welder Wildlife Foundation, Caesar Kleberg Wildlife Research Institute, Borderlands Research Institute, and Gulf Coast Bird Observatory conduct essential research. Nonprofit organizations also play a role. Groups like Ducks Unlimited, Audubon Texas, the Texas Wildlife Association, the Nature Conservancy of Texas, the Native Prairies Association of Texas, Texas Conservation Alliance, Texas Land Conservancy, and many others collaborate to identify conservation priorities, educate young people, and safeguard wildlife habitats.

HIGHLIGHT 2.1 - WILDLIFE CONSERVATION MILESTONES IN TEXAS AND THE U.S.

- 1872 – Yellowstone National Park is formed.
- 1891 – Forest Reserves, later known as National Forests, are established by the Forest Reserve Act.
- 1900 – Lacey Act prohibits interstate and international trade in illegally-killed wildlife.
- 1903 – First NWR, Pelican Island in Florida, is set aside.
- 1909 – Texas Game, Fish & Oyster Commission begins requiring purchase of hunting licenses in Texas.
- 1918 – Migratory Bird Treaty Act provides protection for migratory birds and sets up collaborative conservation of migratory birds with neighboring countries.
- 1933 – *Game Management* is published by Aldo Leopold, ushering in the era of wildlife management.
- 1934 – Migratory Bird Hunting Stamp Act requires waterfowl hunters to purchase a stamp to benefit wetland conservation.
- 1937 – Pittman-Robertson Act places an excise tax on firearms and ammunition that provides revenue to states for wildlife conservation.
- 1938 - First degree in Wildlife Research and Game Management in Texas is awarded to Day Lay.
- 1945 – TPWD predecessor produces its first publication, "Principal Game Birds and Mammals of Texas: Their Distribution and Management," written by Phil D. Goodrum.
- 1945 – First WMA, Sierra Diablo, is established in Texas.
- 1968 – The Texas Committee on Natural Resources emerges as one of the first Texas conservation organizations. They advocate for protection of forest and water resources.
- 1972 - TPWD establishes a program to assist private landowners.
- 1973 – The current version of the Endangered Species Act provides protection for a federal list of threatened and endangered species.
- 1973 –TPWD hires its first nongame biologists (now called wildlife conservation biologists).
- 1983 – Wildlife Conservation Act places regulatory authority for wildlife statewide with TPWD.
- 1985 – The Farm Bill establishes the Conservation Reserve Program to assist farmers in protecting habitats.
- 1995 – Texas Legislature allows properties with an approved wildlife management plan to qualify for Open-Space property tax valuation.
- 2000 – States begin to receive funding to prepare and implement Wildlife Action Plans. A list of Species of Greatest Conservation Need is prepared.

Figure 2.11. Biologist Dan Lay continued a long career at TPWD that witnessed a dramatic evolution in wildlife management across the state. Although he worked in many parts of the state, his first love was the forests of East Texas. Photo: Brent Ortego, TPWD.

THE NORTH AMERICAN MODEL OF WILDLIFE CONSERVATION

In the European Middle Ages, wildlife belonged to royalty; an average citizen had little access to hunting or wildlife use. As settlement of the United States and Canada occurred, a new tradition surrounding wildlife resources emerged. This system, known as the "North American Model of Wildlife Conservation," is based on seven key principles:

1. **Shared Ownership** – "Wildlife resources are conserved and held in trust for all citizens." – In contrast to earlier European traditions where ruling classes owned the wildlife, in North America wildlife belongs to the public at large.

2. **Restricted Sales** – "Commerce in dead wildlife is eliminated." – Market hunting allowed the sale of public wildlife for private gain and resulted in the overharvest of many species; therefore, laws have been implemented to prevent unregulated trade in wildlife species.

3. **Democratic Laws** – "Wildlife is allocated according to democratic rule of law." – Landowners control access to the land, but governments have the legitimate authority to offer fair regulations designed to conserve wildlife for the public and to provide opportunities for all members of the public to enjoy this resource.

4. **Respect for the Resource** – "Wildlife may only be killed for a legitimate, non-frivolous purpose." – Inherent in all wildlife policy is the principle that species are killed to provide resources such as meat or fur, to protect property, or to manage ecosystems. Wasteful killing without a purpose is discouraged.

5. **International Cooperation** – "Wildlife is an international resource." – Conservation of many species requires collaboration across boundaries to protect habitats and prevent unregulated trade.

6. **Equal Access** – "Every person has an equal opportunity under the law to participate in hunting and fishing." – States strive to provide opportunity to all people, regardless of race, age, gender, physical disability, or socioeconomic status.

7. **Scientific Management** – "Scientific management is the proper means for wildlife conservation." – Since the 1930s wildlife scientists have attempted to apply scientific principles to wildlife management, with a focus on ecosystems and habitats, rather than the production of one species.

Figure 2.12. Duck stamps allow hunters to play a role in conservation. The first federal duck stamp was drawn by Ding Darling, Source: U.S. Biological Survey (public domain).

These principles provide a sustainable template for wildlife science and have been widely admired around the world. In recent years, historians have noted that many of these principals are based on values of respect for the natural world associated with Native American traditions, leading to an acknowledgement that indigenous Americans also have much to contribute to our understanding of wildlife management.

HIGHLIGHT 2.2 - LEGAL CATEGORIES OF WILDLIFE IN TEXAS

Most species of wildlife found in Texas receive some level of regulation or protection, either from TPWD regulations, the Migratory Bird Treaty Act, or the Endangered Species Act. The following categories are recognized by TPWD in its state laws and regulations:

Game Animals – Refers to animals (usually mammals) for which hunting regulations, including **bag limits**, seasons, and methods of harvest, have been adopted. The Texas Legislature grants authority to TPWD to adopt regulations, subject to a public comment period. Game animals in Texas include white-tailed deer, mule deer, pronghorn, desert bighorn sheep, javelina, squirrels, and alligators.

Furbearing Animals – Furbearers are those species that may be legally taken, with the proper license and within the proper regulations, for the sale of their hide or pelt. Some species also require a tag to be issued to help control international trade in similar-looking species. Furbearers in Texas include badger, beaver, foxes, mink, muskrat, nutria, opossum, otter, raccoon, ringtail, and skunks.

Game Birds – Refers to birds for which hunting regulations, including bag limits, seasons, and methods of harvest, have been adopted. Games birds are usually divided into two categories—migratory and upland. Migratory birds include waterfowl (ducks, geese, coots, rails, gallinules & moorhens), doves, sandhill cranes, snipe, and woodcock. State hunting regulations for migratory game birds must conform to guidelines offered by USFWS. Upland game birds include quail species, turkey, chachalaca, and the introduced ring-necked pheasant. There are a few game birds, such as prairie-chickens and the Montezuma quail, for which all seasons are closed, essentially making them protected species.

Threatened and Endangered Species – Animals listed as threatened or endangered (T&E) are vulnerable or declining and may not be killed or collected without a special permit. Species designated as T&E by either USFWS or by TPWD are protected in the state. Plants may also be listed as T&E, but do not receive the same level of protection. Species may be added or removed from threatened and endangered lists when changes in status occur.

Nongame Species – These are species of vertebrate and invertebrate wildlife indigenous to Texas that are not classified as any of the above categories. In general, anyone who has a hunting license may collect or harvest nongame species at any time; however, a permit is required to possess more than 25 specimens or to sell these species. Birds receive more protection. The Migratory Bird Treaty Act protects almost all species of birds, preventing killing or collection unless a permit is issued.

Exotic - An exotic animal is any animal that is not indigenous to Texas or regulated by TPWD. Examples include feral hogs, aoudad sheep, axis deer, elk, sika deer, fallow deer, red deer, blackbuck, nilgai antelope, and birds not indigenous to Texas or protected by the federal Migratory Bird Treaty Act, such as emus, European starlings, house sparrows, Eurasian collared-doves, and rock pigeons. There are no bag limits or closed seasons on exotic species; however, landowner permission and a hunting license are usually required.

CHAPTER 3 – BASICS OF ECOLOGY

"It really boils down to this: that all life is interrelated. We are all caught in an inescapable network of mutuality, tied into a single garment of destiny. Whatever affects one destiny, affects all indirectly."
- Dr. Martin Luther King, Jr., 1967 Christmas sermon

"When we try to pick out anything by itself, we find it hitched to everything else in the Universe."
- John Muir, founder of the Sierra Club

To understand wildlife management, one must understand ecology. In the original Greek ***ecology*** means study (*logos*) of the home (*oikos*). Today, the term can be defined as the study of the relationships that living organisms have with each other and with their environment, including both **biotic** (living) and **abiotic** (non-living) components. Ecology is both wildlife's foundation—its home—and the key that links it all together—its relationships.

WILDLIFE ECOLOGY HAPPENS AT MANY SCALES

For wildlife conservation to be successful it must focus on both the "little picture" and the "big picture"—studying both individual animals and their homes or their ecology. There are several levels of ecological classification:

- **Species** – A species is a group of living organisms consisting of similar individuals capable of interbreeding.

- **Population** – A population is a group of organisms belonging to the same species occupying a particular area at the same time.

- **Community** – A community is made up of interacting populations in a natural environment.

- **Ecosystem** – An ecosystem consists of the biological communities of a given area and the physical environment with which they interact. Ecosystems can be large, such as a forest, or small, such as the collection of organisms that exist within a rotting log. It is composed of both living (biotic) and non-living (abiotic) parts.

- **Ecoregion** – An ecoregion is a major ecosystem defined by distinctive geography, climate patterns, and characteristic species of plants and animals. Most wildlife scientists recognize 10-12 major ecoregions in Texas.

- **Biome** – A biome is a larger region stretching across a continent or several continents where abiotic factors produce similar plant and animal communities. Major terrestrial biomes include tropical rainforest, temperate forest, taiga (cold climate forests), tundra, desert, grassland, and **savanna**.

- **Biosphere** - The biosphere includes all the portions of the earth where life occurs, including the abiotic factors that life depends upon.

(opposite) Figure 3.1. Decomposers, such as this inky cap fungus, are important components of energy flow in ecosystems.

Table 3.1. An example of how conservation of pronghorn can happen at several ecological levels.

Ecological Level	Example	Example of Wildlife Conservation Activity
Species	Pronghorn	Gather data on captive pronghorns to determine gestation period.
Population	Pronghorn herd in the Marfa Grasslands	Conduct annual surveys to assess population fluctuation in the Marfa herd
Community	Pronghorns, bobcats, & coyotes in the Marfa Grasslands	Assess predation rates on pronghorn fawns
Ecosystem	Pronghorns, predators, grassland birds, reptiles, and other vertebrates and invertebrates in the Marfa Grasslands occupy an ecosystem characterized by grasses and shrubs, low rainfall, higher elevation, clayey loam soils, basin topography, and occasional droughts and fires.	Assess animal movement and survival in relation to rainfall patterns and grass density
Ecoregion	Trans-Pecos Mountains & Basins – a region with many habitats that have low rainfall and high insolation (levels of sunlight)	Assess whether habitat exists so that pronghorn in the Marfa Grasslands are able to interact and interbreed with herds in other West Texas grasslands
Biome	Great Plains – grasslands that stretch from Canada to Mexico	Work with other states to assist landowners in protecting arid grasslands and pronghorn herds

Figure 3.2. A biologist working to conserve pronghorns in West Texas might have to think about issues at many levels (Table 3.1).

A PERFECT FIT

One of the beautiful attributes of ecosystems is that every species is special in its own way. Different species in ecosystems have unique roles called **ecological niches**. An ecological niche includes an organism's habitat, food source, shelter, activity patterns, predators, and "job" or the role it fulfills. Even within the same habitat, there can be many niches. In a Texas grassland or savanna, quail feed on seeds and plants on the ground, while roadrunners chase down lizards and insects on the ground. Species with similar diets still have different niches - such as owls hunting at night while hawks hunt during the day. Other factors like nesting habits and seasonal presence can also create unique niches. Although species may compete in their niches, no two animals have the exact same niche—each one is distinct!

Some species have very specific needs, whereas others can adapt to variable conditions. **Specialists** are "picky," requiring specific types of food, habitats, or environments. For example, golden-cheeked warblers only nest in Central Texas, almost always using mature Ashe juniper bark for their nests. Texas horned lizards are usually only found where harvester ants are present, as ants make up 90% of their diet. On the other hand, **generalists** can thrive in different conditions and use a variety of resources. Raccoons, coyotes, and northern mockingbirds are examples of generalists. They can adapt to different environments and may be found in a large geographic area, sometimes even becoming pests in some places.

Species adapt to survive in their native environments through a process called **natural selection**. This means that the environment favors individuals with traits that help them survive. For example, cricket frogs in East Texas are predominantly dark brown and green, allowing them to blend in with the dark waters and vegetation. In Central Texas, they are primarily gray to match the limestone streams. Having a diverse set of genes is important for species to adapt well. When populations become very small, genetic diversity can decrease, making it harder for species to cope with changes. Protecting the remaining genetic diversity is crucial for endangered species facing this challenge. The Attwater's prairie-chicken, a species that needs vast open **prairies**, has been affected by low genetic diversity due to a small surviving population. Efforts to conserve this unique prairie-dwelling species along the Texas coast focus on habitat conservation and maintaining a genetically diverse population in zoos.

INTERCONNECTIONS IN ECOSYSTEMS

In ecosystems, species rely on each other for survival. One important aspect is energy flow. Organisms need energy to carry out their biological processes. Energy in most ecosystems starts with the sun. **Producers**, such as plants, can capture sunlight through **photosynthesis**. They convert sunlight, water, and carbon dioxide into glucose and oxygen. Some microorganisms can also make sugars like glucose using chemicals instead of sunlight.

Producers make their own food and store it in their bodies. Other organisms called **consumers** eat the producers to get energy. Some consumers may eat other consumers, resulting in primary, secondary, or even higher levels of consumers. Energy transfer is not perfectly efficient, and about 90% of the energy is lost as heat or waste when it is transferred to the next **trophic** (or feeding) **level**. This means that the **biomass** (or weight) of producers is greater than the biomass of primary consumers, there are more primary consumers than secondary consumers, secondary consumers outnumber tertiary consumers, and so on. This pattern is shown in an **ecological pyramid** (Fig. 3.3).

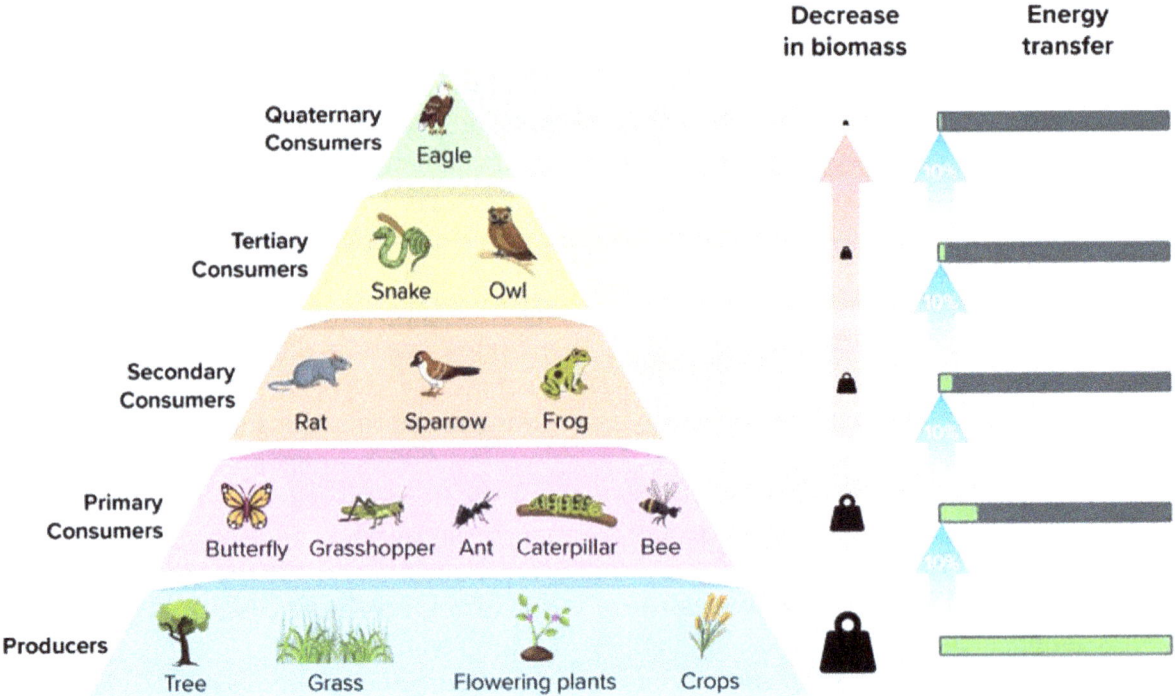

Figure 3.3. Ecological (or trophic) pyramid. Source: CK-12 Foundation, CC 4.0.

Sometimes animals are labeled by what they eat. Animals that only eat plants are called **herbivores.** Herbivores are considered primary consumers. **Carnivores,** on the other hand, eat only other animals and are secondary or later consumers. **Omnivores**, who eat both plants and animals, can act as both primary and secondary consumers. **Decomposers** eat or break down dead organisms. It is important for wildlife scientists to understand what different animals eat, and this is often depicted in a **food chain**. In a Texas savanna habitat, for example, leafhoppers feed on native plants like western ragweed, bobwhite quail chicks eat leafhoppers, Cooper's hawks can eat quail, raccoons may eat Cooper's hawk eggs, golden eagles may feed on raccoons, and when the golden eagle dies, a turkey vulture will feed on the carcass. In a food chain, the animal being eaten is called the **prey**; the animal hunting and eating it is the predator.

In any ecosystem, there are many food chains because there are many different species, and most species eat more than one type of food. For example, raccoons and snakes can also eat bobwhite quail eggs, and golden eagles can eat snakes. Bobwhite quail and mourning doves eat seeds from western ragweed. Cooper's hawks prey on mourning doves, and great-horned owls often hunt raccoons. Turkey vultures and black vultures eat almost any dead animal. All these connections create a complex network called a food web (Fig. 3.5), showing how animals' diets and actions can affect many other animals and plants in the ecosystem.

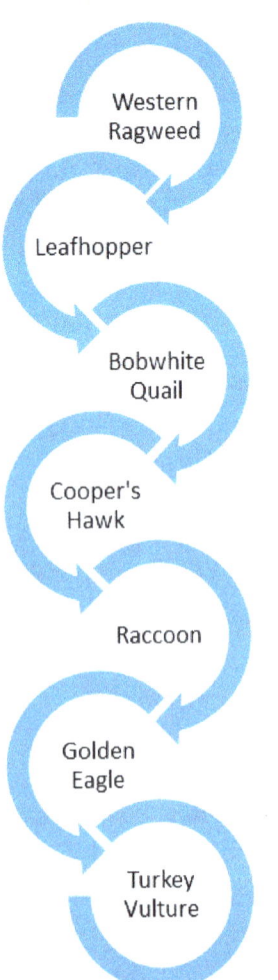

Figure. 3.4. Example of a Texas savanna food chain.

28 | Chapter 3 – Basics of Ecology

Figure 3.5. A comparison of a food chain versus a simplified food web. Source: Shutterstock.

Trophic pyramids, food chains, and food webs are types of ecological models. A **model** is a representation or simulation of a real-life phenomenon, organism, or process. Ecological models, often called concept-process models because they illustrate processes such as energy flow, can help scientists predict how changes in one part of a system may affect other parts of the system. In some cases, mathematical formulas can be applied to models so that specific outcomes can be predicted.

In addition to food relationships, species interact in other ways. In some circumstances, it is a fight for existence. In other relationships, the connection is so specific that one species cannot exist without the other. When the life history of two organisms is linked by their location and interaction, ecologists refer to the relationship as **symbiosis**. Symbiosis may be beneficial to both species, but it is not always. Some examples of symbiosis include:

- **Mutualism** is a partnership between two different species that benefits both. There are many examples of mutualism in ecosystems, such as pollinators like insects and hummingbirds that visit flowers for food and help plants reproduce at the same time. Similarly, certain animals, like white-tailed deer and desert bighorn, have helpful bacteria in their stomachs that aid digestion. Many plants rely on special fungi called mycorrhizae to enhance the absorption of water and nutrients from the soil. In return, these fungi receive nutrients from the plant roots. Mutualism is a win-win situation in nature.

- **Commensalism** – Commensalism is an association between two species in which one species benefits, and the other is neither benefited nor harmed. Brown-headed cowbirds are a species of songbird that once associated with bison, feeding on insects stirred up by the moving grazers. The cowbirds were benefitted by bison, but the bison did not receive any benefit. In modern Texas, cowbirds, true to their name, are often found around cattle.

Figure 3.6. Biologists have recorded mutualistic relationships between badgers and coyotes hunting together in some habitats. The coyotes chase down prey in open areas, whereas the badgers dig for prey such as ground squirrels that take shelter in burrows. Both predators may then share the meal. Photos: USFWS (public domain).

- **Competition** occurs when two or more organisms need similar limited resources such as nutrients, space, or light. Competition can be between individuals within the same species (intraspecific) or between different species (interspecific). Examples of intraspecific competition include male bullfrogs defending their territory from other male bullfrogs. Woodpeckers competing with squirrels for acorns is interspecific competition. Plants also compete with other plant species for space, nutrients, and light, sometimes using chemical warfare. Competition can lead to the dominance of individuals or species with better adaptations.

- **Exploitation** refers to a relationship where one organism benefits while another is harmed. Examples include predators hunting prey, animals consuming plants, and parasites feeding on other organisms. These relationships can affect the balance of species in an ecosystem. However, it is interesting to note that natural exploitation usually does not lead to species extinction, because both species have adaptations that help them survive. For instance, birds of prey like red-tailed hawks have excellent eyesight to spot their prey from a distance, but potential prey like cottontail rabbits have features that help them blend into their surroundings and evade detection. Many Texas plants have developed toxic substances to defend against herbivores. Freshwater mussel larvae act as parasites on fish, but the infected fish can become more resistant to future parasite infections. Cowbirds engage in an exploitative relationship called nest parasitism. Chapter 12 describes this interesting interaction between cowbirds and other songbirds, along with how both the parasite and hosts have adapted.

SPECIES TO WATCH FOR

Certain species are especially important in setting conservation goals for ecosystems—these are species that wildlife managers want to "watch for." **Indicator species** are one such group. Indicator species reflect specific environmental conditions, providing cues to the wildlife scientist. For example, bushy bluestem grass is an indicator of wetland soil; four-wing saltbush indicates salty soils. The Houston toad and the large-fruit sand-verbena, both endemic to Texas, can only be found in deep sands habitats. Conservation of these soil types in the Post Oak Savanna may therefore be a high priority. Additionally, some species serve as sentinel species, like the fountain darter and Texas wildrice. The presence of these species indicates good water quality. Amphibians and freshwater mussels are also considered sentinel species due to their sensitivity to water quality.

Keystone species are also important to identify in ecosystems. Keystone species are species that significantly influence ecosystem structure, composition, and function through their activities. The term comes from the name for the stone placed at the top of an arch. The keystone keeps all the other stones in place and supports the arch structure. If a keystone species is removed, the ecosystem could change drastically or even collapse. These changes are often referred to as "cascade effects"—a rippling impact on other species from the decline or disappearance of one species. Several types of keystone species have been recognized.

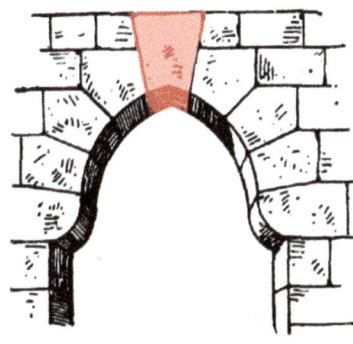

Figure 3.7. A keystone in a doorway arch. Source: Otto Lueger, WC (public domain).

- **Apex predators** (predators at the top of the food chain) have an important role in maintaining balance in ecosystems. For example, when gray wolves were reintroduced into Yellowstone Park, they helped reduce coyotes and overpopulated elk herds, allowing streamside habitat and species like trout, songbirds, beavers, and scavengers to thrive. In other habitats, coyotes can have an important role. In some habitats, when coyotes are removed the number of middle-level predators (**mesopredators**) like raccoons and foxes can increase, negatively impacting ground-nesting birds like quail.

- **Pollinators** can play a vital keystone role in ecosystems. These mutualists help in the transfer of pollen from one plant to another, allowing plants to reproduce. In addition to insects and hummingbirds, some mammals, including the Mexican long-nosed bat, also act as pollinators. In West Texas, these bats use their unique bristled tongues to collect nectar from agave plants. This process helps pollinate the agave, which benefits many other animals and even humans who rely on it as a resource.

- **Ecosystem engineers** create, modify, maintain, or destroy habitats, thus affecting other species. Examples include American beavers building dams that create pools used by many wetland species. Pileated woodpeckers create large holes in dead and dying trees, aiding nutrient cycling and providing shelter for many species. Harvester ants disperse seeds and improve soil conditions. The burrows of prairie dogs also have a significant impact on soils and other species (Highlight 3.1).

Other roles of keystone species have been recognized. Herbivores, such as bison, white-tailed deer, or muskrats, may change the vegetative structure of communities. Large populations of Brazilian free-tailed bats deposit guano that provides nutrients for a wide variety of cave organisms. Scavengers such as vultures help to remove dead animals from the environment and facilitate the recycling of nutrients.

Finally, focusing on **umbrella species** can also be an effective conservation strategy. Umbrella species have extensive habitat needs; conserving them helps protect many other species in the ecosystem. For example, in West Texas, expansive habitats that support black bears and mountain lions support a higher diversity of species than areas without these large carnivores. In South Texas, conserving tropical woodland habitat for the endangered ocelot also benefits other rare species like chachalacas, indigo snakes, Texas tortoises, and the more common javelina. In East Texas, managing fire-maintained forests for the endangered red-cockaded woodpecker provides habitat for a variety of bird species, the Louisiana pinesnake, and Texas trailing phlox, a rare plant. Even plants like Texas wildrice can be considered an umbrella species, as maintaining river flow for its conservation supports other rare species like the fountain darter, San Marcos salamander, Cagle's map turtle, and even whooping cranes downstream.

HIGHLIGHT 3.1 - PRAIRIE DOGS – KEYSTONE SPECIES OF THE PLAINS

Five species of prairie dogs were once widespread throughout the Great Plains of the United States, even extending into the deserts of the Southwest. These burrowing rodents live in colonies or "towns" that are made up of family groups called coteries. Within the towns, the coteries dig extensive burrow systems that may extend as deep as 4.5 meters (15 feet). Burrows contain many "rooms" or separate areas for raising young, listening for predators, and even toileting.

Prairie dogs are primarily grazers, with grass and other leafy vegetation making up 98% of their diet. Within the town, prairie dogs graze grass to a height of less than 15 centimeters (6 inches) to give them a better lookout for predators. When a sentinel—a lookout prairie dog—spots a predator, it gives a series of sharp whistles, or "barks," to warn the prairie dogs to return to their burrows.

Prairie dogs have a strong influence on both plants and animals in the environment, with more than 160 species depending on prairie dog towns. Their grazing activity provides open habitat for species such as mountain plover and limits the spread of invasive woody vegetation like mesquite. Their burrowing activity increases nutrients in the soil, benefiting pollinators and grazers such as bison and pronghorn. Numerous animals, like burrowing owls, badgers, reptiles, amphibians, and other rodents, seek shelter in prairie dog burrows. Prairie dogs are prey for coyotes, swift foxes, ferruginous hawks, and especially for the black-footed ferret, which feeds almost exclusively on prairie dogs.

Black-tailed prairie dogs once inhabited a large area in the western United States, including vast acreages in Texas. However, they were considered pests and were reduced by hunting, poisoning, and disease. As a result, their population decreased by 99%. This decline also caused other species like the black-footed ferret, mountain plover, and burrowing owl to be at risk. Wildlife scientists in Texas are now working with landowners to study the remaining prairie dog populations and explore the possibility of reintroducing black-footed ferrets to restore this ecosystem.

Figure 3.8. Prairie dogs, a keystone species of the plains, giving warning calls to other colony members.

A RENEWABLE SYSTEM

The Law of Conservation of Mass states that the matter that makes up the earth can neither be created nor destroyed. However, the atoms and molecules that make up matter can be rearranged, resulting in some amazing cycles that shift and renew ecosystems in an ultimate recycling system. During this process, matter can end up in four primary reservoirs or "spheres." The biosphere is made up of all the living organisms on Earth. The **geosphere** is composed of all the rocks and minerals making up the Earth's crust. The **hydrosphere** is composed of the water in the Earth's oceans, rivers and lakes. The **atmosphere** is composed of all the gases making up the Earth's air. Because these transformations move chemicals through both living things such as plants and non-living components such as rocks, they are called **biogeochemical cycles.** The water cycle, carbon cycle, nitrogen cycle, and phosphorous cycle are some of the models that have been developed to describe these processes.

THE WATER CYCLE

Water is key to life on Earth. It transports nutrients and gases in plants and animals. It helps maintain internal body temperatures. It is key to reproduction in animals and photosynthesis in plants. It supports the body structure of organisms like earthworms, jellyfish, and plants and makes up 60% of the human body. It provides habitat for fish and many other organisms. It covers almost three-quarters of the globe's surface.

Earth contains about 185 quadrillion gallons of water. That's a lot! At any one time, the vast majority (97%) is salt water in the oceans. Of the 3% that is freshwater, more than two-thirds is found in glaciers and ice caps. About one-third is found in groundwater or stored in the soil. Only a miniscule amount of water on earth is found in rivers or lakes. An even smaller percentage can be found in the atmosphere or in the living organisms of the biosphere. Although these percentages represent the average distribution of water, water molecules can move from one place to another in a process represented by the water cycle.

One can start to examine the water cycle with evaporation, the process whereby the sun warms water on the Earth's surface and changes it into water vapor in the atmosphere. Living organisms can also move water to the atmosphere. In respiration animals exhale water vapor. In transpiration

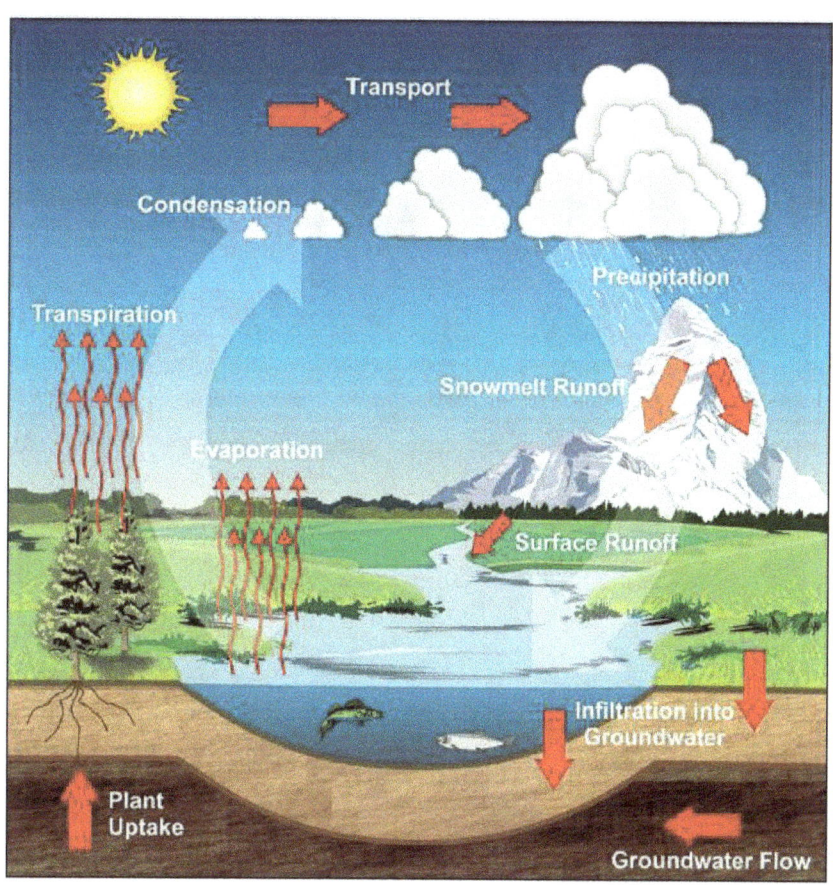

Figure 3.9. The water cycle. Source: Atmospheric Infrared Sounder, CC 2.0.

plants release water vapor into the atmosphere through microscopic pores in the plant's leaves and stems.

As water vapor rises in the atmosphere, it cools and changes back into liquid, forming clouds or fog through the process of condensation. When water vapor condenses onto small particles in the atmosphere, such as dust specks, it can fall from the sky as precipitation. Condensation and precipitation move water from the atmosphere to the hydrosphere and geosphere.

Precipitation falling on land can take several routes through an ecosystem. It can be taken up by organisms. It can percolate through soil and other porous surfaces. In the soil it can be held by the soil, absorbed by the roots of plants, or collect in aquifers—underground pools in the pores and crevices of bedrock. If precipitation falls to the ground faster than it can soak in, the water becomes runoff. Gravity pulls the runoff downhill, where it can collect in streams, rivers, lakes or oceans. Some of that water eventually returns to the atmosphere, and the whole cycle begins again.

THE CARBON CYCLE

Carbon is the fourth most abundant **element** in the universe and the second most abundant element in human bodies (after oxygen). Some parts of the carbon cycle are very slow. Limestone rocks containing calcium carbonate slowly form from ocean creatures. Only the extreme temperatures associated with volcanoes can release carbon from rocks, forming carbon dioxide (CO_2) in the atmosphere. Other parts of the carbon

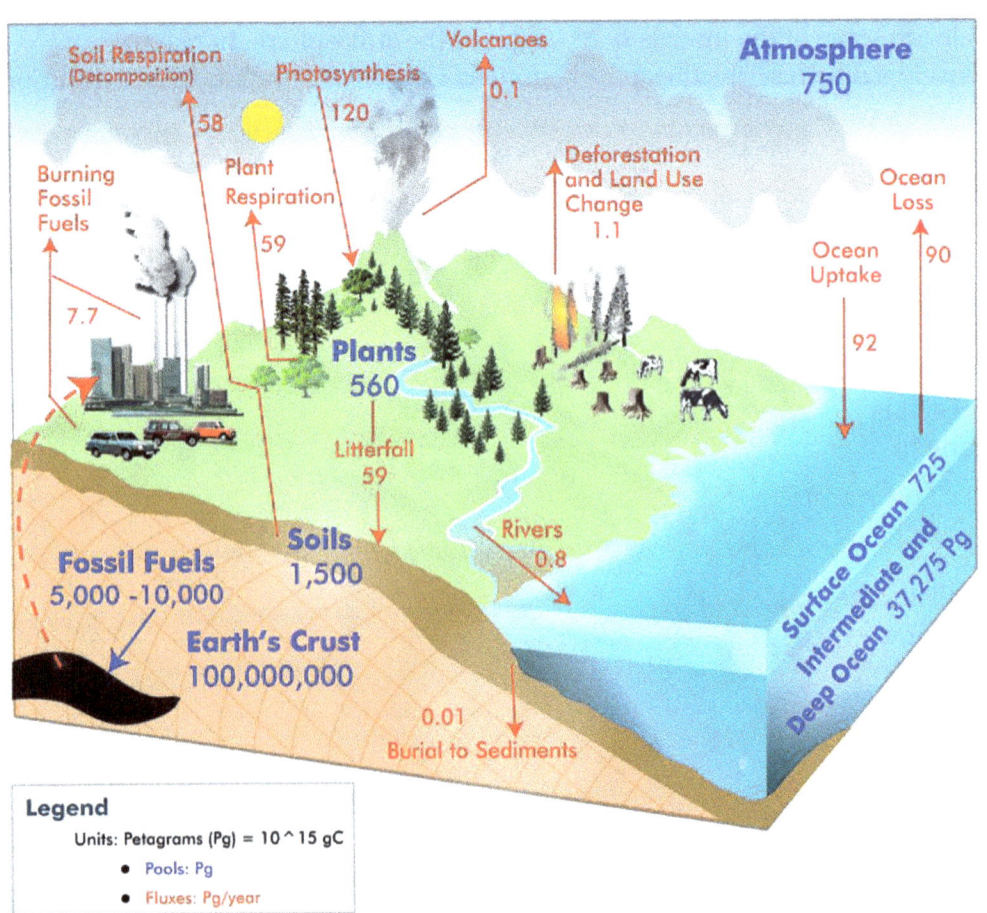

Figure 3.10. Most of the carbon on earth is stored in the earth's crust and oceans (blue numbers); however, the movement of carbon between other sources (red numbers) greatly affects life on earth. Source: NASA (public domain).

cycle are very dynamic. This biological portion of the carbon cycle is closely tied to the flow of energy through ecosystems. Plants and other producers absorb CO_2 from the air or water and convert it into carbohydrates through the process of photosynthesis. When a consumer eats a plant, the carbon in carbohydrates begins to be transferred through food chains. Carbon leaves the biosphere through cellular respiration, when organisms breathe out CO_2 or excrete methane into the atmosphere. Carbon can also enter the atmosphere through the burning or decomposition of organisms. Then the cycle begins again.

THE NITROGEN CYCLE

Nitrogen is abundant on Earth. It makes up 78% of our atmosphere. Nitrogen is also an essential component of living tissues, and farmers know how important nitrogen is to plant growth. Unfortunately, atmospheric nitrogen (N_2) is not usable by plants. Instead, to enter the biosphere N_2 must be converted into other nitrogen **compounds** such as ammonia or nitrates. This process, called nitrogen fixation, can happen in three ways:

- Bacteria living in the soil or in symbiosis with plants convert atmospheric nitrogen into ammonia.

- High-energy natural events, such as lightning, fire and volcanic eruptions, convert small amounts of nitrogen gas into nitrates.

- Humans combine nitrogen gas with hydrogen to make nitrogen-based fertilizers.

Once N_2 has been fixed into ammonia or nitrates, plants can use those compounds to build their tissues. From there, nitrogen can travel through food chains to other organisms. Organisms excrete ammonia in their waste products back into the environment. When organisms die, decomposers also convert the nitrogen compounds in their bodies back into ammonia. Ammonia sticks to soil particles and stays put. In contrast, denitrifying bacteria in water or the soil can convert nitrates into nitrogen gas. This process, denitrification, removes nitrogen from the biosphere, geosphere, and hydrosphere and returns it to the atmosphere.

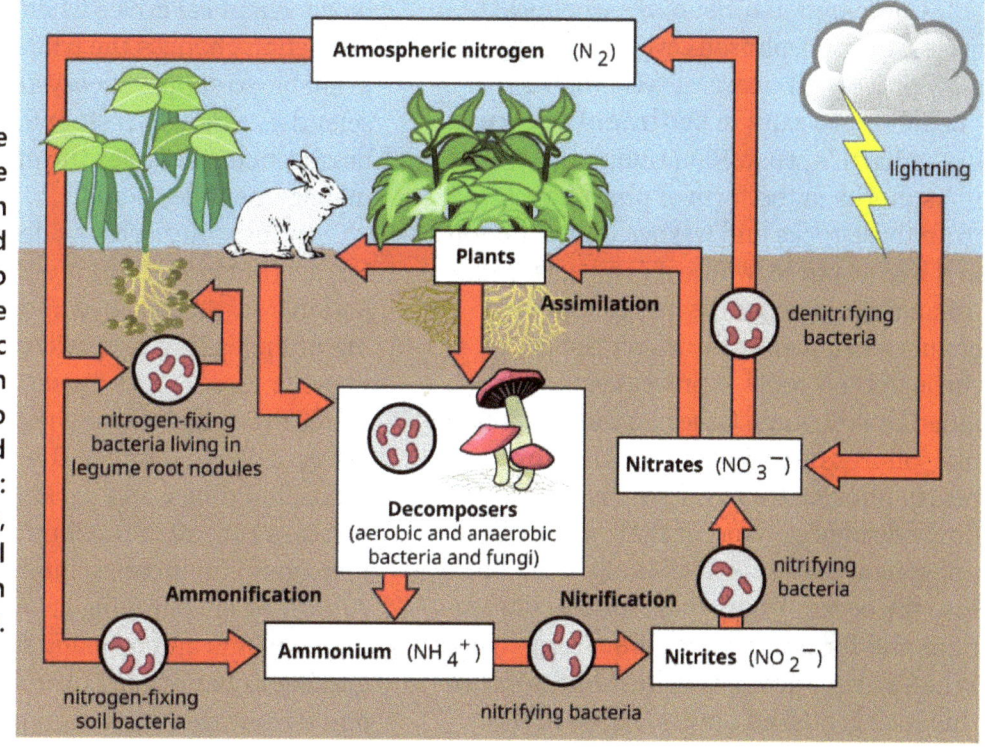

Figure 3.11. The nitrogen cycle relies upon lightning and bacteria to make atmospheric nitrogen available to plants and animals. Source: Johann Dréo, Environmental Protection Agency, CC 3.0.

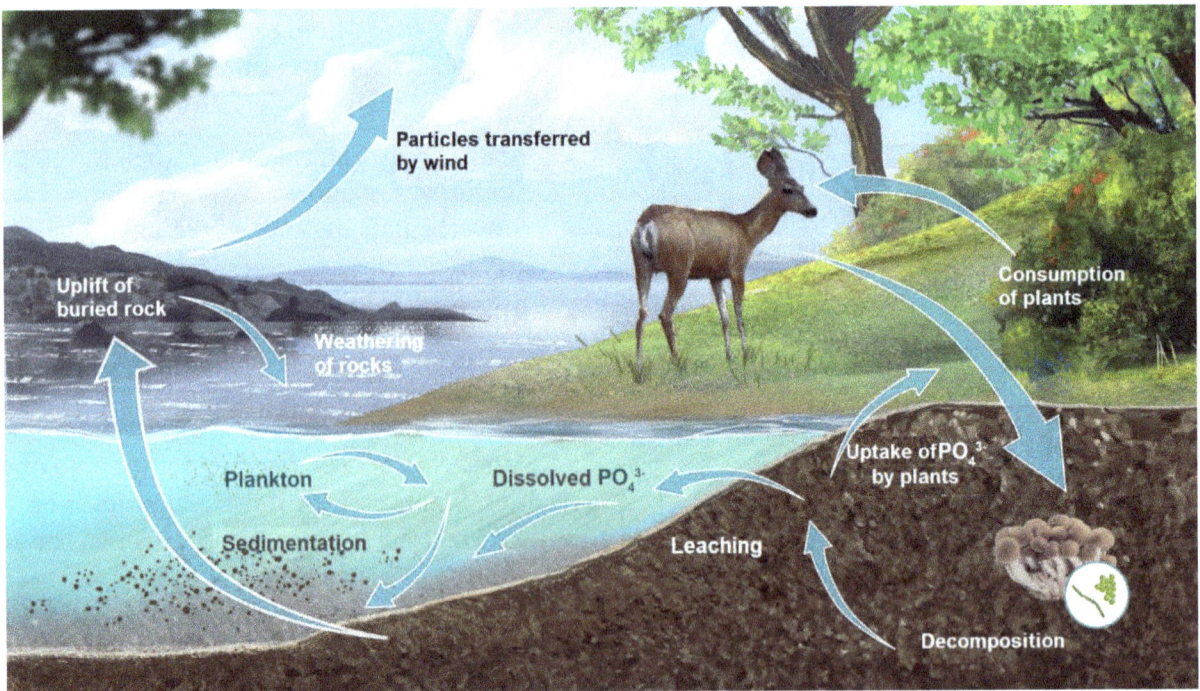

Figure 3.12. The phosphorus cycle. Source: Englishsquare, CC 3.0.

PHOSPHOROUS CYCLE

Although phosphorus makes up less than one percent of most organisms, life could not exist without it. Phosphorus is a part of DNA, RNA, and the energy molecule ATP. It is a structural component of cell membranes, bones, and teeth.

Compared to other biogeochemical cycles, the phosphorous cycle is a relatively slow process. The largest reservoir of phosphorus occurs in **sedimentary rocks** in the Earth's crust. Rain removes phosphorus in the form of phosphate (PO_4^{3-}) from these rocks and washes it into the soil and hydrosphere, where it can be used by organisms. Phosphorus moves into food chains when plants absorb phosphate through their roots, a process sometimes aided by soil fungi. Other organisms obtain phosphorus by eating plants. Decomposers return phosphates to the soil by breaking down dead organisms or their waste. All phosphate molecules eventually wash back into the ocean, where they settle to the sea floor and form layers of sedimentary rock. When Earth's geologic plates shift, some of this rock is pushed upward. Rock at the surface undergoes weathering and erosion, and the cycle begins again.

RESTORING BIOGEOCHEMICAL CYCLES

Although the elements in these cycles are never destroyed, human activities can disrupt biogeochemical cycles when pollution or extraction changes the balance of compounds available to support ecosystems. Wildlife scientists can help restore the balance of biogeochemical cycles through their land management decisions.

In the water cycle, widespread cutting of forests (deforestation) reduces transpiration and increases the speed and amount of run-off, often leading to erosion and pollution of water bodies. Forest conservation and restoration can reverse these effects. Overgrazing and impervious cover, such as roads, buildings, and parking lots, also increase run-off and pollution. Wildlife managers can increase infiltration and slow run-off by encouraging growth of deep-rooted native plants and by using fire and moderate grazing to remove or control brushy growth and encourage **perennial** grasses.

Humans also impact the carbon cycle. Hydrocarbons are fuels such as oil, coal, and natural gas that are formed from the living organisms buried beneath the Earth's surface. When these "fossil fuels" are burned, carbon that has been stored underground for many years is suddenly released into the atmosphere as carbon dioxide. Though CO_2 makes up only 0.04% of the Earth's atmosphere, it has a profound impact on the Earth's climate. Large gas molecules such as CO_2 naturally trap the sun's radiant energy at the earth's surface, an important life-sustaining process known as the **greenhouse effect**. However, climatologists worry that rapidly increasing levels of CO_2 and other human-generated gases are causing rapid changes in the heat-trapping capability of Earth's atmosphere, resulting in a set of influences referred to as **climate change**.

Wildlife scientists can play a beneficial role in influencing the carbon cycle. When carbon is trapped in soils and native vegetation, it helps offset increases in the amount of carbon dioxide in the atmosphere. Some environments, such as **grasslands** and forests, are especially important in serving as a storage area, or "sink," for carbon compounds. Forests store carbon within tree biomass above ground, and grasslands provide stable storage of carbon underground in root systems and deep soils. Other habitats, such as wetlands, are especially prone to release greenhouse gases when destroyed; therefore, wildlife scientists can work to promote their stability.

Similarly, many human activities affect nitrogen and phosphorus cycles. Fertilizers contain high levels of nitrogen and phosphorus. When fertilizers are over-applied in residential areas or agricultural systems, nutrients enter water bodies where they can cause rapid growth (blooms) of algae and plants, a process called **eutrophication**. When these aquatic producers die, their decomposition removes oxygen from the water, often resulting in oxygen shortages for other organisms like fish. Extremely high levels of nutrients have even created "dead zones" in major water bodies, such as the Gulf of Mexico, also called the Gulf of America (the Gulf). These impacts can be lessened when wildlife managers restore native grasslands and wetlands to capture and slow run-off or when homeowners make choices to promote native plants that do not require fertilizers.

Chapter 3 – Basics of Ecology | 37

HIGHLIGHT 3.2 - BREAKING THE LEAD CYCLE

Why would TPWD biologists sit around a big table at the J.D. Murphree WMA with knives in hand and plates full of raw duck gizzards? No, it wasn't a somewhat unsavory feast. It was an investigation into a human-influenced chemical cycle, and the biologists were searching inside the gizzards for tiny flecks of lead.

Lead is an element that naturally occurs at low levels in the Earth's crust. It is not usually recognized as a biogeochemical cycle; however, humans have introduced lead into the soil, water, and air in a variety of ways. Lead was once added to gasoline and emitted into the atmosphere during the burning of fuel. Paints used to contain lead that could enter the soil through paint flecks and dust. Lead was once used in plumbing, leading to instances of lead entering water supplies. Over time, scientists recognized that this increased exposure to lead could cause brain damage and other serious health effects in humans, especially children, and many uses of lead were phased out.

Ironically, hunters, who enjoy the natural environment and support habitat conservation, contributed to the lead cycle. Until 1991 nearly all shotgun shells were filled with pellets made of lead. When hunters shoot at ducks, geese, doves, or other birds, the small pellets from the shotgun shells fall onto the ground, building up in heavily hunted areas. It was estimated that 700 tons of lead shot fell on hunting areas on the Texas coast from the early to mid-1900s. Concentrations of lead shot in some heavily hunted areas were as high as 1.5 million lead pellets per acre, and some scientists estimated that 1.9 billion lead pellets could be found in the soils of the Chenier Plain NWR Complex on the upper Texas coast alone.

These lead pellets are often accidentally eaten by birds. When waterfowl, doves, and many other birds forage for seeds they also swallow pieces of grit and gravel. The seeds and pebbles end up in the gizzard—a muscular pre-stomach that has the job of using grit to grind up tough food items. To a bird, a lead pellet feels like a seed or pebble, so the birds swallow the pellets as well.

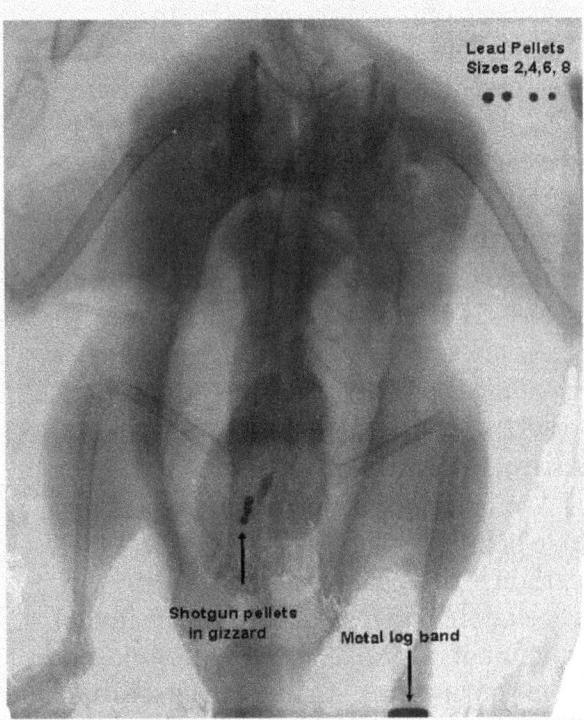

Figure 3.13. An x-ray reveals lead pellets in the gizzard of a spectacled eider duck. Source: P.L. Flint, WC (public domain).

As the lead pellets are ground in the gizzard, the lead is dissolved and absorbed into the body. Then the toxicity begins. Even a small amount of lead in the bloodstream can cause irreparable bone, nerve, and muscle damage, almost always leading to death. One of the first reports of lead poisoning in waterfowl came from Texas in 1894. Studies in the 1960s found that over 95% of mallards ingesting lead died from lead poisoning. By the 1980s those TPWD biologists examining gizzards found that 15-20% of the duck gizzards contained lead. Surveys by USFWS during that time estimated that one to three million ducks and geese died each year due to lead poisoning.

Lead poisoning is a slow and painful way to die. Birds become anemic and weakened, lose weight, and suffer paralysis and convulsions. Birds with lead poisoning often appear very emaciated and have greenish diarrhea. Many birds starve to death or may drown because they cannot hold their head above water. Weakened birds become easy prey for predators. Waterfowl are not the only species affected by lead poisoning. Up to 130 species have been found to be impacted by lead ammunition in the environment. Lead can cycle its way through the food chain and has been found in raccoons, mink, and birds of prey.

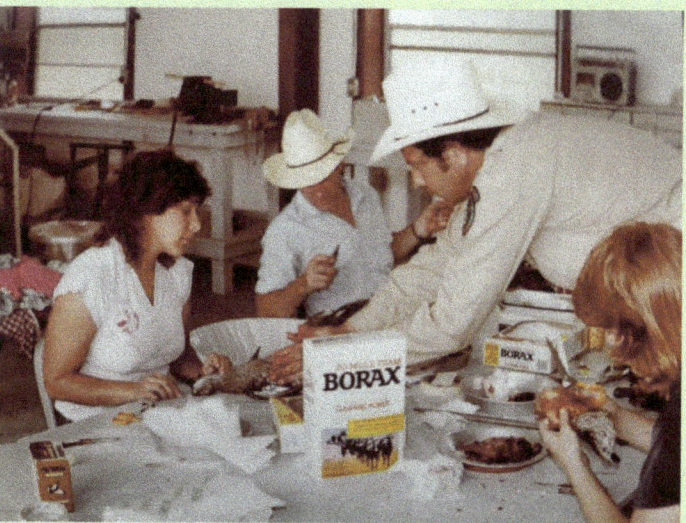

Figure 3.14. TPWD biologists examining duck gizzards. Photo: Charles D. Stutzenbaker.

Texas was a leader in the national response to reduce lead poisoning in waterfowl. TPWD began by sponsoring research regarding the effectiveness of non-toxic shot, such as shotgun shells with steel, copper, or tungsten pellets. Then, in the late 1970s, lead shot was banned for use in some waterfowl hunting areas along the upper Texas coast. Non-toxic-only waterfowl hunting zones were expanded statewide in the 1980s. Finally, in 1991, federal regulations imposed a nationwide prohibition on the use of lead shot to hunt ducks or geese. The changes made a huge difference for waterfowl. Overall, lead ingestion rates dropped by one-half to two-thirds. In Texas, ingestion rates in mottled ducks—a species whose stronghold is the Texas coast—dropped from nearly 50% to less than 10%.

Figure 3.15. Mottled ducks live year-round on the coasts of Texas and Louisiana, making them vulnerable to consuming lead pellets that were deposited over many hunting seasons.

Challenges still remain in the lead cycle. Lead persists in the environment from all those years of hunting, and, as a result, waterfowl continue to find decades-old lead pellets in wetlands. Outside of wetlands, Texas wildlife management areas and national wildlife refuges have begun to require dove hunters to use non-toxic shot, but dove hunting at other sites still deposits 35 billion lead pellets into the environment each year. Hunters also use lead slugs in deer hunting, and fishermen use lead weights in fishing. These sources of lead can also cycle through the ecosystem. In a recent nationwide study, lead fragments were found in 14% of bald eagle carcasses and 12% of golden eagle carcasses. A 10-year study in Wisconsin found that 12% of bald eagles found dead had died from ingesting lead. It's going to take more work to "get *all* the lead out."

CHAPTER 4 – THE PHYSICAL ENVIRONMENT OF TEXAS

A PHYSIOGRAPHIC PLATFORM FOR DIVERSITY

Texas is huge, encompassing 261,232 square miles of land and 7,365 square miles of water within its land boundaries. It stretches 770 miles from the extreme westward point of the Rio Grande near El Paso to the extreme eastern point of the Sabine River in Newton County. North to south, it stretches 801 miles from the northern edge of the Panhandle to the extreme southern turn of the Rio Grande on the border between Cameron County and Mexico.

Texas is diverse, too. Altitude ranges from broad expanses of sea-level wetlands along the Gulf Coast to 8,749 feet on the top of Guadalupe Peak in West Texas. Rugged mountains and desert basins are found in West Texas. Flat wind-swept high plains bounded by the spectacular canyons of the Caprock Escarpment dominate the Panhandle. Rolling prairies and scattered woodlands dominate the central part of the state, uplifted and eroded along the Balcones Fault that traces just west of Interstate 35. East and south of the fault, grassland, forest, and savanna landscapes are gently rolling as they slope and flatten to the **coastal plain** along the Gulf and the subtropical climates of the Lower Rio Grande Valley. The physical geography, or **physiography,** of Texas supports an abiotic diversity of geology, climate, and soils that produce a biotic diversity of plants and animals reflected in Texas's ecoregions.

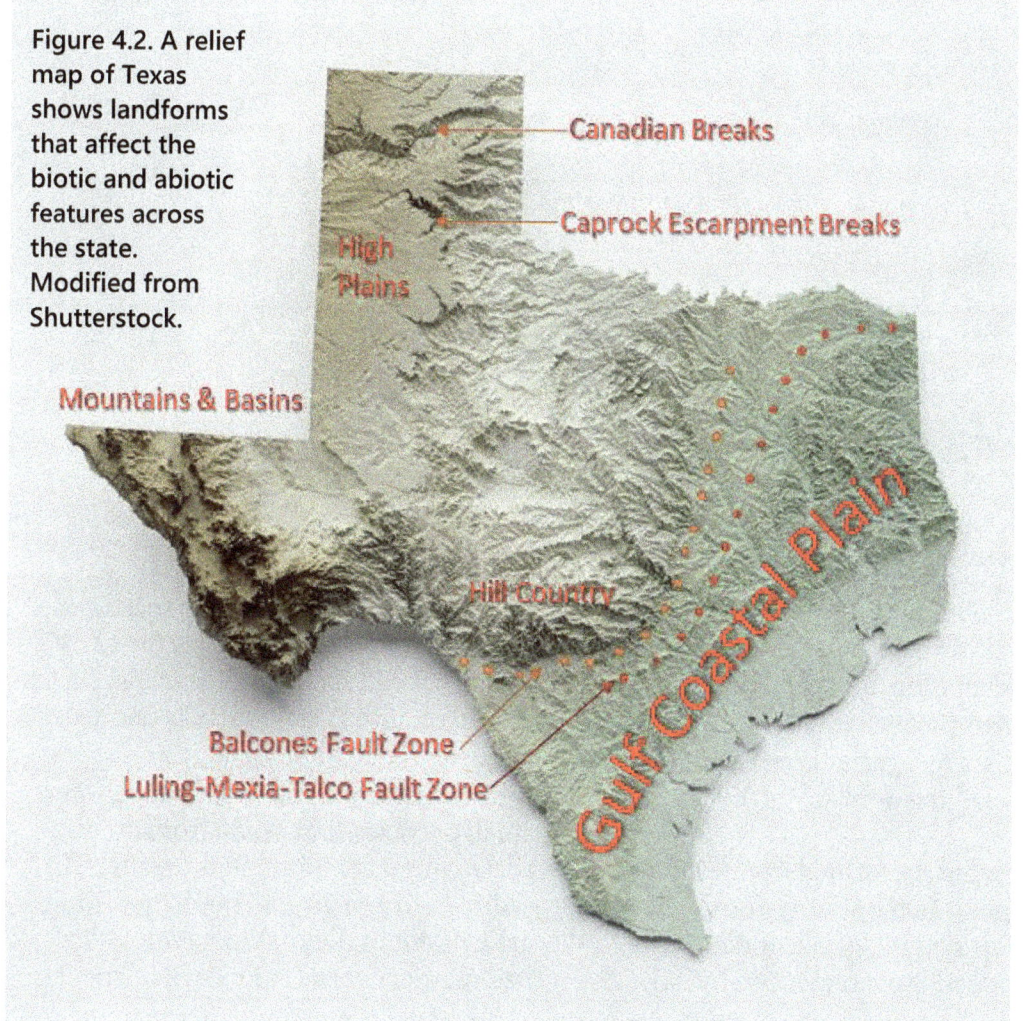

Figure 4.2. A relief map of Texas shows landforms that affect the biotic and abiotic features across the state. Modified from Shutterstock.

(opposite) Figure 4.1. The colorful formations of Palo Duro Canyon were shaped by erosion along the Caprock Escarpment. The various layers reveal at least four periods in the geologic history of Texas. Photo courtesy TPWD © 2025, (Earl Nottingham, TPWD).

Chapter 4 – The Physical Environment of Texas | 41

HIGHLIGHT 4.1 - TEACHING TECHNIQUES - TOPOGRAPHIC MAPS

Topography is the arrangement of the natural and man-made features of an area. Relief maps, as shown in Figure 4.2 above, create an interesting display of the various elevations of the land, but greater details can be found in topographic maps. The topographic map centering on Del Rio and Val Verde County shown below provides information about roads, streams, man-made structures, vegetation cover, and elevation. Each brown line represents a different elevation level. The closer the lines are together, the steeper the topography is. Topographic maps at different scales are available for the entire United States from the U.S. Geological Survey.

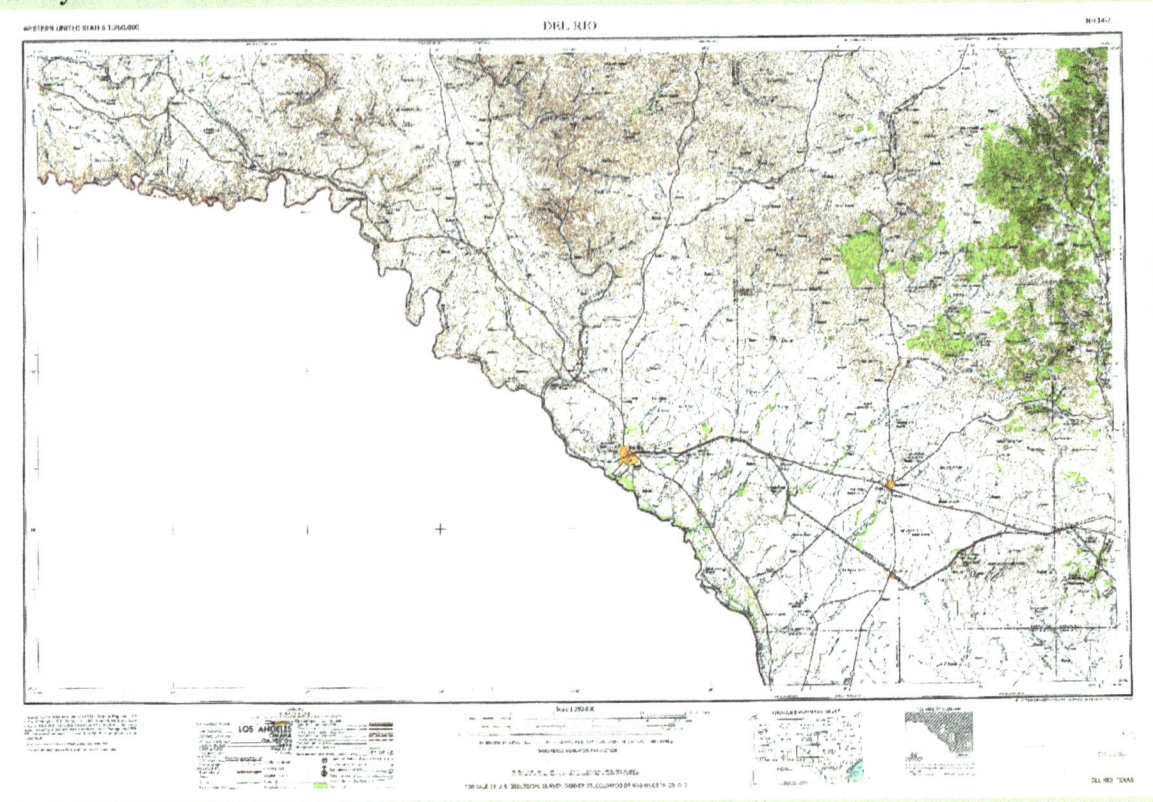

Figure 4.3. Topographic map for Del Rio, Texas. Source: US Geological Survey (public domain).

TEXAS GEOLOGY

Minerals in the Earth's crust (its outer layer) are constantly being shaped by erosion, deposition, cementation, heat, and pressure in a process called the rock cycle (Fig. 4.4). As a result, geologists recognize three main rock types that make up the **geology** of Texas:

- **Igneous rocks** are formed from cooling molten magma. **Intrusive igneous rock**, such as granite, forms under the earth's crust and is characterized by large crystals. **Extrusive** igneous rock forms when magma or lava reaches the earth's surface. It can have small crystals, such as basalt, or air bubbles, such as pumice.

- **Sedimentary rocks** are formed when small particles that erode from other rocks compact and cement together. In sedimentary rocks like sandstone, grains are evident in the rock structure. Sedimentary rocks composed of smaller

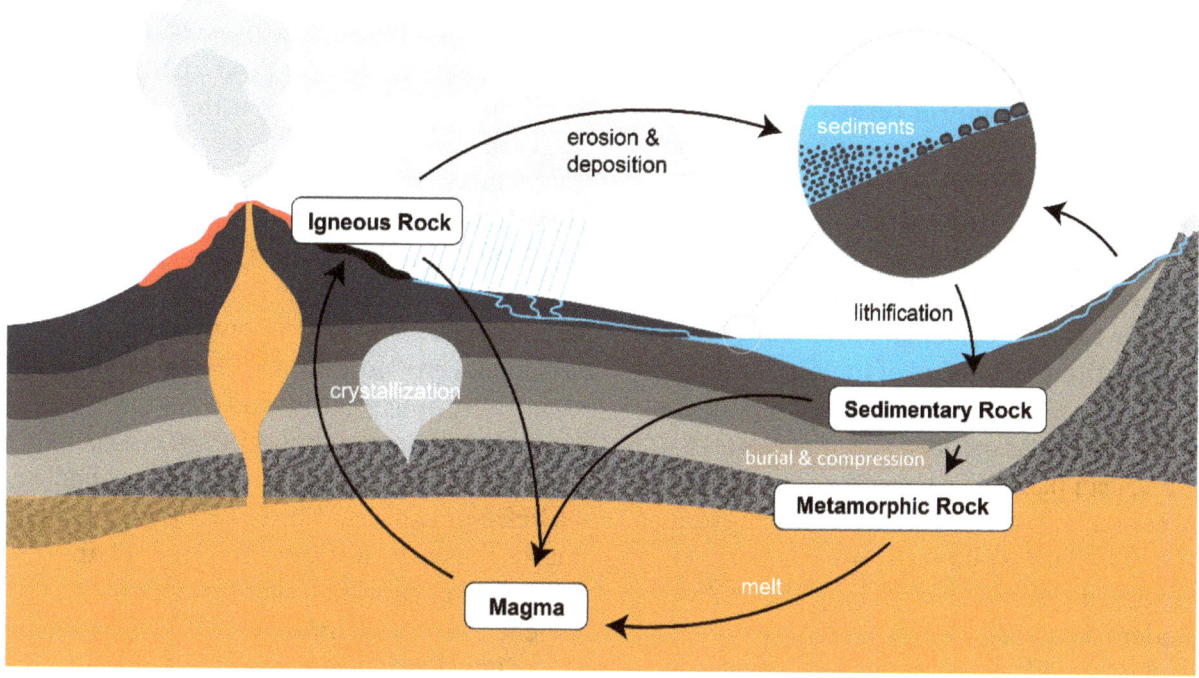

Figure 4.4. The rock cycle. Source: Carie Frantz, CC 4.0.

particles, including shale or limestone, have a smooth, solid appearance. Sedimentary rocks can be porous and make up some important aquifer areas in Texas.

- **Metamorphic rocks** are formed when heat or pressure transforms other types of rock. This process can result in rocks that are dense with flattening or distortion of crystals. Familiar transformations include the formation of marble from limestone, the formation of slate from shale, and the formation of gneiss from granite.

The rock cycle allows these three types of rocks to transform and reform over time. Rocks can be worn down by erosion, but sediments may form back into rocks in a process called lithification. Pressure and heat transform sedimentary and igneous rocks into metamorphic rocks. As rocks are buried further under the earth they may even melt back into lava, which in turn can cool into igneous rock, allowing the cycle to continue.

The geology of Texas is diverse and reflects a dynamic history of lithification, metamorphism, tectonic movements, sedimentation in ancient seas, igneous volcanic eruptions, earthquakes, and erosion by water and wind. This dynamic geological history produced a variety of minerals, laid the foundation for a range of ecosystems from deserts to forests, and contributed to the state's rich mineral and fossil resources.

The earth's crust is divided into large plates that rest on top of the semi-molten magma below. Texas is underlain by the Texas Craton, a portion of the original North American plate. Ancient igneous and metamorphic rocks of the craton are exposed

Figure 4.5. Enchanted Rock is a large batholith of granite that is now part of a state park. Photo: G. Lamar, CC 2.0.

Chapter 4 – The Physical Environment of Texas | 43

by erosion in a few places in the state, notably, in parts of West Texas near Van Horn and the granite outcroppings of the Llano Uplift in Central Texas.

The earth's plates do not remain static over time. They shift and move on top of the mantle below, in a process called **plate tectonics**. When these plates collide or separate, they can produce dramatic landforms. The Ouachita Mountain Range, the southern extension of the Appalachian Mountains, were formed when the South American plate collided with the North American plate. The Ouachitas once arced across much of Texas, roughly tracing the path of Highway 90 from West Texas eastward and northward along IH-35. Now under the surface in most of Texas, remnants of the Ouachitas are visible only in a few outcroppings near Marathon in West Texas.

Geologic faults—cracks in the Earth's crust caused by stress and movements—also create changes in the Earth's surface. Movement along the Balcones Fault, which lies along the old Ouachita Range, led to the uplift of the Hill Country, also known as the Edwards Plateau, in Central Texas. West Texas has folded mountains from plate collisions, limestone mountains formed in seas and uplifted by tectonics, igneous mountains formed by volcanic activity, and basins formed by mid-continent rifting, where the land is spreading apart. The Gulf was formed in a rift that developed when the North American and South American plates moved apart from each other.

Water is also a powerful force in shaping geology and topography. Shallow seas have covered Texas during several periods, first covering much of West Texas, laying down rich sediments, some of which would become the oil-rich sedimentary rocks of the Permian Basin. Reefs formed during that period would later be uplifted to form the highest peaks in Texas (in the Guadalupe Mountains). During later periods the entire state was covered by shallow seas, producing much of the limestone now evident in Central Texas. Skeletal fossils of aquatic plesiosaurs have been uncovered from El Paso to Austin and north to the Dallas-Fort Worth area. The edges of these ancient seas bear the fossilized footprints of dinosaurs preserved in limestone.

The flat plains in the Texas Panhandle are underlain by ancient Permian seabed deposits but were subsequently overlain by sediments eroded from the Rocky Mountains. Over time, precipitation and accumulation of calcium carbonate within the sediments of the High Plains formed a hard layer called **caliche** or caprock. The Caprock Escarpment marks the eastern edge of the High Plains where water runoff has cut through the caliche to create dramatic landforms, such as Palo Duro Canyon.

The southeastern part of the state, called the Western Gulf Coastal Plain, is relatively flat terrain formed along the edge of the continental plate adjacent to the Gulf. As North America and other continental plates drifted apart, a shallow basin was formed. This basin was intermittently filled with seawater and sediments. Oil, gas, and coal formed in those sediments. Vast salt deposits formed during dry periods when the seawater evaporated. These salt beds would ultimately be covered in sediments and serve as traps for oil and gas along the current Texas coast. Rising and falling sea levels added to the erosion and deposition, creating various features like shorelines, alluvial plains, and wetlands. Texas rivers that originate in the higher ground of West Texas carry water, sand, and gravel to the sea, forming the deltas, lagoons, beaches, and barrier islands of the Western Gulf Coastal Plain.

Modern Texas showcases a diverse assortment of geology because of these past and current geological processes. The surface geology of Texas (Fig. 4.6) is mainly composed of sedimentary rocks, such as sandstones and mudstones in the Panhandle and Gulf Coastal Plain and limestones and

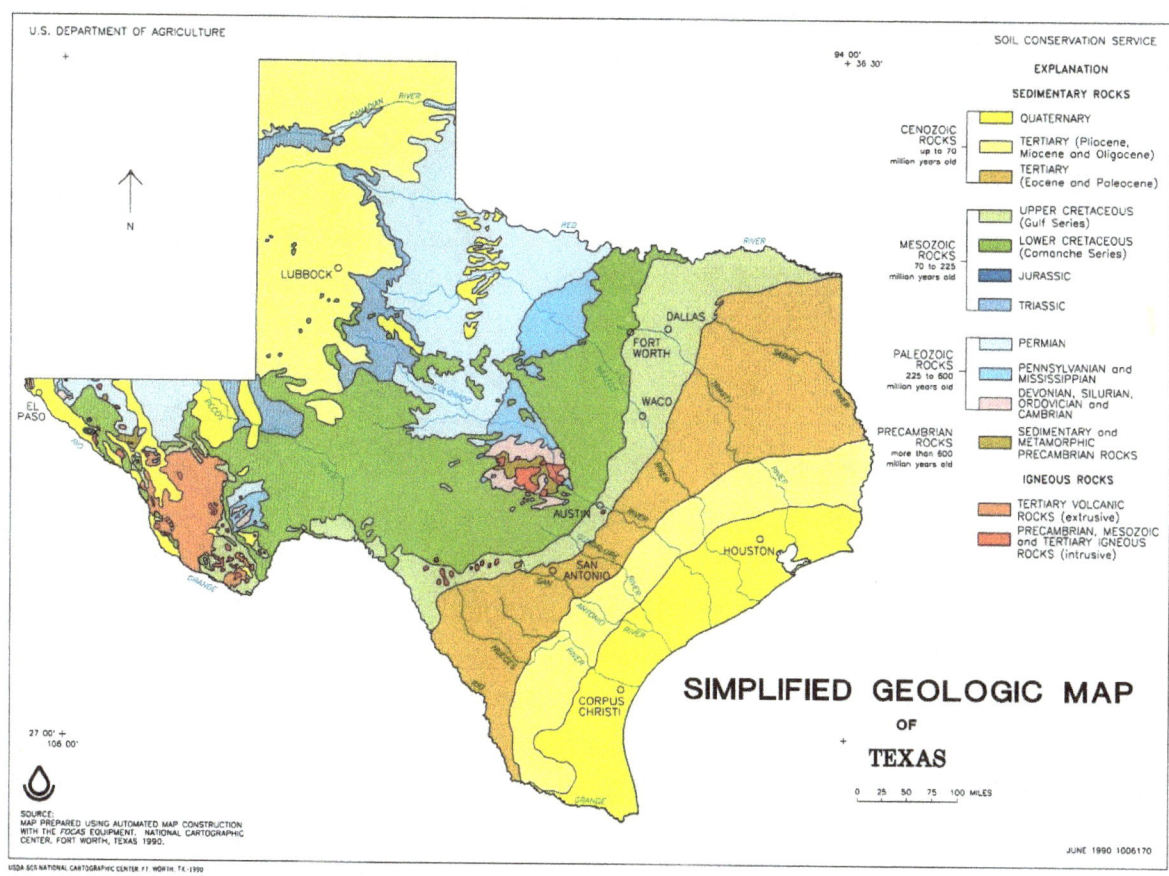

Figure 4.6. Geologic map of Texas. Source: USDA (public domain).

carbonates in Central and West Texas. In the Llano uplift of Central Texas, ancient igneous granite and metamorphic rocks like gneiss are evident. Evidence of more recent volcanic activity is found in West Texas, with numerous **calderas** and **lava domes**. Extrusive igneous rock, such as basalt formed from lava flows and tuff formed from volcanic ash deposits, is common in Big Bend National Park and the Chinati Mountains. Intrusive igneous rock is also evident in a variety of rock formations exposed by erosion, especially in the Davis Mountains and in the fantastic Solitario formation in Big Bend Ranch State Park (SP).

TEXAS CLIMATE

Climate is the weather conditions prevailing in an area over a long period. It includes patterns in temperature, humidity, atmospheric pressure, wind, and rainfall. Most of Texas falls into two major climate zones, humid subtropical and semi-arid (Fig. 4.7).

The eastern two-thirds of Texas experiences humid subtropical climate, with warm temperatures and year-round precipitation. The Texas Panhandle and portions of South and West Texas experience semi-arid climates where rainfall is lower, and temperatures can fluctuate dramatically between day and night. Desert climates occur in West Texas and are characterized by very low precipitation that occurs mainly in the late spring and summer.

Texas's climate is affected by latitude, humidity, and topography. The amount of rain increases from west to east, with averages ranging from 8 inches per year in El Paso to 56 inches per year in Orange (Fig. 4.8). The mountains in western North America block moisture from the Pacific, making the western part of Texas arid. However, during the summer, Gulf moisture can reach the West along with moisture from tropical storms in the Pacific, bringing heavy rains and causing deserts to bloom.

The timing and amount of rainfall influence the types of vegetation found in different areas, from forests and wetlands in the east to deserts in the west, with grasslands, savannas, and **shrublands** in between.

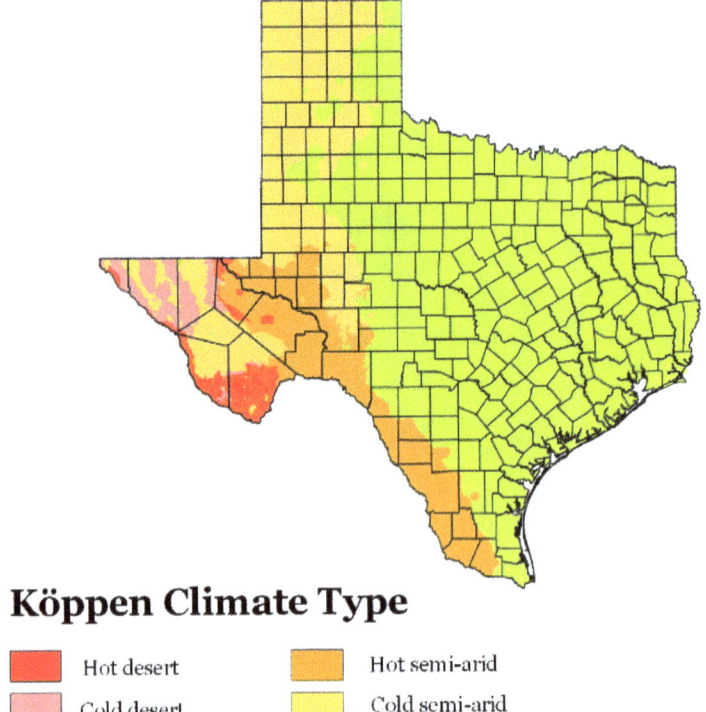

Figure 4.7. Climate types in Texas. Source: Icy98, CC 4.0.

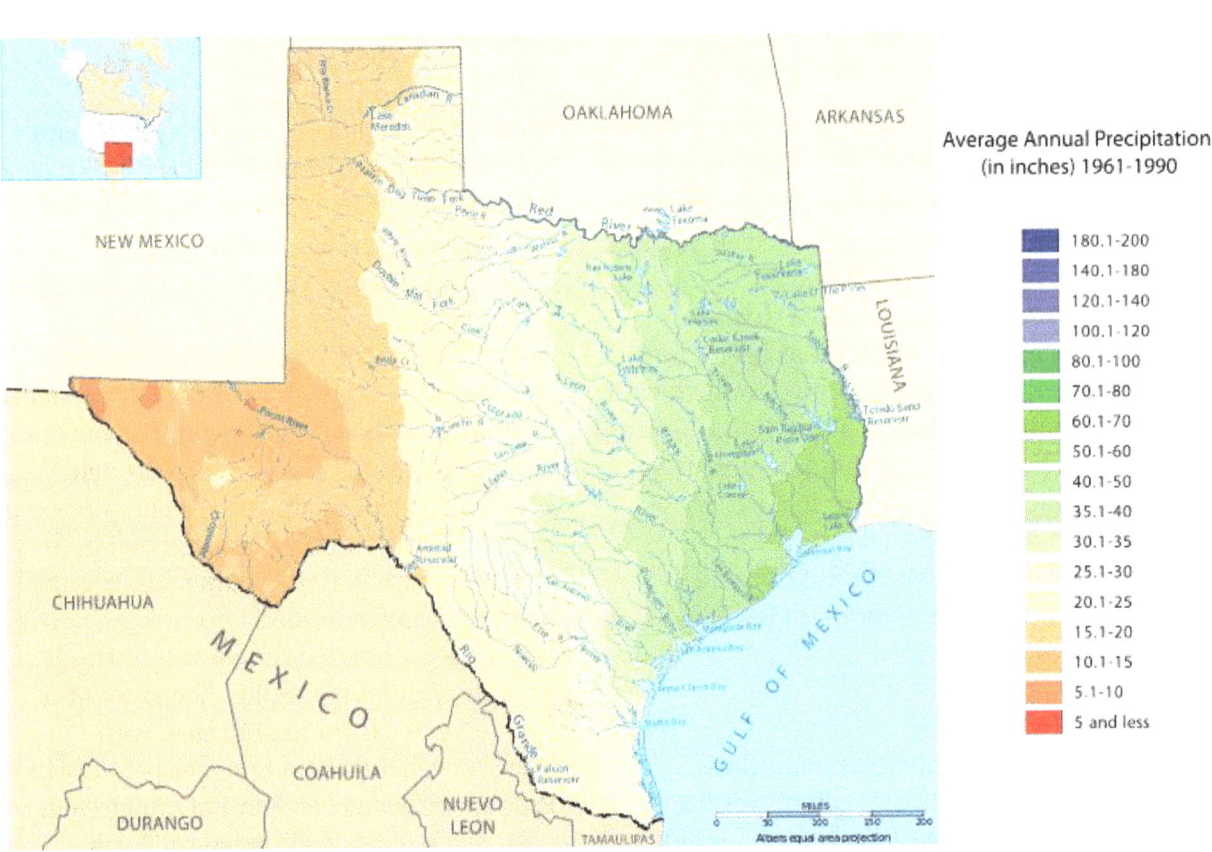

Figure 4.8. Annual rainfall in Texas. Source: Wikimedia Commons (public domain).

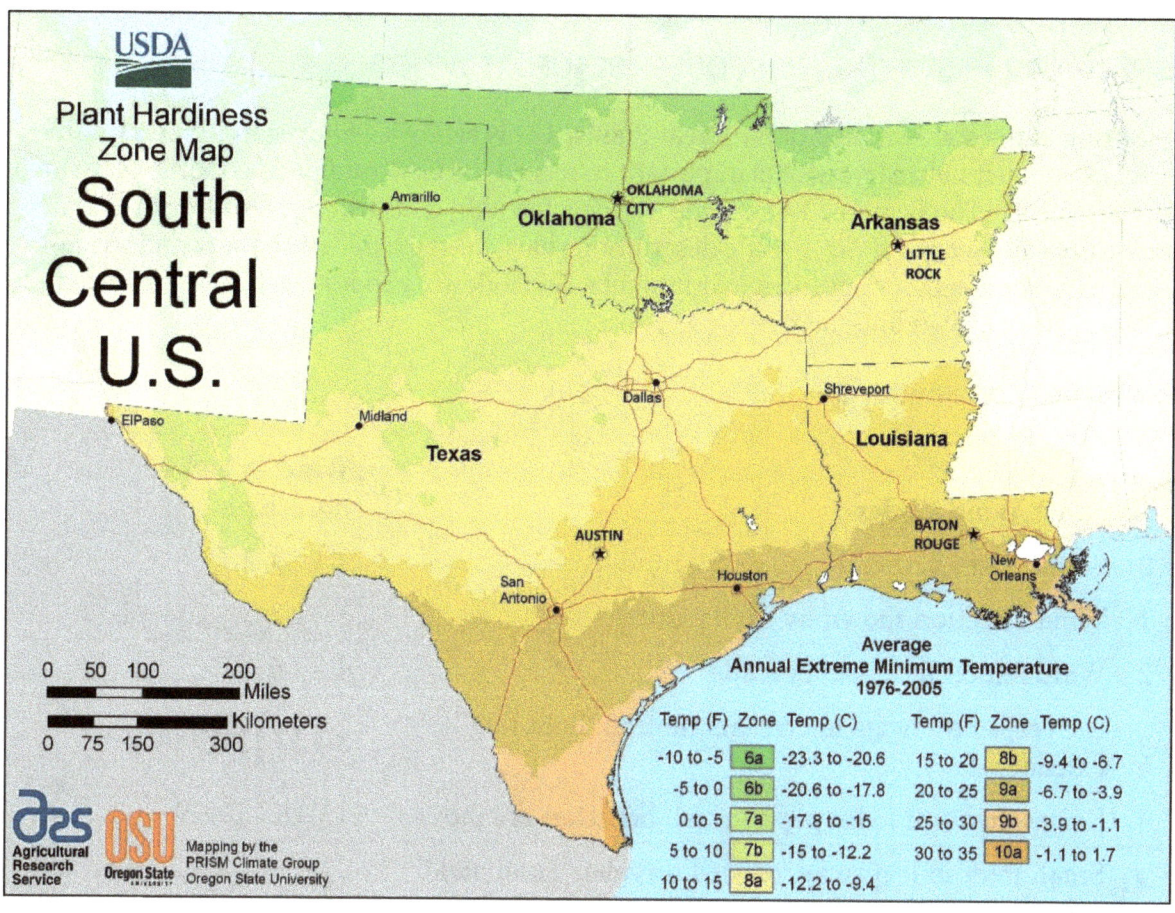

Figure 4.9. Hardiness zones reflect variations in extreme minimum temperatures in Texas. Source: USDA (public domain).

As Figure 4.9 shows, the Texas panhandle has the coldest winters; however, winter cold fronts from Canada or the Pacific can bring sudden temperature drops throughout the state. Perhaps surprisingly, the highest average July temperatures are in the Dallas-Fort Worth area and along the Rio Grande near Laredo, not at the southern tip or in the desert areas. Coastal areas are cooler due to ocean heat absorption and sea breezes. The deserts of West Texas are cooler due to altitude and lower humidity. Humidity is highest in the coastal regions, even producing subtropical woodlands in South Texas.

Weather in Texas can fluctuate significantly from year to year. The El Niño-La Niña cycle is a weather pattern created by ocean currents in the Pacific Ocean. During El Niño years, Texas experiences more winter rainfall in the eastern part of the state. During La Niña years, Texas experiences lower winter rainfall. However, even outside this cycle, much of the state is subject to drought—periods of very low or absent rainfall. Extended droughts create problems for farmers, ranchers, and water users and can result in degradation of native habitats. There is concern that increasing ocean and atmospheric temperatures, a trend commonly called global warming, will result in climate changes that increase the severity of drought, heat waves, and storms in the state. Higher global temperatures also contribute to the melting of the world's glaciers and pack ice, which, in turn, contributes to rising sea levels that can impact coastal wetlands and the species and economies that depend on them.

HIGHLIGHT 4.2 - TEACHING TECHNIQUES - BEAUFORT WIND SCALE

Sitting at the southern end of the Great Plains, Texas can be a windy state! When wildlife biologists conduct field surveys, it is important to record the environmental conditions that may affect the survey. The Beaufort Wind Scale was developed by a British naval officer to standardize sailing conditions, but the descriptions have been expanded so that wind speeds may be estimated in other outdoor environments. The following codes can be used to document wind speeds during wildlife surveys

Table 4.1 Beaufort Wind Scale

Beaufort Wind Codes		Wind Speed in mph/Kph
0	Smoke rises vertically	< 1 / < 2
1	Wind direction shown by smoke drift	1-3 / 2-5
2	Wind felt on face; leaves rustle	4-7 / 6-12
3	Leaves, small twigs in constant motion; light flag extended	8-12 / 13-19
4	Raises dust and loose paper; small branches are moved	13-18 / 20-29
5	Small trees in leaf sway; crested wavelets on inland waters	19-24 / 30-38

TEXAS SOILS

Soil is a part of an ecosystem that is both abiotic and biotic. It is made up primarily of minerals weathered from rocks, water, air, and **organic matter** derived from the decay of organisms. However, soil also contains a vast array of many microscopic and macroscopic organisms. For this reason, many biologists describe soil as "living."

Geology and climate work together to produce soil, and soils in one area can differ dramatically from soils in another area. The primary influences on soil characteristics are the rocks that composed the parent material, the slope of the landscape, the climate, the age of the soil, and the types of plants and animals that are present. As soil forms, it develops recognizable layers, called **soil horizons**. Though often hidden underground, when exposed, the horizons are revealed in a **soil profile** that can provide clues about the characteristics and history of the soil (Fig. 4.10).

Table 4.2. Layers in a soil profile:

O horizon - contains a high amount of organic matter; it is absent from many soils.

A horizon - topsoil; is dark-colored with a mix of **humus** (decayed material) and minerals

B horizon - subsoil; contains minerals and clay particles that have washed down from above

C horizon - no organic matter; contains rock fragments

R horizon - solid rock (or bedrock)

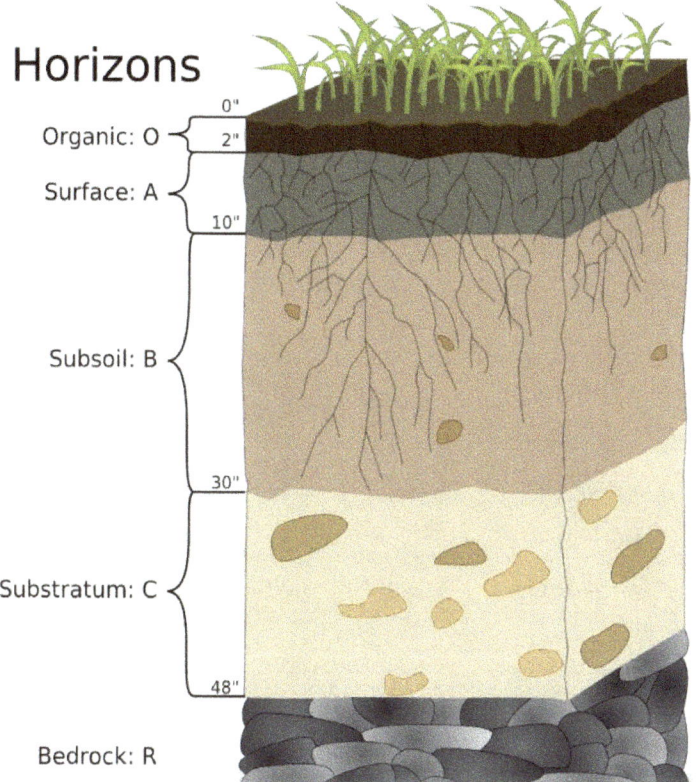

Figure 4.10 Horizons in a soil profile. Source: EssensStrassen, CC 4.0.

A number of characteristics are used to classify soil. These variations have a direct impact on the type of vegetation that can grow in an area and can influence the animal community as well. Soil characteristics include:

Color – Soil color is affected by organic matter, moisture, and the minerals that make up the soil. Soils that are high in organic matter are dark brown or almost black. Wet soils sometimes have a mottled pattern of grays, reds, and yellows. Soils high in iron can oxidize (rust) and appear deep orange-brown to yellowish-brown.

Structure – Depending on its composition, soil tends to form clumps (or peds) of predictable size and shape. Structure affects the pore space in the soil—influencing root growth, air and water movement, and ultimately the type of vegetation found in an area. Abuse of soil, such as compaction and erosion, can break down soil structure.

Chemistry – Soil chemistry affects soil fertility. Major soil nutrients needed by plants include nitrogen, phosphorus, potassium, magnesium, and calcium. Soil pH measures whether a soil is acidic (pH < 7) or alkaline (pH > 7, also called basic). The pH can affect how well plants can absorb these nutrients. Parent material and climate affect pH. Limestone-based soils are alkaline. Soils in higher rainfall areas tend to be more acidic.

Texture – Soils have different textures based on the size of the particles in the soil. Sand particles are the largest and feel gritty to the touch. Sandy soils drain rapidly, holding little water for vegetation. Clay particles are the smallest and are sticky when wet. Clay soils often swell when wet and shrink when dry. They are slow to absorb water and do not give up water or air easily to vegetation. They are also difficult for burrowing animals to penetrate. Silt particles are intermediate in size and are slippery to the touch. The ideal soil for many plants and animals is a combination of all three particle sizes—a type of soil called a "loam." A soil texture triangle (Fig. 4.11) depicts the many different combinations of particle sizes.

Texas has a variety of soils due to its climate, geology, topography, and size. The state has 12 primary soil groups, called soil orders (Fig. 4.12), that are categorized based on their physical, chemical, and biological properties. Forested areas in East Texas have acidic and less fertile soils called Ultisols. Other woodlands and some grasslands across the state have less acidic soils called Alfisols. Coastal areas and many prairies have fertile soils called Vertisols that are dominated by clay. Central Texas and the northern Texas Panhandle have alkaline prairie soils called Mollisols that are rich in basic minerals. In

West Texas, dry climates have led to unweathered soils called Aridisols that may have high salt, gypsum, or carbonate levels. These soil types, along with other factors like topography and climate, contribute to the diversity of plant and animal life in Texas, a diversity explored in the ecoregions of Texas in Chapter 5.

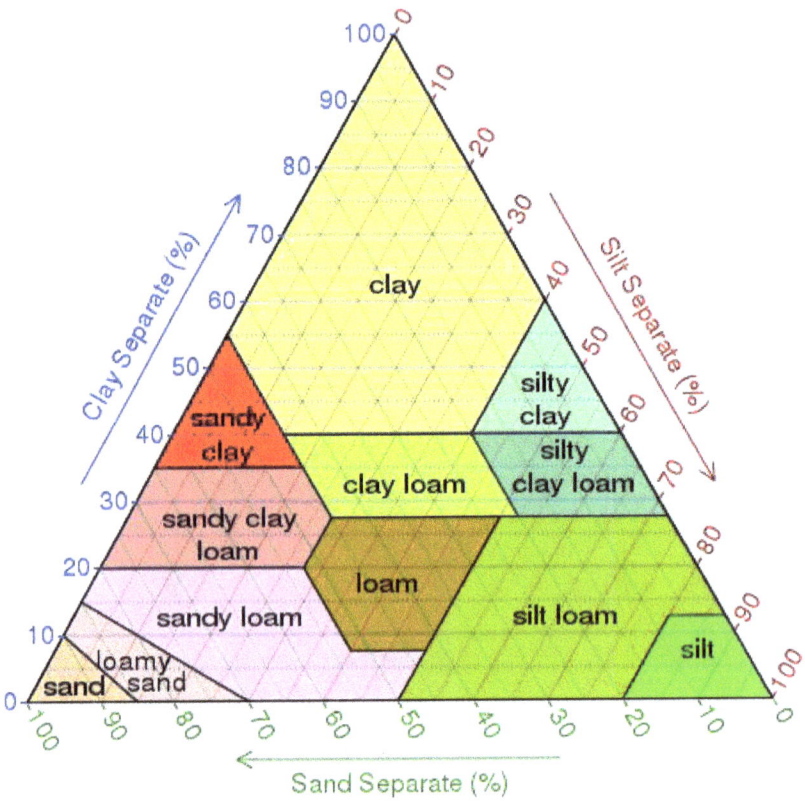

Figure 4.11. Soil texture triangle. Source: Apouzin, CC 4.0.

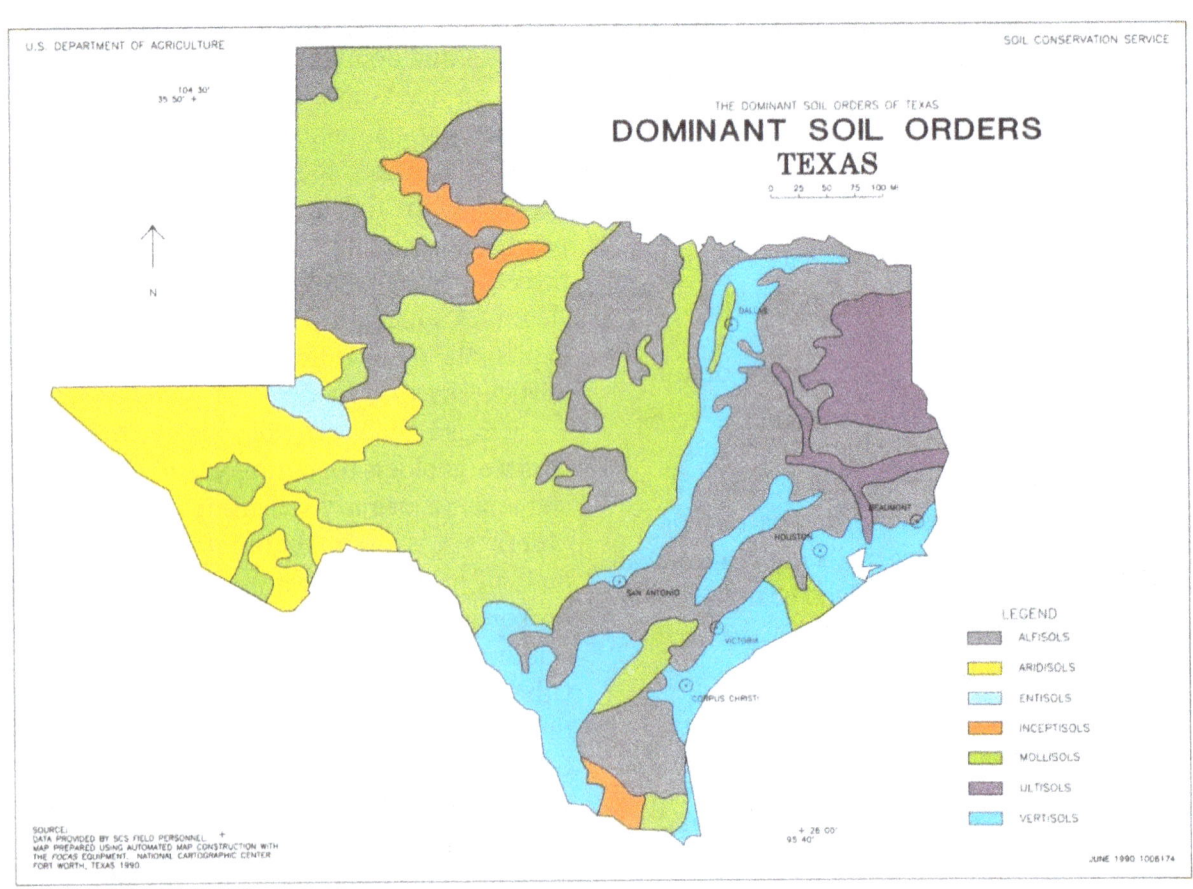

Figure 4.12. Dominant soil orders in Texas. Source: US Department of Agriculture (public domain).

CHAPTER 5 – ECOREGIONS OF TEXAS

Texas lies at a crossroads for abiotic and biotic diversity from all parts of the United States. To the east, it is part of the pine forests of the southeastern U.S. In its northeastern corner, the **hardwood** forest belt enters the state. The Great Plains of the central United States run through the center of Texas, the Mexican subtropics jut into the South, and Rocky Mountain habitats and southwestern desert habitats can be found in West Texas. The patterns in Texas's topography, geology, climate, and soils lead to a complexity of vegetation communities and wildlife populations.

With all this diversity, a map of all the vegetation types in Texas can look very complicated (Fig. 5.2).

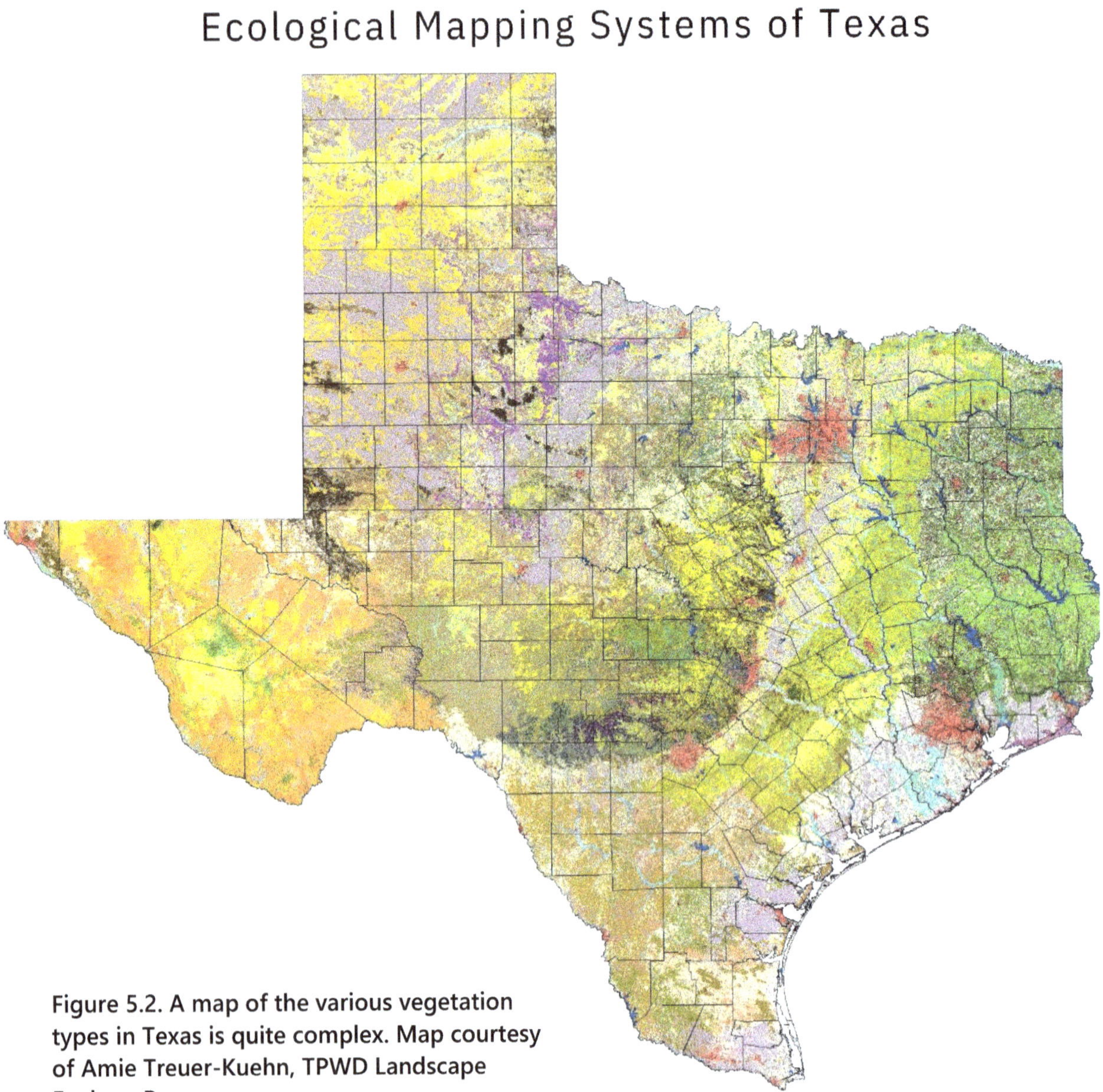

Ecological Mapping Systems of Texas

Figure 5.2. A map of the various vegetation types in Texas is quite complex. Map courtesy of Amie Treuer-Kuehn, TPWD Landscape Ecology Program.

(overleaf) Figure 5.1. Soils and geology, such as the granite substrate of the Llano Uplift, shape the plant and animal communities within the ecosystem.

Despite the vegetation complexity, ecologists have noticed patterns in ecosystem distribution and have proposed several maps to summarize these patterns. These "ecoregion" maps represent major ecosystems defined by distinctive geography, soils, rainfall and climate patterns, and characteristic species of plants and animals.

Many different, but similar, ecoregion maps for Texas exist (Fig. 5.4). This book relies on a Natural Regions of Texas map produced by the Lyndon B. Johnson School of Public Policy as part of a document entitled "Preserving Texas's Natural Heritage" published in 1978 (Fig. 5.3). The Natural Regions map divides Texas into 11 ecoregions.

Figure 5.3. The Natural Regions of Texas. Source: LBJ School of Public Affairs. Modified by Paul Daugherty, TPWD IT-GIS.

Chapter 5 – Ecoregions of Texas | 53

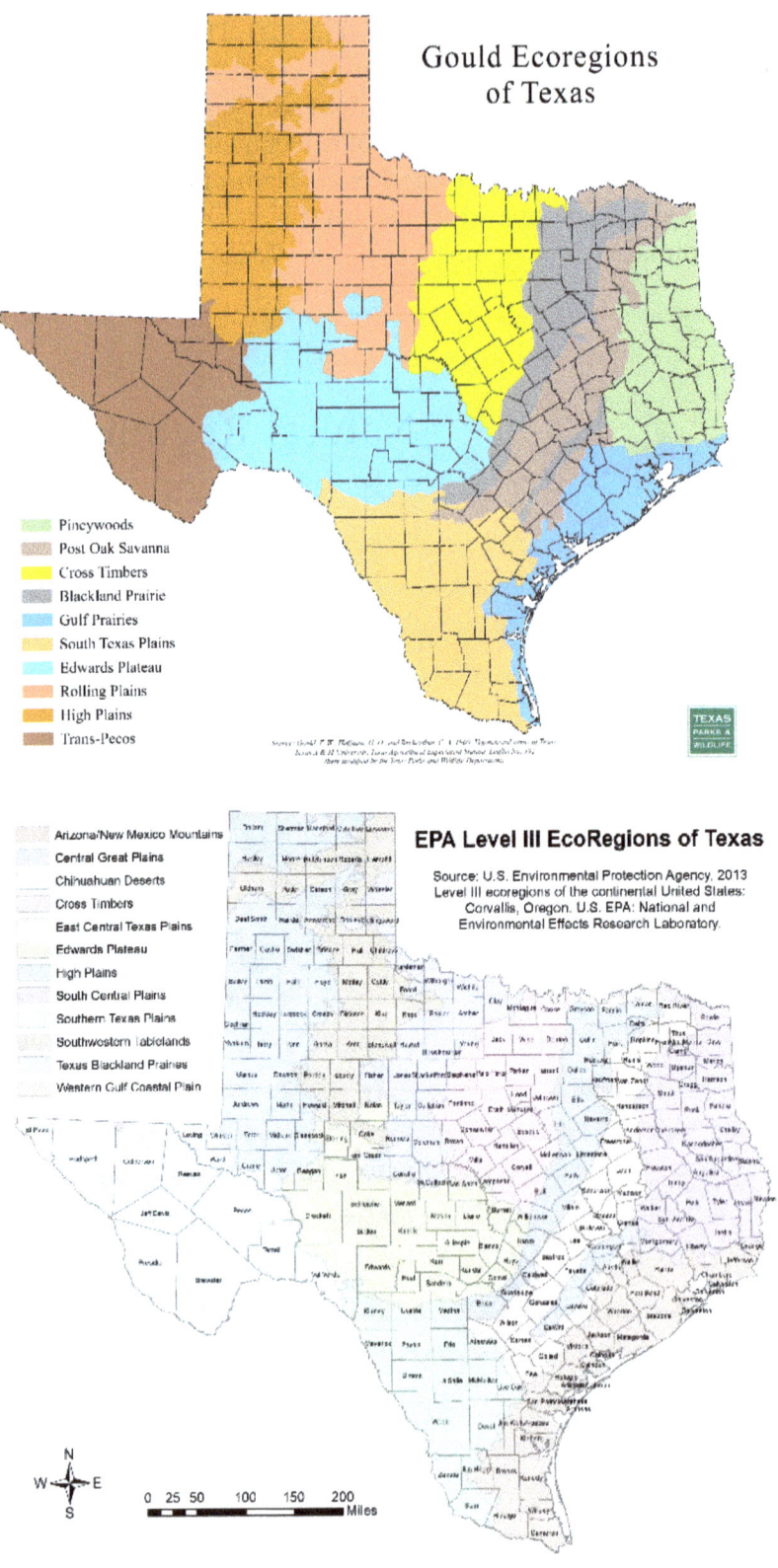

Figure 5.4. Examples of some other variations of ecoregion maps for Texas. Gould ecoregions map prepared by Paul Daugherty, TPWD IT-GIS. EPA map prepared by Environmental Protection Agency (public domain).

REGION 1 - Pineywoods

Figure 5.5. The Pineywoods ecoregion hosts a variety of both pine and hardwood forests.

Located on the eastern edge of the state, the Pineywoods ecoregion encompasses over 23,000 square miles of gently rolling hills dominated by forest land. The Pineywoods are warm and wet. It is, on average, the wettest ecoregion in the state, ranging from 46 to 54 inches of rain per year. High temperatures in the summer average 93°F; winter lows average 36°F. Humidity is high. Soils in the Pineywoods ecoregion are predominantly deep sands and sandy loams and are generally acidic. Elevations range from about 100 feet up to 500 feet, with a few higher points on ancient ridgelines. Two subregions are recognized in the Pineywoods—Longleaf Pine Forest in the southeast and Pine-Hardwood Forest throughout the rest of the region.

The region is part of a much larger area of pine-hardwood forest that extends into Louisiana, Arkansas, and Oklahoma and once covered most of eastern North America. Sandy soils and fires, started either by lightning or Native Americans, play a role in the dominance of pines in the southern pine forests; meanwhile, soils and elevation affect the composition of the pine forests. Loblolly pine historically grew in the lower slopes and bottoms; longleaf pine grew in the very sandy soils of southeast Texas; and shortleaf grew in association with longleaf in the middle Pineywoods, transitioning to a shortleaf-oak-hickory association in northeast Texas. In more recent years, slash pine has also been widely introduced in the Pineywoods.

Although pine forests dominate most of the region, there is a natural mosaic of hilly uplands, mid-slopes, bottomlands, stream and river drainages, and **swamps** that create a diversity of forest types. Slopes support mixed pine-hardwoods forests containing dogwoods, red and white oaks, beech, and magnolia. Wetter, low-lying areas with little to no fire occurrence support biologically diverse ecosystems dominated by hardwoods such as sweetgum, magnolia, tupelo, elm, ash, pecan, and overcup oak. Swamps comprised of bald cypress and water tupelo are common. **Bog** habitats are scattered among woodlands. Even small prairies and glades (open grassy areas often associated with unique shallow soils) are inserted into the forests.

Several rare species are associated with Pineywoods habitats. Fire-maintained mature pine forests in some locations host the endangered red-cockaded woodpecker and once hosted the threatened Louisiana pine snake. Once extirpated, black bears are making a comeback in forested habitats in Northeast Texas, but the ivory-billed woodpecker, jaguar, and red wolf are now gone. Rare plants in the region include Texas trailing phlox found in sandy habitats, white bladderpod and golden gladecress found in wet glades, and southern lady's-slipper in slope forests.

Dominant land use in the Pineywoods includes lumber and cattle production, although the region is also home to the largest oil field in Texas (discovered in 1930, it became the most productive oil field in the contiguous United States). There are four national forests in the Pineywoods, but the majority of forested land in Texas is privately-owned.

REGION 2 - OAK WOODS AND PRAIRIES

Figure 5.6. The Engeling Wildlife Management Area hosts oak forests and savannas typical of the Oak Woods and Prairies ecoregion. Photo: William L. Farr, CC 4.0.

To the west of the Pineywoods lies the Oak Woods and Prairies ecoregion. The region is composed of several belts of hardwood forest covering 19,500 square miles of gently rolling terrain. Rainfall ranges from about 28 inches per year in the west to about 45 inches per year in the more easterly portions. Upland soils in the Oak Woods and Prairies ecoregion are dominated by acidic sandy loams and sands that overlie a dense layer of clay below. Indeed, the distribution of these sand claypan soils predict the distribution of the oak woods. Elevations climb slowly from 300 feet above sea level in the southeastern portions to more than 1700 feet above sea level in the west. Three belts of woodlands are recognized in this ecoregion. From east to west, they are the Post Oak Savanna, the Eastern Cross Timbers, and the Western Cross Timbers.

The Oak Woods and Prairies are a transition zone between more forested areas to the east and grassland biomes to the west, and the three belts of woodland are separated by belts of prairie. The dominant vegetation structure includes oak-dominated forests and oak savanna habitats, where trees are more widely spaced and interspersed with bluestem grasslands. Dominant plant species are post oak and blackjack oak, along with black hickory. Fires help to maintain the open savannas; when they are excluded a thick understory of yaupon, winged elm, and eastern red cedar can develop. Some unique plant communities include peat bogs and dwarf palmetto wetlands that form in stream meanders, depressional areas, and topographic breaks. Even a patch of isolated pine forest—the Lost Pines—occurs in the Oak Woods and Prairies in the Bastrop area.

Endemic species in the Oak Woods and Prairies are strongly associated with the sandy soils of this ecoregion. The large-fruit sand-verbena is an endangered plant restricted to deep sandy soils in only three counties. Navasota ladies'-tresses is an orchid almost exclusively found in the Post Oak Savanna. The Houston toad is an endangered amphibian that was once more widespread but is now only found in deep sandy soils of the Oak Woods.

Cattle ranching is the major agricultural industry in the Oak Woods and Prairies, although significant areas have been lost to urbanization in the Dallas-Fort Worth Metroplex. Introduced grasses such as bermudagrass have been widely introduced. In addition, fire suppression has allowed understory plants to overtake the oak savannas in many areas.

REGION 3 – BLACKLAND PRAIRIE

Figure 5.7. Texas blackland prairie is preserved in only a few small tracts, such as Clymer Meadow Preserve, owned by the Texas Nature Conservancy. Photo: Wilafa, CC 4.0.

Lying between belts of the Oak Woods ecoregion, there were once large swaths of grasslands. Rainfall (28-40 inches of rain/year) and the flat to rolling topography in the grasslands are similar to the Oak Woods. Elevations are also similar, ranging from 300 to 800 feet above sea level. But the soils are quite different. These prairie soils are heavy soils dominated by dark-colored alkaline clays, interspersed with patches of gray acidic sandy loams. The dark clays are often called "black gumbo," and the Blackland Prairies ecoregion is named for the deep, fertile black soils that characterize the area. The ecoregion is divided by belts of the Oak Woods into several segments, including the Fayette Prairie to the southeast and the Grand Prairie to the west in addition to the primary Blackland Prairie.

Covering 25,500 square miles, the Blackland Prairie encompasses the southern end of the "True Prairie," a region of tallgrass grassland hosting grasses up to nine feet tall that once extended all the way to southern Canada. Dominant plants in Texas include big bluestem, little bluestem, and Indiangrass, along with an abundant diversity of herbaceous flowering plants. Switchgrass and eastern gamagrass are common in prairies in wetter sites, and dropseeds are more common in sandier soils. In the Grand Prairie, shorter grasses such as little bluestem and sideoats grama (the State Grass of Texas) predominate.

Few species in the Blackland Prairies are listed as endangered, but several unique species are found there. The Parkhill Prairie crayfish is known only from two counties in North Texas. It doesn't associate with wetland areas but instead burrows down deep to the water table underground. The Grand Prairie hosts two yucca species found nowhere else—the Glen Rose yucca and the pale yucca—along with an endemic species of harvester ant named for former human inhabitants of the area, the Comanche harvester ant.

In truth, the native grasslands of the Blackland Prairie are its most endangered feature. Because of the fertile soils, much of the original prairie has been plowed to produce food, fiber, and forage crops. According to TPWD, less than 1% of the original prairie remains. In addition to agriculture, hay production, and cattle grazing, much of the ecoregion, which lies along IH-35 from Austin to Dallas-Fort Worth, is now highly urbanized.

REGION 4 - GULF COAST PRAIRIES AND MARSHES

Figure 5.8. The Gulf Coast Prairies and Marshes ecoregion provides grassland, wetland, and forest habitats along the Texas coast.

Bordering the Gulf from the border with Louisiana south to the Rio Grande, the Gulf Coast Prairies and Marshes ecoregion is remarkably flat, but remarkably diverse. The entire region is less than 150 feet in elevation, with an average rainfall that may be more than 50 inches per year in the east to about 25 inches in the south. Soils are acidic sands and sandy loams, with clays in the upland prairies and the river bottoms. Soils near coastal edges may have high salt content. As one moves from inland to more coastal reaches, distinctive subregions include the Upland Prairies and Woods, the Estuarine Zone, and the Dunes and Barrier Islands.

The 21,000 square-mile region includes remnant tallgrass prairies and salty prairies, scattered patches of live oaks, extensive low-lying wetlands, river bottom forests, and sandy barrier islands. Brownseed paspalum is an indicator species for intact upland prairies, while moister, saltier prairies are dominated by gulf cordgrass. Live oak patches (or live oak "**mottes**"), often on ancient dune ridges, provide habitat for migrating songbirds and numerous other wildlife species. Coastal marshes, grading from freshwater to saltwater, are essential habitats for waterbirds, alligators, and commercial fisheries. Strings of **riparian** forest made up of cedar elm, sugar hackberry, live oak, and other species border the many rivers that dissect the marsh, providing habitat for forest-dwelling birds. (Jaguars and black bears once roamed these forests as well).

Barrier islands protect mainland habitats from storms, offering beaches and tidal flats for shorebird foraging and grasslands for rodents, badgers, and birds.

Biodiversity is rich in the Gulf Coast Prairies and Marshes ecoregion, and many threatened species rely on its varied habitats. Endangered Attwater's prairie-chickens and endemic plants such as the slender rushpea, prairie dawn, and black lace cactus are found in the prairies. Coastal marshes are essential winter habitat for the endangered whooping crane and for the threatened diamondback terrapin. Bald eagles nest in riparian woodlands along coastal rivers. Peregrine falcons and threatened piping plovers feed on barrier islands. Once-endangered, now-recovered brown pelicans nest on small islands in the bays, and endangered Kemp's ridley sea turtles are expanding their nesting sites on Texas beaches.

Many acres of the Gulf Coast Prairies have been converted to agriculture, including sorghum, cotton, corn, and rice. Brush species such as mesquite, acacias, and introduced Chinese tallow and Macartney rose are more common now than in the past due to fire suppression. Urbanization has also swallowed up prairie and marsh habitats. Coastal marshes suffer from decreased freshwater inflows and increased saltwater intrusion, while construction of housing and condos has impacted some of the barrier islands.

REGION 5 - COASTAL SAND PLAIN

Figure 5.9. Little bluestem in a tract of native prairie on the Coastal Sand Plain. Photo: Bill Carr.

Sometimes considered to be part of the Gulf Prairies to the north or the Brush Country to the West, the Coastal Sand Plain is a smaller ecoregion about 60 miles long by about 60 miles wide located along the Texas coast just south of Kingsville. Home to the famous Kenedy and King Ranches, among others, this unique section of coastal prairie receives 20 to 30 inches of rain annually. Like the rest of the coastal prairie, its topography is low and flat.

Despite its small size, the soils and unique origin of the region cause ecologists to give it its own distinct designation. Formally known as the South Texas Eolian Sand Plain, the region is believed to have formed from an ancient barrier island. Over time, southeasterly winds swept sands from the island and from nearby Padre Island over the mainland, creating a sandy mosaic of dunes, swales, flats, and wetlands.

Seacoast bluestem once dominated the grassland communities that formed in the sandy soils, with camphor weed on the ridges and gulfdune paspalum in the swales. A total of 14 species of plants are found only in this small ecoregion. Oak mottes host additional unique species, such as Bailey's ball-moss, a gigantic version of the more common species found elsewhere in Texas. Oak mottes also provide habitat for the ferruginous pygmy-owl, a species popular with birders. Wetlands that form in the swales underlain by clay are inhabited by the threatened black-spotted newt and Rio Grande lesser siren.

The region is also known as the "Wild Horse Desert," reflecting an observation of large mustang herds in the mid-1800s and the lack of permanent surface water. Large ranches primarily managed for wildlife and cattle occupy most of the Coastal Sand Plain, lessening threats to the region. However, earlier overgrazing is evidenced by the encroachment of South Texas brush species.

REGION 6 - SOUTH TEXAS BRUSH COUNTRY

Figure 5.10. The South Texas Brush Country hosts an assortment of diverse brushland habitats. Photo: Joe Herrera, TPWD.

The South Texas Brush Country covers 28,000 square miles at the far southern tip of Texas. Annual rainfall is highly variable from year-to-year in this sub-tropical region, increasing from about 19 inches in the west to about 32 inches in the east. Summer temperatures are high, with very high evaporation rates. Soils of the region are alkaline to slightly acidic clays and clay loams. Topography is flat to gently rolling with elevations generally ranging from sea level to about 1,000 feet. The Bordas Escarpment is an elevated ridge covered with shrubby vegetation that runs north-south. Streams to the west of the escarpment flow to the Nueces River; streams to the east flow toward the Gulf.

Sometimes called the South Texas Plains, a variety of vegetation communities occur in the Texas Brush Country. Ecologists suggest that originally the South Texas Plains supported a mixture of grassland and brushland communities, but that overgrazing, especially by sheep, resulted in a decline in grass coverage and an increase in woody coverage. Today, most of the region is covered with a mixed brush community often described as Taumalipan Thornscrub, because of the similarity to similar communities in Mexico. Dominant species include mesquite, blackbrush, guajillo, granjeno, and Texas persimmon. Along rivers and ancient streambeds a more robust forest may form, including anacua and Texas ebony, among others. In the Lower Rio Grande Valley where the river approaches the Gulf, sabal palms join the riparian community, creating a unique subtropical habitat.

The South Texas Brush Country is an incredibly important wildlife area and is thought to host some of the greatest wildlife diversity in the state. Much of the region is in large cattle ranches that make most of their income from hunting leases for white-tailed deer, mourning doves, and bobwhite quail. At least six rare plant species are found only in the South Texas Plains, including the endangered Zapata bladderpod that stays dormant until moisture is available and the tiny star cactus. The region also hosts a number of rare wildlife species, including the endangered ocelot, a beautiful wild cat dependent on thick **brushlands** and forests in the Lower Rio Grande Valley. Though rarely sighted, jaguarundi and white-nosed coati are also closely linked with Texas Brush Country habitats that overlap into Mexico.

Although many land changes have occurred in this region, the Brush Country remains rich in wildlife and a haven for many rare species of plants and animals. Hunting, livestock grazing, and crop production are the principal agricultural land uses, with crop production contributing to widespread loss of woodland habitats along the lower Rio Grande. Nevertheless, the cities of that area receive significant economic benefit from birders who flock to the area to spot tropical specialties such as green jays, chachalacas, Aplomado falcons, and a variety of species that occasionally wander north into South Texas.

REGION 7 - EDWARDS PLATEAU

Figure 5.11. The Edwards Plateau hosts grasslands, savannas, and oak-juniper forests on calcareous limestone soils.

The 31,000 square-mile Edwards Plateau ecoregion is defined by its geology. It is an uplifted region in the center of the state with shallow alkaline clay soils overlying limestone rocks. It is bounded on the west by the Pecos River and on the eastern and southern edge by the Balcones Escarpment, an uplifted area formed along the Balcones Fault. In the west, rainfall averages about 15 inches annually. In the east where the Edwards meets the Blackland Prairies, rainfall averages up to 35 inches per year. Elevations range from about

1,000 feet above sea level in the east to about 3,000 feet in the west. Subregions include the Live Oak-Mesquite Savanna, the Lampasas Cut Plain that lies adjacent to the Grand Prairie, and the Balcones Canyonlands along the eastern edge.

Vegetation communities in the Edwards Plateau (the region and several other features in Texas are named for Haden Edwards, a settler who advocated for Texas independence from Mexico) also vary from west to east. In the west, the Edwards Plateau blends into the shortgrasses of the High Plains, the arid scrub vegetation of the Chihuahuan Desert, and the Brush Country of South Texas. Vegetation here is short and sparse. Moving eastward, the central part of the Edwards Plateau was once a savanna, with scattered groves of oaks and Ashe juniper (often called cedar) scattered among bluestem grasslands. Finally, in the Balcones Canyonlands, an area of hills and canyons eroded along the Balcones Escarpment, one finds the true "Hill Country" (Balcones is a Spanish word that translates to "balconies," perhaps reflecting the multiple elevations in this steep region). Vegetation in this wetter, steeper region is protected from fire and fed by springs. It includes mature forests of red oaks, Texas live oaks, Ashe juniper, cedar elm, and numerous understory trees. Riparian stands of bald cypress and sycamore trees line rivers and streams.

Many rare and endemic species occur in the Edwards Plateau. Cliff chirping frogs and barking frogs are found only in the canyons of the Texas Hill Country, and spring-fed streams host numerous endemic salamanders. The limestone geology is honeycombed with thousands of caves occupied by endemic invertebrates and millions of Brazilian free-tailed bats. Golden-cheeked warblers nest only in the Edwards Plateau, and it hosts some of the highest populations of nesting black-capped vireos and painted buntings in the country. More than 100 species of plants are found only in this ecoregion.

Numerous changes have occurred in the Edwards Plateau. Western areas have less grass cover due to historic overgrazing. Oak savanna habitats in many places are overgrown with Ashe juniper due to overgrazing and fire suppression. Rapid urban growth has taken place along the eastern edge of the Hill Country. Hunting and ranching are the primary land uses in the more rural areas (white-tailed deer populations are extremely high in this region). Outdoor recreation, camping, and birding also contribute to the economies of the Balcones Canyonlands.

REGION 8 - LLANO UPLIFT

Figure 5.12. The Llano uplift, set within the Edwards Plateau, is characterized by granite formations instead of limestone rock. Photo: Jeff Forman, TPWD.

Sometimes considered part of the Edwards Plateau, this small 5,000 square-mile region is distinctive because of its geology. The Llano Uplift is characterized by granite and other igneous and metamorphic rock that is believed to represent some of the oldest rocks from the Texas Craton. Elevation ranges from 825 to 2,250 feet above sea level. Rainfall is about 30 inches per year. In contrast to the clay limestone soils of the Edwards Plateau, soils in the Llano Uplift are coarse sands derived from weathered granite. It is sometimes called the Central Mineral Region because of the deposits of gold, copper, iron, tourmaline, smoky quartz, and Texas blue topaz (the official Texas State Stone) that are exposed from the ancient rocks.

The most distinctive topographic features of the Llano Uplift are its large, exposed granite domes, including Enchanted Rock. These exposed batholiths "exfoliate" through weathering, producing granite boulders that accumulate in lower-lying areas. The domes also erode as lichens grow on their surface and as water collects in low spots, disintegrating the stone and producing springtime pools of water called vernal pools. These pools may support Rock Quillwort, an endangered endemic plant. In the sandy soils between domes one can find woodlands and grasslands containing blackjack oak, mesquite, cedar elm, black hickory, little bluestem, and basin bellflower, an endemic plant species. As many as eight additional endemic plant species may occur in this small region.

Bird, mammal, and reptile species in the Llano Uplift are similar to the Edwards Plateau. Hunting, ranching, and nature tourism are the predominant industries.

REGION 9 - ROLLING PLAINS

Figure 5.13. Red soils and rolling topography typify the Rolling Plains ecoregion. Photo: Mark Mitchell, TPWD.

As one moves west from the Blackland Prairies and Cross Timbers, rainfall decreases, and forests and tallgrasses give way to the mid-height grasses of the Rolling Plains. Rainfall in this region ranges from 20 to 30 inches per year, but its occurrence is sporadic, and evaporation is high. Sometimes called the Rolling Red Plains, the soils are reddish neutral to alkaline sands, loams, and clays, tinted by the color of their shale, mudstone, and sandstone sources. Elevation varies from 800 to 3,000 feet above sea level. Topography is gently rolling throughout much of the region, but the Caprock Breaks on the west are rugged canyons that are the source of the Red, Brazos, and Colorado Rivers. Another subregion, the Canadian Breaks, is a broader valley along the Canadian River lying 500 to 800 feet below the High Plains.

The original grasslands of the Rolling Plains would have been a mixed-grass prairie composed of sideoats grama, little bluestem, and blue grama grazed by prairie dogs and nomadic herds of bison and frequently burned by wildfires and fires set by Native Americans. However, as domestic grazing increased and fires diminished, grasslands came to be dominated by lower-value **increaser grasses** and woody species such as honey mesquite. Stream floodplains host various hardwood species. Steep slopes, cliffs, and canyons in the Escarpment Breaks support an open woodland dominated by juniper species.

Although some wildlife species, such as bison and black-footed ferrets, are gone from the Rolling Plains, the ecoregion still provides important wildlife habitats. Quail and horned lizard populations are higher here than in many parts of the state. The Rolling Plains are the only place where the endemic Texas kangaroo rat and Palo Duro mouse can be found. Grasslands on managed areas provide important grassland bird habitat, and interior least terns nest along rivers of the region. The Concho watersnake and Brazos River watersnake are endemic reptiles found along their respective rivers. The Texas poppymallow is an endemic plant.

Crop and livestock production are the major agricultural industries in this region. About one-third of its area is converted to crops such as wheat, cotton, and sorghum. The Escarpment Breaks compose the second-longest canyon system in the United States, and many tourists are drawn to the colorful walls of Palo Duro Canyon SP.

REGION 10 - HIGH PLAINS

Figure 5.14. The High Plains once hosted a shortgrass prairie dotted with playa lakes that supported wetland vegetation. Photo: Jeff Bonner, TPWD.

The open vistas of the High Plains, the southern end of the Great Plains, lie on a level high plateau in the Texas panhandle 3,000 to 4,500 feet above sea level. The soils are dominated by clays underlain by caliche, a hardened calcium carbonate layer called the caprock. At its eastern edge, the limestone-like caprock is cut by erosion to create the Escarpment Breaks, forming the border with the Rolling Plains. Annual rainfall is only 15 to 22 inches, and extended droughts are common and notorious. The Canadian River divides the 34,500 square mile region into two segments—the Canadian-Cimarron High

66 | Chapter 5 – Ecoregions of Texas

Plains north of the Canadian River and the Llano Estacado, or Staked Plains, south of the river.

The original vegetation of the High Plains was shortgrass prairie dominated by buffalograss and blue grama. These grasses form sods of thickly packed roots that help the grasses withstand both drought and the herds of bison that once passed through. Once scattered all across the prairie were shallow round wetlands called **playa lakes**. These wetlands provided wildlife habitat and recharge to the massive Ogallala Aquifer that underlies the region. Along the western edge of the Texas panhandle sandy habitats occur. Some support taller grass communities made up of bluestems and sideoats grama. Other sites support woody species such as shinnery oak and sand sagebrush.

Immense herds of buffalo once passed through the High Plains. Today prairie dogs and pronghorns, though reduced in numbers, are the primary grazers. Black-footed ferrets, once predators of prairie dogs, are extirpated, but the endangered lesser prairie-chicken hangs on in shinnery oak habitats and in the eastern Panhandle. Shinnery oak also provide habitat for the endangered dunes sagebrush lizard.

About 90% of the High Plains is in agricultural use. Much of the High Plains is grazed, but many acres of grasslands and even some playas have been converted to cropland, utilizing irrigation drawn from the Ogallala Aquifer. Corn, wheat, cotton, and sorghum, along with silage, hay, and soybeans are the primary products.

REGION 11 - TRANS-PECOS

Figure 5.15. A glimpse into the Solitario in Big Bend Ranch SP reveals desert vegetation in a rugged volcanic terrain. Photo: Mark Mitchell, TPWD.

Chapter 5 – Ecoregions of Texas | 67

The most arid of Texas's ecoregions, the Trans-Pecos is perhaps also the most complex. Over 38,000 square miles in size, it is sometimes called the Trans-Pecos Mountains and Basins to reflect the variation in topography. Elevations range from 2,500 feet above sea level on the desert floor to 8,749 feet on Guadalupe Peak. Rainfall ranges from less than 9 inches per year in El Paso to more than 20 inches on mountain tops. **Insolation** (sunshine levels) and evaporation are high, adding to water deficits. Some mountains are characterized by volcanic rocks, but others are founded on limestone. The variation in geology, terrain, and climate produce a wide variety of soil types. Not surprisingly, the region is divided into multiple different subregions, all with distinctive vegetation types.

The Desert Scrub subregion occurs in the flat basin areas of the Trans-Pecos and includes drought-tolerant species typical of the Chihuahuan Desert, such as creosote bush, yuccas, ocotillo, and lechuguilla. Basin areas where water collects and evaporates can leave behind salty soils dominated by salt-tolerant plants that define the Salt Basin subregion. The Desert Grassland subregion, which supports short- and mid-height grasses, occurs in deeper soils at slightly higher elevations, such as mesas or lower mountain slopes. Higher up mountain slopes, the Mountain Ranges subregion can support forests of Mexican pinyon pine, oak, and other alpine species. To the east, the Stockton Plateau supports a short shrub savanna interspersed with springs similar to the western Edwards Plateau. The Sand Hills overlap the shinnery oak habitats of the southern High Plains.

With high vegetation diversity, the Trans-Pecos supports high wildlife diversity as well. Seventy species of mammal occur here, including the only breeding populations of black bear in Texas and the largest number of mountain lions in the state. The mountains of the Trans-Pecos funnel many Mexican birds into Texas. It is the only region in Texas where birders can reliably find the Montezuma quail, common black hawk, acorn woodpecker, Colima warbler, and numerous others. Unique bats and moths pollinate desert plants. Desert springs support endemic populations of pupfish found nowhere else. Almost half the plant species found in Texas occur in the Trans-Pecos. Many are endemic, including highly sought-after cacti.

The Trans-Pecos contains Texas's two largest national parks, Big Bend and Guadalupe Mountains, and several large TPWD and Nature Conservancy properties, drawing tourism to the region. However, much of the region is in large private ranches. These ranches graze cattle while also providing critical wildlife habitat.

As noted above, ecoregions have distinctive characteristics, but they also have a diversity of vegetation communities within them. Figure 5.16 illustrates that each natural region can be divided into several subregions. Chapters 7-10 explore the characteristics, conservation, and management of the forests, grasslands, desert scrublands, and wetlands scattered across the various subregions.

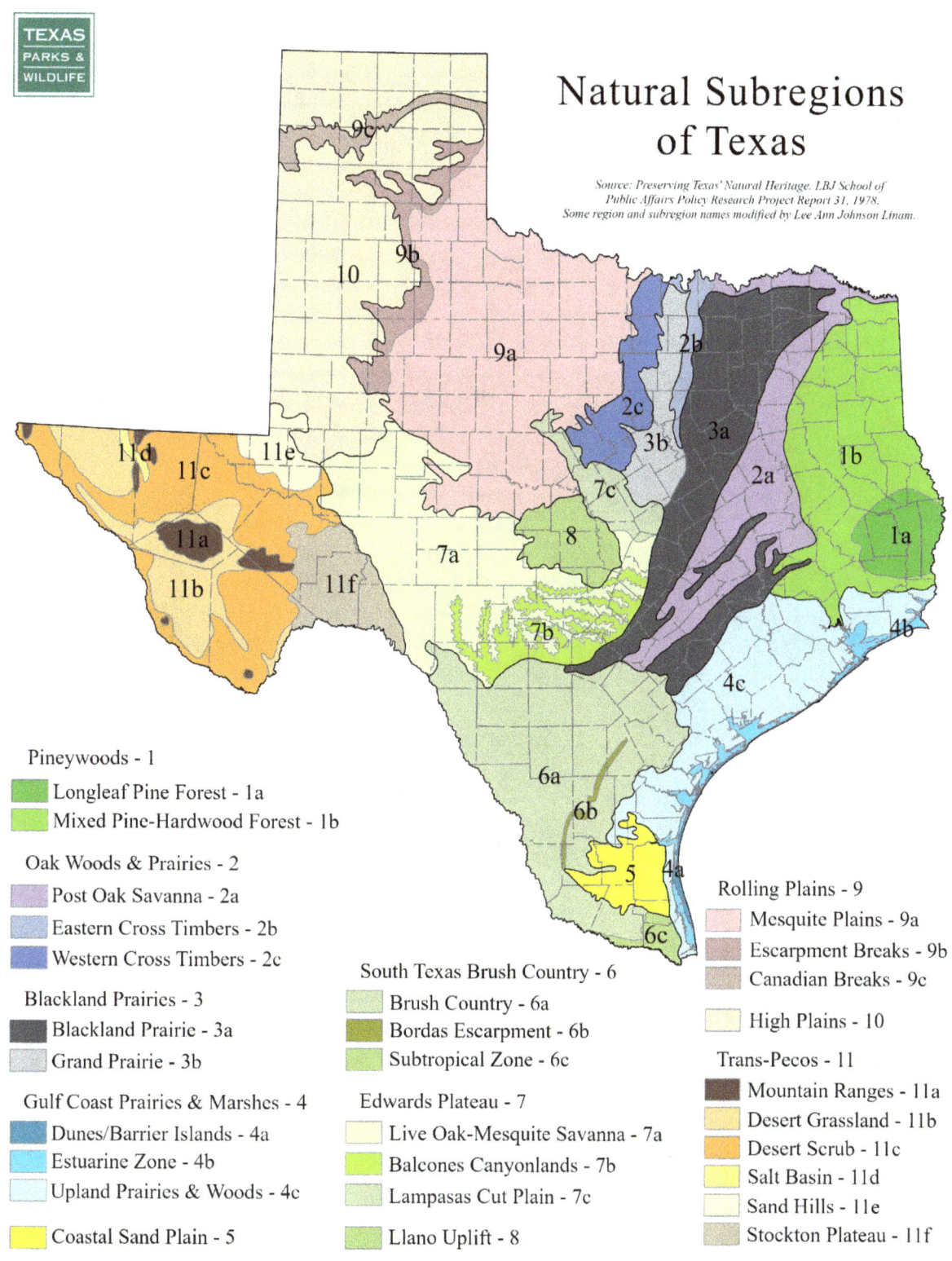

Figure 5.16. Subregions found in the natural regions of Texas reflect variations in topography, soils, and vegetation. Source: LBJ School of Public Affairs. Modified by Paul Daugherty, TPWD IT-GIS.

CHAPTER 6 – HABITAT MANAGEMENT

Wildlife depends on habitat!

Conserving wildlife isn't just a matter of saving animals; it means saving the places they live—their habitats. **Habitat** is the natural environment that provides all the essential elements a species needs. Because each species' needs differ, habitats differ for different types of animals. For example, thrashers and woodpeckers are both birds found in forests. Thrashers require thick, brushy habitat near the forest floor, whereas woodpeckers are mostly found along trunks that emerge in open forest layers. Sometimes habitat differs depending on the sex of the animal. Female Cagle's map turtles spend most of their time in quiet pools in rivers, but males spend more time in riffles and transition areas. Understanding habitat and how to create it is a critical component of wildlife science.

PLANTS AS A FOUNDATION

Topography, geology, soils, and climate shape habitat. However, habitat is often defined by the plants, or vegetation, associated with an animal. Texas is home to about 5,000 species of plants. Of these, 367 are **endemic** to Texas—found nowhere else outside the state. Plant identification is important to wildlife scientists for several reasons. Some wildlife species may depend on a specific type of plant, such as the association of Mexican long-nosed bats with agave and monarch butterflies with milkweeds. Other types of plants are significant because they benefit a large variety of wildlife. One example is *Croton*, a genus of plants often called doveweed because it produces large seeds favored by doves and other seed-eating birds.

(opposite) Figure 6.1. Land management activities, such as grazing or prescribed fire, can affect wildlife habitat. Source: Shutterstock.

IDENTIFICATION OF PLANTS

The ability to identify plants helps wildlife managers assess important shelter or food plants (favorite wildlife food plants are called "ice cream plants!") and remove **invasive species**. Biologists often use a tool called a dichotomous key as they learn to identify different plant or animal species (Highlight 6.1). Dichotomous keys and reference books called field guides use characteristic parts of plants to help identify species. Some distinguishing parts include:

- *Roots* – Roots absorb water and nutrients for the plant. They are usually classified either as tap roots, with one primary well-developed root, or fibrous roots with a network of fine, branched hair-like filaments.

- *Stems* – Stems provide the main body structure of a plant. In some species they are herbaceous (green and flexible). Herbaceous stems can be hollow or solid. In other plants, stems may be woody, such as the trunk of a tree. The bark of woody stems, especially tree trunks, may also aid in identification.

- *Leaves* – Leaves are the primary part of the plant where photosynthesis occurs. They derive their green color from chlorophyll, a catalyst used in photosynthesis. Leaf characteristics are very helpful in identification. Botanists examine the shape of the leaf, the arrangement of the leaves on the stem, whether the leaf is one simple leaf or a compound leaf composed of many leaflets, and the pattern of the veins in the leaves (Figs. 6.2 and 6.3). Some evergreen plants have scale-like or needle-like leaves.

- *Flowers* – Flowers are the reproductive structure of the plant. Some species have well-developed, showy flowers to attract pollinators, but wind-pollinated plants may have small inconspicuous flowers. Some flowers have both male (stamen) and female (pistil) parts, but some plants produce separate male and female flowers. **Monoecious** species bear both male and female flowers on one plant. In contrast, in **dioecious** species the male and female flowers occur on separate plants.

Fruits – After flowers are pollinated, the seeds develop within structures called fruits or cones. Fruits can vary from the small, hard grains produced by grasses to larger, fleshy structures designed to attract wildlife to eat the fruit and disperse the seeds. **Mast** is a broad term that refers to the various nuts and fruits produced by woody plants.

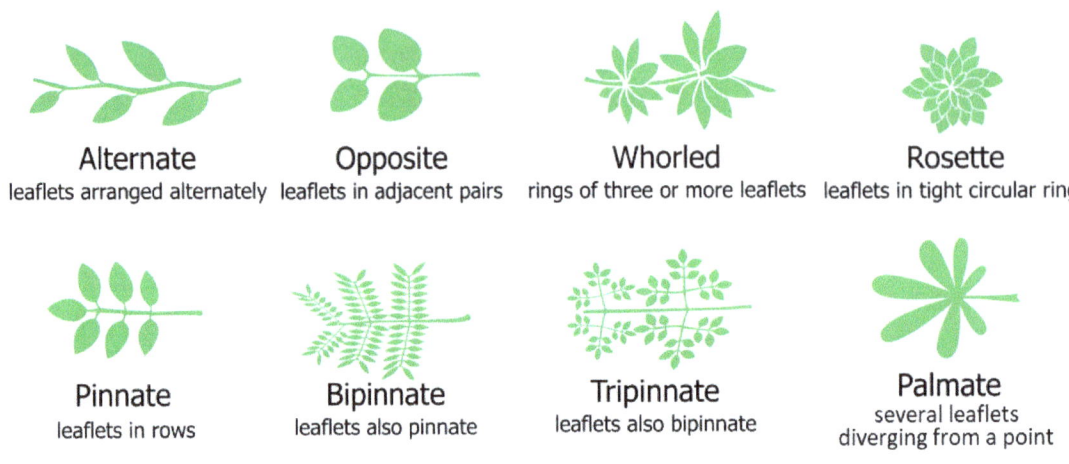

Illustrations modified from Leaf_morphology_no_title.png: User: Debivort. Wikimedia Commons. CC BY-SA 3.0

Figure 6.2. Various leaf arrangements.

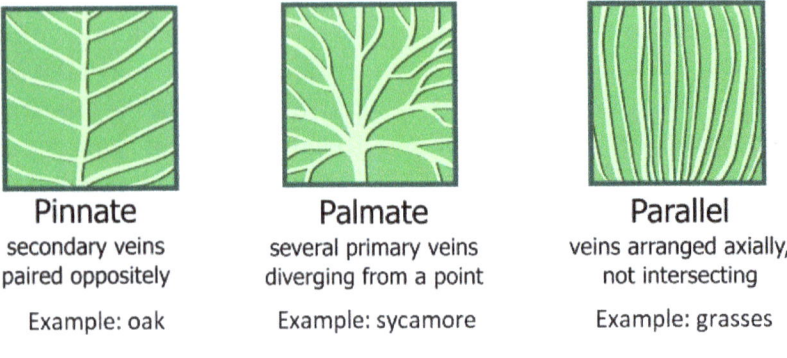

Illustrations modified from Leaf_morphology_no_title.png: User: Debivort. Wikimedia Commons. CC BY-SA 3.0

Figure 6.3. Examples of leaf venation.

HIGHLIGHT 6.1 - SAMPLE DICHOTOMOUS KEY

Dichotomous keys consist of pairs of questions. The user begins with the first pair, or couplet, and then uses the results to obtain an answer or proceed to another pair of questions. The example below examines the characteristics of the leaves of a few common **evergreen** plants in Central Texas.

Dichotomous Key for Selected Central Texas Evergreen Trees and Shrubs

1. Has scale-like leaves...Ashe juniper
1. Leaves are not scale-like..2 (go to question pair #2)

2. Leaves are compound..3
2. Leaves are simple...5

3. Leaflets are in palmate groups of 3, tipped with spines............................Agarita
3. Leaflets are arranged pinnately..4

4. Leaflets are shiny dark green, arranged in pinnate groups of 7-11............Texas mountain laurel
4. Small leaflets with pointed tips, usually in pinnate groups of 5................Texas ash

5. Leaves small, stems branch out at right angles...Elbow bush
5. Larger simple leaves, stems not at right-angles...6

6. Leaves are opposite, velvety texture on bottom..Lindheimer's silktassel
6. Leaves are alternate and leathery..Texas live oak

STRUCTURAL PLANT GROUPS

Sometimes, wildlife managers only need to understand the structure of the plants in a habitat rather than their specific identifications. In these circumstances, plants may be categorized into four basic plant groups:

Grasses – Grasses, classified in the plant family Poaceae, are characterized by jointed stems and small, inconspicuous flowers pollinated by the wind. As members of the Class Monocotyledoneae (the monocots), they have leaves with parallel veins and non-woody stems. Texas has more than 700 grass species, including many species introduced for livestock grazing. Native grasses provide habitat structure for wildlife and food for grazers such as bison, prairie dogs, and pronghorn.

Grass-like Plants – These monocot species may look like grasses but do not have jointed stems like grasses. They are often associated with moist or wet soils, such as sedges ("sedges have edges") and rushes ("rushes are round"), and provide seed sources and cover for waterfowl and other wetland species.

Forbs – Forbs are plants with broad leaves that grow new green stems every year. They can be annual, biennial, or perennial. Annual forbs grow from seeds and die each year. Biennial forbs take two years to grow and develop before they die. Perennial forbs have a living part that can survive several years. Some people call such plants weeds, but forbs are valuable plants for wildlife. They provide nutritious and tender green tissue, seeds, and fruit that can be critical protein and energy sources.

Woody plants – Woody plants are perennial and include trees and shrubs. Trees form a dominant main stem called a trunk; shrubs often form multiple stems near the base of the plant. Woody plants can provide important habitat structure for wildlife, such as nesting sites, and wildlife species that are **browsers** feed on new, tender growth of woody plants.

Grass – Bushy Bluestem 	Grass-like – Cedar Sedge 	Forbs – Paintbrushes & bluebonnets 	Woody – American Beauty Berry 

Figure 6.4. Examples from different Structural Plant Groups.

VEGETATION COMMUNITIES

Climate, topography, and soils all affect the species of plants found in a given region. Although individual species can differ across the state in different plant communities, in general the abiotic factors of Texas lead to the formation of four primary vegetation types:

Forests and Woodlands – Forest communities are areas with trees grouped in a way so their leaves, or foliage, shade the ground. Forest habitats in Texas are typically found in the Pineywoods of East Texas, the oak belts of Central Texas, the mountain regions of West Texas, the subtropical areas of the Lower Rio Grande in the South Texas Brush Country, and along river floodplains throughout the state. Forested areas often receive higher rainfall and have looser soil, supporting woody plant growth. Forests provide nesting, shelter, and food sources for wildlife, including mast production from trees. Forest floors may be covered with a thick layer of leaf litter, providing habitat for various moisture-loving species. Texas forests are explored further in Chapter 7.

Deserts - Deserts have low rainfall and much sunshine. In Texas, the Chihuahuan Desert lies in the Trans-Pecos ecoregion of West Texas. The average rainfall is about 10 inches per year and mainly occurs during a limited time in late spring and summer. Desert plants and animals are adapted to survive in high temperatures and low moisture. Reptiles are common, but amphibians are rare. Many mammals in the desert burrow or are active at night. Chapter 9 explores Texas desert habitats.

Grasslands – Grasslands occur at intermediate rainfall levels. Also called prairies, plains, or **rangelands,** they once dominated the central half of Texas. Prairies have minimal woody vegetation; however, due to fire suppression and periods of overgrazing, woody species have encroached on many grasslands, creating brushlands in some areas. The soils in grasslands are deep and rich, formed by years of plant growth and held together by the deep roots of prairie plants. Grassland wildlife includes large grazers like bison and pronghorn and small mammals and birds that rely on the shelter and nesting cover provided by native grasses. Various Texas grassland ecosystems are discussed in Chapter 8.

Wetlands – Wetlands occur where soils hold moisture for much of the year. Grass-like sedges, rushes, or trees adapted to flooded conditions dominate the vegetation. The largest expanse of wetlands in Texas occurs in the Gulf Coast Prairies and Marshes ecoregion, where vast marshes border the Gulf, and in the forested wetlands of the Pineywoods. Wetlands are an especially critical habitat for waterfowl, wading birds, and amphibians. Chapter 10 presents more information on Texas wetlands.

PLANT COMMUNITY DYNAMICS

Left undisturbed, landscapes in Texas tend to reach one of the vegetation types described above as a **climax community**. A climax community is a stable vegetation type that develops over time and includes the plants and animals best adapted to an area's abiotic factors.

Vegetation communities are often affected by disturbances that can alter their characteristics. These disturbances, such as strong winds in a forest, fires in grasslands, droughts in wetlands, heavy rainfall in the desert, or even human impacts like clearing, can change the appearance and nature of the plant communities. Over time, if nature is allowed to take its course, the local climate, soils, and topography tend to bring the vegetation back to its original state in a process known as **ecological succession**. Sometimes, the disturbance is severe, like a volcanic eruption or a landslide that covers the ground completely. In that case, succession must start to rebuild the soil using species like lichens and fungi to break down rocks and add organic matter. This is called **primary succession.** Often, however, the disturbance only affects the dominant plant cover. In those cases, pioneer plant species can initiate **secondary succession** using the remaining soil and organic matter. (Fig. 6.5).

Succession can require a decade to several centuries, depending on the level of disturbance and type of climax community. As the vegetation communities change, the wildlife communities may also vary (Fig. 6.6). Some vegetation communities stabilize in a subclimax condition. For instance, grassland vegetation may have naturally developed into woodlands in the Post Oak Savanna. However, frequent wildfires limit the growth of woody plants, making grassland savannas the dominant habitat. When fire is suppressed, thick understory invades post oak savannas and woodlands spread into the grassland areas.

Figure 6.5. Process of secondary succession following a fire. The habitat begins as a forest community (box 1). The fire removes all vegetation but leaves the soil intact (box 4). As secondary succession begins, pioneering forbs arrive carried by wind and animals and dominate the landscape (box 5), followed by shrubs (box 6) and immature trees (box 7). Finally, after many years, the forest returns to its original structure (box 8). Source: Katelyn Murphy, CC 3.0.

Different stages of forest succession provide different types of habitat. Young forests provide habitat for a wide variety of creatures at multiple key times in their lives (i.e. breeding, brooding, migration)

Figure 6.6. Wildlife communities shift as plant succession occurs. Source: Jessica Miller Mecaskey, "Wildlife Diversity in Successional Habitat," 2022.

Another dynamic in vegetation communities is non-native or exotic plants. Non-native plants that harm vegetation communities by outcompeting or excluding native species are called invasive species. The Federal Invasive Species Advisory Committee defines invasive species as "alien species whose introduction causes or is likely to cause economic or environmental harm or harm to human health." Invasives are the second most significant threat to native species diversity after habitat loss. Examples of non-native plants that have taken over habitats in Texas include:

- Macartney rose and Chinese tallow in Texas coastal prairies
- Kudzu in East Texas
- Saltcedar (also called tamarisk) in the Trans-Pecos
- King Ranch bluestem (also called yellow bluestem) in the Edwards Plateau
- Buffelgrass and Guineagrass in the South Texas Plains
- Giant reed and non-native phragmites along waterways
- Purple loosestrife in prairies
- Japanese honeysuckle, privet (*Ligustrum*) species, and chinaberry trees near urban areas

Invasive species can rapidly increase in population since they usually lack natural biological control. Landowners and homeowners can help by using native plants in landscaping and habitat restoration.

WILDLIFE HABITAT REQUIREMENTS

Plants are an important part of wildlife habitat, but they are not the only components of habitat. Habitat must provide four basic requirements for animals: food, water, shelter, and adequate space. Any of these factors that are lacking or in short supply are said to be **limiting factors**. If limiting factors are present, they can reduce the habitat's **carrying capacity**. Carrying capacity is the maximum number of individuals of a species that a habitat can support indefinitely. By addressing limiting factors in these four areas, wildlife managers can increase the habitat quality and its carrying capacity.

FOOD[2]

Animals vary in their food requirements. Some species are dietary specialists, such as monarch butterfly larvae that only feed on milkweed. In contrast, other species like raccoons and ravens are dietary generalists. Managing dietary specialists can be challenging. For example, one obstacle for biologists attempting to reintroduce endangered black-footed ferrets is that their specialty prey, prairie dogs, have also dramatically declined in number.

The diets and energy needs of species can change throughout the year. For instance, migratory birds require high-energy food in the weeks leading up to **migration**. Javelinas can use prickly pear cactus pads as a vital food source during drought and stress but need fruits, seed pods, and tubers with higher amounts of protein and fiber to sustain themselves long-term.

Diets can also vary during different stages of the life cycle. Adult gallinaceous birds, like quail, prairie-chickens, and turkeys, primarily eat plants as adults but require insects as chicks. Therefore, suitable habitat for these birds should include plants that produce seeds and an abundance of insects. In contrast, many frog tadpoles are herbivorous, whereas adult frogs primarily eat insects. Because of these diverse dietary requirements, managing healthy habitats is more effective than directly feeding wildlife.

WATER

Water is essential for life, although wildlife species vary in the amount of water they require. Some have remarkable adaptations for minimizing their water needs. Kangaroo rats can extract water from the seeds they eat. Desert tortoises can concentrate their urine. Texas horned lizards can harvest even tiny amounts of rainfall by collecting it with their scales. Other species, such as otters, beavers, waterfowl, wading birds, and most amphibians and turtles, are inextricably tied to water. These species require both abundant water and the plant and animal life it supports. In the middle, most wildlife species thrive best when water is readily available in their habitats.

Wildlife can access water from various sources, and habitat management can maximize its availability and quality. One important goal is to protect water quality and infiltration by reducing erosion and siltation. Some valuable strategies include:

- Maintain riparian vegetation along streams
- Prevent overgrazing
- Use erosion barriers, such as creating brush fences on slopes and maintaining healthy native grass cover in watersheds.
- Install small check dams in drainages to maintain pools of water during dry conditions. (Note: Dams on larger streams may require a permit.)
- Use **prescribed fire** and selective brush clearing to maintain healthy native grasslands, which feed underwater aquifers and enhance spring flow.
- Construct **wildlife water guzzlers** to capture, store, and release rainwater. These structures can provide water during dry periods when wildlife can be under stress (Fig. 6.7).

---

[2] *Clip-art from RawPixel Public Domain Collection.*

Figure 6.7. A wildlife guzzler collects water during rainfall and directs it into reservoirs where wildlife can drink from it. Photo 1: Mark Mitchell, TPWD. Photo 2: Lee Ann Johnson Linam.

SHELTER

Shelter is the vegetation or physical structure that provides protection from environmental threats. Wildlife requires shelter for nesting, resting, avoiding weather extremes, and escaping predators. Shelter may come from vegetation (sometimes called "cover") or the habitat's physical structure. For example, screech owls require cavities in trees for nesting cover; peregrine falcons use cliff edges. Similarly, some bat species, such as Brazilian free-tailed bats, use caves for roosting. Others require hollow trees, loose bark, or even palm fronds. To avoid weather extremes, desert animals use vegetation as shade to escape the midday sun. Other species, such as spadefoot toads, burrow themselves in loose sandy soil. Prairie dogs also use soil to construct extensive burrows to escape predators. Cottontail rabbits may dash into brush thickets. All these biotic and abiotic components can help meet shelter needs for wildlife.

ADEQUATE SPACE

Habitat managers must be concerned with issues of both quality and quantity. When habitat is managed for any species, it is important to understand how much habitat is needed. An individual animal usually has a range of familiar areas that it passes through regularly to obtain food, seek shelter, or find mates. This area is called its **home range**. The home range may be less than one acre for some species, such as hispid cotton rats. Home ranges may be as large as 100-150 square miles (64,000-96,000 acres) for other species, such as wolves and mountain lions. Habitat areas must be large enough to provide home ranges for many members of a species so that a viable population can exist.

In addition, some species are territorial, meaning they seek to exclude other members of the same species from their home range. Wintering whooping crane pairs and families in coastal Texas claim and defend wetland **territories** that average 425 acres in size. Wildlife scientists can use that knowledge to estimate how many acres of wetlands must be conserved on the Texas coast to meet a recovery goal of 1,000 whooping cranes. On the other hand, sometimes species overlap and share home ranges. Whooping cranes sometimes leave their marsh territories to feed in uplands, often sharing these habitats with other whoopers. The home range size for the northern bobwhite quail is about 40 acres; however, several coveys (family groups) of quail might overlap in their use of habitat during winter.

The arrangement of space is also relevant. If habitats are broken into small pieces or "fragmented," then wildlife may be unable to travel safely between different areas of habitat they need (Fig. 6.8). **Fragmentation** occurs frequently when human development, such as roads or housing areas, is inserted into wildlife habitat. In such circumstances, wildlife scientists may create **habitat corridors** connecting different habitat patches so that wildlife can safely move around.

Other corridors may be more artificial. For example, only two populations of endangered ocelots, a beautiful spotted cat more common in Mexico and Central America, occur in South Texas. However, some of the few remaining fragments of thick woodland habitat near Laguna Atascosa NWR are separated by busy roads. After several ocelots were killed by cars (up to 10% of the population in some years), the Texas Department of Transportation constructed tunnels under the roads, allowing ocelots to safely move between habitat fragments. Biologists were delighted when remote game cameras detected the ocelots and more than one dozen additional wildlife species using the tunnels (Fig. 6.9). Similar tunnels have been installed in Bastrop County to aid the endangered Houston toad as it moves between habitat patches.

Finally, it is essential to understand species' preferences for habitat diversity. Some species are **edge species** or species that prefer an intersection of different vegetation types. Many game species are edge species. White-tailed deer prefer habitats that mix open areas for foraging with wooded habitats for escape cover. On the other hand, some species prefer expanses of uniform vegetation. Prairie-chickens avoid grasslands interrupted by woodlands that can provide perches for predators. Golden-cheeked warblers are more likely to be found in unbroken forests because forest edges offer easier access for nest parasites like brown-headed cowbirds.

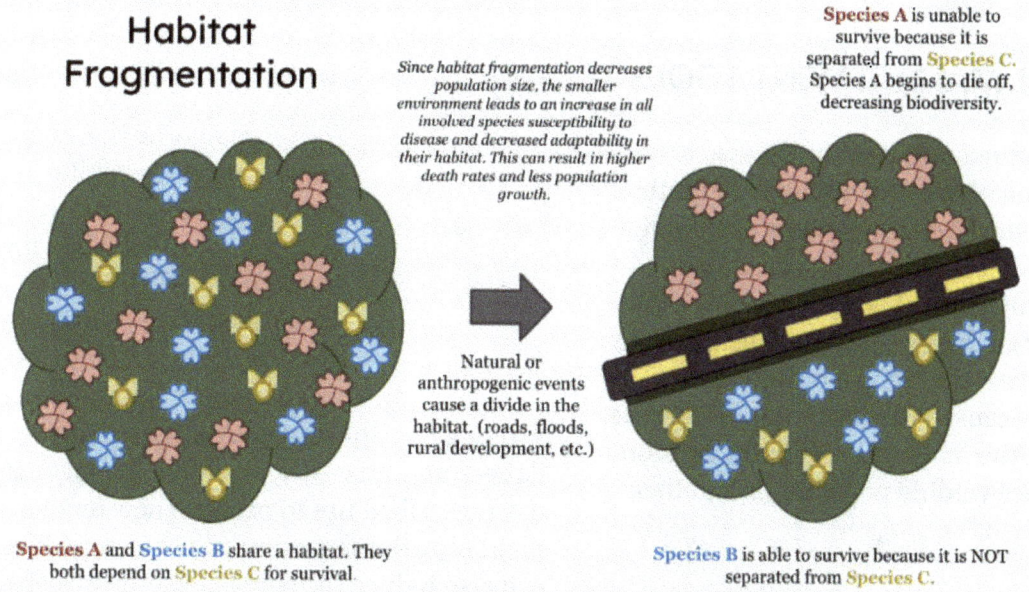

Figure 6.8. Habitat fragmentation can affect species survival. Source: LlacaUrbden1, CC 4.0.

The ocelot is a beautiful, endangered cat that occurs in the U.S. only in a few habitat patches in far South Texas. Photo: Steve Sinclair, USFWS (public domain).

Ocelot habitat in South Texas. Photo: USFWS Headquarters, CC 2.0.

Ocelot with radio-transmitter collar using highway tunnel between habitats patches. Photo: TxDOT/USFWS (public domain).

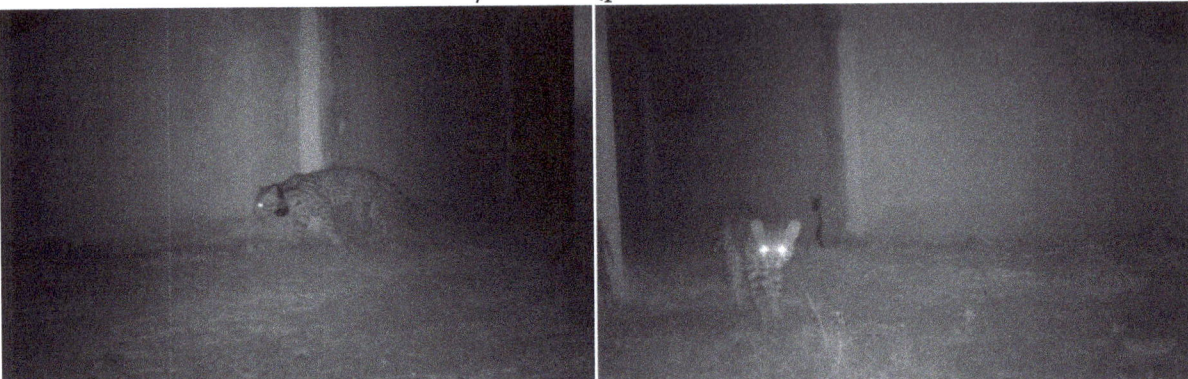

Figure 6.9. The Texas Department of Transportation (TxDOT) and USFWS worked together to create safe corridors between patches of ocelot habitat.

HABITAT MANAGEMENT TOOLS

Habitat management involves a combination of disturbance and restoration to provide the four components of wildlife habitat. Aldo Leopold emphasized that wildlife management is a purposeful, hands-on activity guided by specific goals. Leopold suggested that the same tools that had historically caused harm to wildlife—the axe, plow, cow, fire, and gun—could help restore and enhance wildlife populations and their habitats.

FIRE

Fire may be the single most important habitat management tool for wildlife scientists. A prescribed fire, sometimes called a controlled burn, is a fire set intentionally for land management purposes. Although out-of-control wildfires can be dangerous and costly, a wildlife manager can use prescribed fire to produce new tender plant growth for grazers and browsers. Fire can also reduce the spread of some invasive plant species. It prevents woody growth in grasslands and maintains open landscapes for pronghorn and grassland birds. In savannas, brushlands, and woodlands, it is used to change the habitat structure, providing more open areas for movement and more edge habitat. Even the red-cockaded woodpecker, which uses stands of mature pines, also relies on controlled fires that prevent hardwood and understory growth in its habitat.

The use of fire requires an understanding of the growth season of different plants so that it can be set at proper times. In addition, because of the risks to both humans and wildlife associated with setting fires, the manager needs to understand safe weather conditions, fuel loads, and fire control methods. The Texas A&M Forest Service and the Texas Department of Agriculture offer training in the use of this essential management tool.

COW

Bison were historically a part of the Texas natural landscape throughout nearly all the state, except perhaps far South Texas. As bison passed through the state, their grazing and hooves impacted the vegetation communities where they roamed. Today, wild bison no longer exist in Texas except for the official State Bison Herd found at Caprock Canyons SP. Still, wildlife managers use grazing by cattle to simulate some of the ecosystem changes the bison brought.

Cattle grazing can disturb the soil in a grassland so that forbs produce more seeds for doves and other seed-eating birds. Grazing can also be used to modify habitat structure, such as creating open bare ground for shorebirds to nest or prairie-chickens to display. In contrast, grazing can be reduced in adjacent prairie areas to retain heavy cover for hens or chicks. Grazing also has the potential to limit woody plants. On the other hand, overgrazing can remove too much wildlife cover, contribute to erosion, and result in invasion by brush, especially in areas where mesquite and Ashe juniper are found.

PLOW

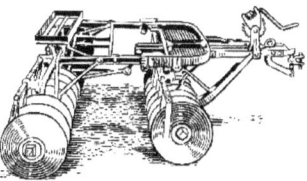

Historically, plowing destroyed wildlife habitats, especially North American grasslands. From 1850 to 1990, around 290 million acres of prairie were lost, mostly converted to cropland. Although the loss of prairies was severe, these agricultural landscapes can be managed to offer some limited wildlife habitat. Native plants along edges and crop residue can provide wildlife food and cover. In some places, planting woody species as windbreaks within the fields or along the edges can provide cover habitat for small mammals and birds. When farmers re-flood rice fields after harvest, they create excellent habitats for waterfowl.

Agricultural tools can be used to restore habitat as well. Tractors, plows, seeders, and planters are used to restore grasslands and forests. Although planting food for wildlife is not usually needed in healthy natural habitats, small food plots can help draw wildlife to certain areas and aid wildlife nutrition during stressful seasons, such as during late winter or fawning.

AXE

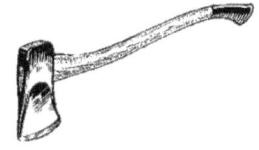

Leopold recognized that timber cutting (represented by the axe) profoundly influenced wildlife habitat in North America. In Texas alone, more than 90% of the original longleaf pine forests were cut. **Bottomland hardwoods** also were cut over, contributing to the extinction of the ivory-billed woodpecker and Carolina parakeet. In modern times, more sustainable timber harvest is often embraced, and selectively removing trees can have a role in habitat management.

Although large tracts of intact forest are desirable for many forest species, such as golden-cheeked warblers, Louisiana waterthrush, and eastern red bat, other species benefit from a mixture of different habitats. Windstorms and fire naturally created a combination of open areas and shelter areas in forests. By harvesting trees in small patches or selectively cutting only certain trees, foresters can mimic natural disturbance while leaving some areas untouched.

In savannas that have been overgrown by woody vegetation and in brushlands that have developed after overgrazing, tools that mimic the axe can help restore open habitats and edges. Tractors can be used to bulldoze, grub, chain, rake, roller-chop, or hydro-axe. However, operators should take care to minimize soil disturbance and compaction. When brush is cleared, it can be stacked into brush piles to provide escape cover for wildlife. Browsing by goats and herbicides can also accomplish brush clearing, although managers should monitor their impact on other plant species.

GUN

Before wildlife harvest laws were adopted, unregulated hunting was one of the primary threats to wildlife in North America. However, Leopold added the gun to his list of tools to emphasize that hunting can help manage wildlife. Chapter 11 discusses the issues related to wildlife overpopulation—one reason why guns can be a tool to protect wildlife habitats. As noted in Chapter 2, hunting also generates income for wildlife conservation.

MONITORING

It is important for the land manager to be able to measure the habitat characteristics and the effect of management strategies. Appendix B presents some methods for monitoring vegetation communities using **quadrats**, **transects**, and point-based methods. Later chapters present methods for monitoring wildlife populations.

LANDSCAPE FOCUS

An individual landowner or wildlife manager can significantly improve wildlife habitat on their property. However, conservation of species is often most effective as a group effort. White-tailed deer and wild turkey have average home ranges of 640 acres (about one square mile), and both can shift seasonal habitat use over several miles, so to manage a population of species such as these, smaller landowners may need to work together. TPWD helps facilitate the development of joint management plans in cooperatives formed of private landowners. These Wildlife Management Associations develop mutual goals and strategies and reap collective benefits.

Some conservation strategies operate at an even larger scale. For endangered species, USFWS offers landowners the chance to protect themselves and the species by participating in Habitat Conservation Plans (HCP). Regional HCPs can balance threats to the species with conservation strategies implemented across a large project area and many landowners. Safe Harbor HCPs help landowners overcome fear of repercussions if habitat improvements serve to attract endangered species to their habitats.

Figure 6.10. The Alum Creek Wildlife Management Association works together to meet habitat management goals. This wildlife cooperative in Bastrop County is also a part of a Habitat Conservation Plan that helps landowners feel comfortable in supporting recovery of the endangered Houston toad. Photo: Tom Hausler.

Finally, some species truly require a landscape-scale perspective for conservation. Many governmental and non-profit organizations now have landscape conservation programs designed to build partnerships across large ecosystem areas or watersheds to conserve wildlife habitat. A ground-breaking effort in this regard is a Lesser Prairie-chicken Range-wide Conservation Plan developed by the Western Association of Fish and Wildlife Agencies with several private and public partners. When pollutants are present in the environment, such as the story of lead poisoning of waterfowl in Chapter 3 and in the story of DDT in Highlight 6.1, habitat management must also be bigger than the local scale—sometimes a national or international scale is needed.

HIGHLIGHT 6.1 - DDT AND SILENT SPRING – CLEANING UP WILDLIFE HABITATS

When the pesticide DDT was introduced during World War II, it was hailed as an answered prayer. The powerful organochloride (a carbon-based compound containing chlorine) was widely sprayed to kill the insects that spread malaria, yellow fever, typhus, and bubonic plague in war zones and then on the home front after the war.

DDT was the first modern synthetic insecticide, and, at the time, environmental chemicals were not widely tested for safety. It was not yet evident that DDT could cause neurological damage in humans and wildlife, and that, once present, the compound would not break down. Because of its persistence, wildlife in environments with DDT residues began to store the chemical in their body tissues. As secondary consumers fed on primary consumers, they concentrated more DDT in their tissue. This **bioaccumulation** continued up the food chain until higher-level predators had dangerously high concentrations of DDT (Fig. 6.11).

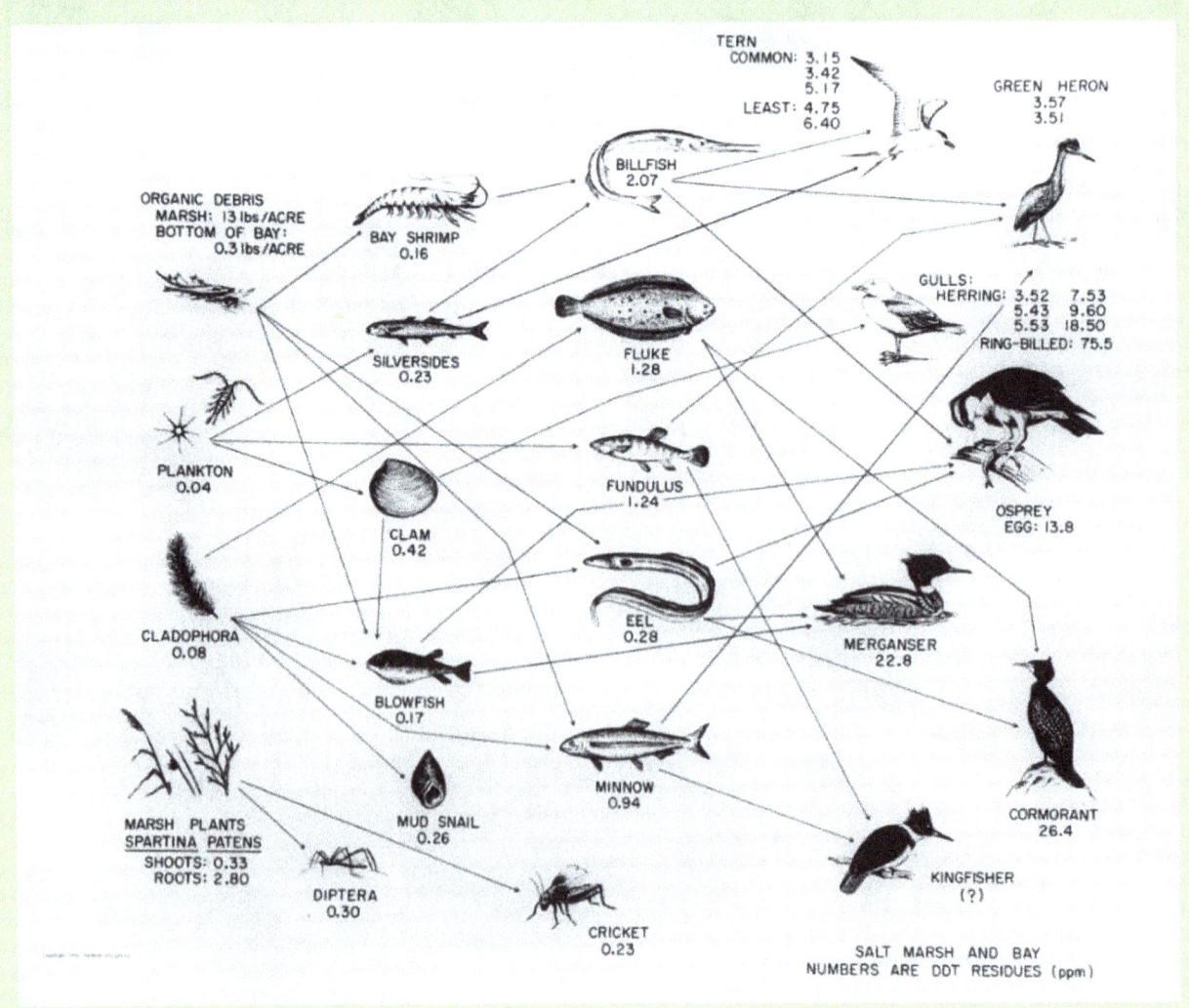

Figure 6.11. Concentration of DDT within an aquatic food web. Source: ENERGY.GOV (public domain).

Some organisms died directly from DDT (biologists reported numerous die-offs of robins feeding on contaminated earthworms), but it had a different indirect devastating effect on birds. The chemical prevented some birds from effectively using calcium to build eggshells. Several species, including ospreys, peregrine falcons, brown pelicans, cormorants, and bald eagles, began experiencing population crashes, as weakened eggshells could not support the weight of the incubating parent birds.

Some biologists noticed the declines, and the U.S. Department of Agriculture began restricting DDT use in the 1950s due to declining effectiveness and the mounting evidence of environmental effects. However, the public finally took notice when a book by marine biologist and nature writer Rachel Carson called *Silent Spring* was published in 1962. In *Silent Spring,* Carson sounded the alarm that increasing levels of pollution, especially DDT, could lead to the collapse of wildlife populations. *Silent Spring* provided impetus to the environmental movement that resulted in the formation of the U.S. Environmental Protection Agency and the passage of the Clean Water Act, the Clean Air Act, and the Endangered Species Act. In 1972, environmental awareness finally led to banning the use of DDT.

Much DDT remained in the environment, but slowly, concentrations began to decrease. As the habitat became less toxic, bird populations began to recover. In the 1970s, there were only four bald eagle nests in Texas. As of 2023, there were nearly 200 eagle nests in Texas. (Texas is also home to hundreds of wintering eagles that nest in the northern U.S.) There were 50 brown pelicans in Texas in 1964; populations are now estimated at over 20,000 in the state. Osprey, cormorant, and peregrine falcon numbers exhibited similar rebounds. Protecting habitats from further contamination offered the opportunity for some impressive endangered species success stories!

Figure 6.12. After DDT levels declined in the landscape, bald eagles and brown pelicans have made a remarkable recovery in the state. Eagle nesting photo by Mark Alan Storey. Brown pelicans photo by Lee Ann Johnson Linam.

CHAPTER 7 - TEXAS WOODLANDS

What a sight the original forests of Texas must have been for settlers arriving here! The Pineywoods had towering pine over 120 feet tall and open longleaf pine flatwoods where travelers could drive their carriage between the ancient trees. There were low-lying woodlands so overgrown with a rich variety of lush vegetation that the area was dubbed the "Big Thicket." In the Cross Timbers, the bands of timber were so intimidating writers described them as "forests of cast iron." Black bears, mountain lions, and Merriam's elk roamed mountaintop islands of pine and Douglas-fir in the Trans-Pecos. In South Texas, 40,000 acres of verdant palm forest greeted Spanish explorers, leading them to christen the Rio Grande "Rio de las Palmas." Over the years, many changes have occurred in Texas forests, but they still provide beautiful landscapes. They also still offer opportunities to manage and conserve.

WHAT IS A FOREST? AND WHAT GOOD IS A FOREST?

There are several types of woodlands (vegetation communities dominated by woody vegetation) in Texas. A **forest** is a habitat with trees grouped in a way so their leaves, or foliage, shade the ground. Forests can be classified as upland (occurring on predominantly dry soils) or lowland (occurring on soils that are often wet). Forests are **deciduous** (hardwood trees that seasonally lose their leaves), evergreen (trees, usually with softer wood, that stay green year-round), or mixed. Several other vegetation types are dominated by woody vegetation. Brushlands are woodlands of shorter species, usually less than 25 feet tall. A **savanna** contains patches of trees scattered within a grassland.

(opposite) Figure 7.1. Riparian forest with an understory of dwarf palmetto in Palmetto SP.

Figure 7.2. A diversity of canopy heights, as provided in this mature tract of hardwood forest in the Davy Crockett National Forest, supports a wide variety of bird species. Photo: William L. Farr, CC 4.0.

Figure 7.3. This educational poster from TPWD illustrates some of the many species associated with forests in the Northern Pineywoods.

The distribution of forests in Texas is strongly associated with rainfall, with higher rainfall areas producing forests. Even within a forested landscape, forests occur in a mosaic influenced by geology, slope, drainage, elevation, and soil types. Forest vegetation often occurs at multiple layers (Fig. 7.2), allowing for many niches and rich biodiversity. In fact, bird species diversity correlates with forest height and the number of canopy layers.

Forests provide many benefits. Forests contribute $23 billion annually to the Texas economy and support more than 144,000 jobs. They are a source of wood for buildings, posts and poles, furniture, paper, and fuel. They also are a source of non-wood forest products, including medicines, cosmetics, and food such as nuts and fruits. In the late 1800s and early 1900s there was a turpentine industry in East Texas that produced medicines, paint, rubber, soap, and varnish, all by tapping the living cambium layer of longleaf pine, much as sugar maples are tapped for syrup in the Northeast.

Forests also provide **ecosystem services**—outputs from nature that benefit human welfare. Intact forests help keep watersheds clean, protect biodiversity, produce oxygen, clean the air, regulate the climate through carbon storage, reduce heat through shading and transpiration, and participate in the rock and nutrient cycles by building soil and aiding decomposition. USFS estimates that the ecosystem services contributed by forests in the U.S. are worth $114 billion in benefits every year!

Finally, forests provide a place for recreation and relaxation. Forests are popular sites for hiking, camping, hunting, and wildlife watching. Over 1 million people visit national forests in Texas yearly; most of them engage in wildlife-related activities. In 2017, these visitors spent over $39 million during their recreation. Perhaps more importantly, recent psychological research suggests that time spent in forests lowers stress, improves mood, decreases depression, and boosts the immune system. Those findings have led mental health experts to prescribe "forest therapy."

TEXAS FORESTS

Woodlands in Texas cover more than 62 million acres, about 38% of the land area of the state. Overall, 94% of Texas forests are on privately-owned land. In East Texas, where most commercial forest land occurs, 69% is family-owned, timber companies own 23%, and 8% is public. There are four national forests in Texas totaling 650,000 acres, including 37,000 acres designated as wilderness areas.

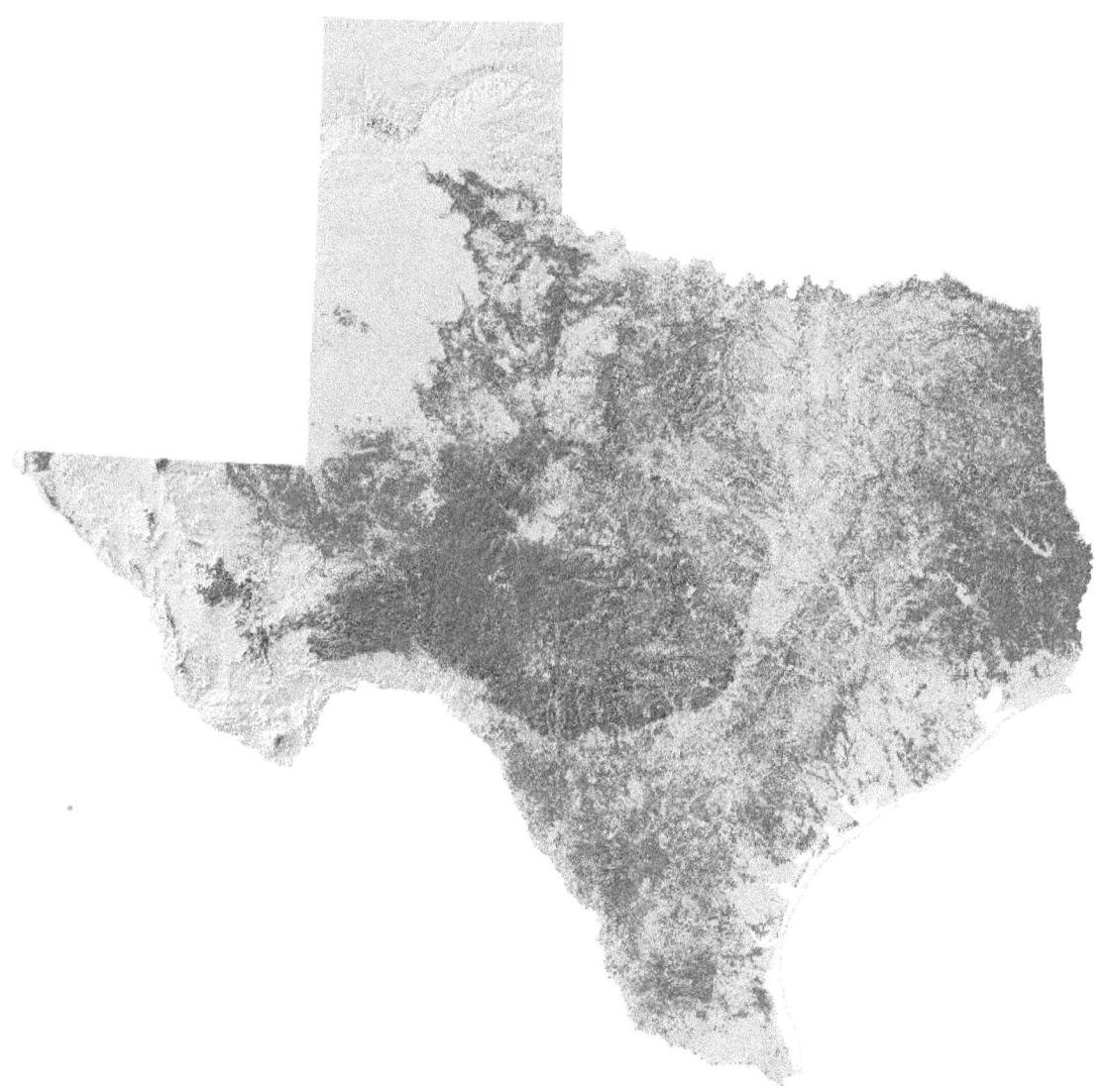

Figure 7.4. Map of Texas woodlands produced in 2017, noting areas where at least 10% of the canopy is shaded by trees and brush. Map courtesy of Texas A&M Forest Service.

There is a remarkable diversity of woodland communities in the state. TPWD recognizes several dozen types of woodlands based on the species that occur in them. At least six of Texas's ecoregions host significant acreages of forest (Fig. 7.5).

Texas is home to more than 250 tree species. More information about the individual tree species that occur in Texas is available from the Texas A&M Forest Service online through their Trees of Texas - List of Trees website.

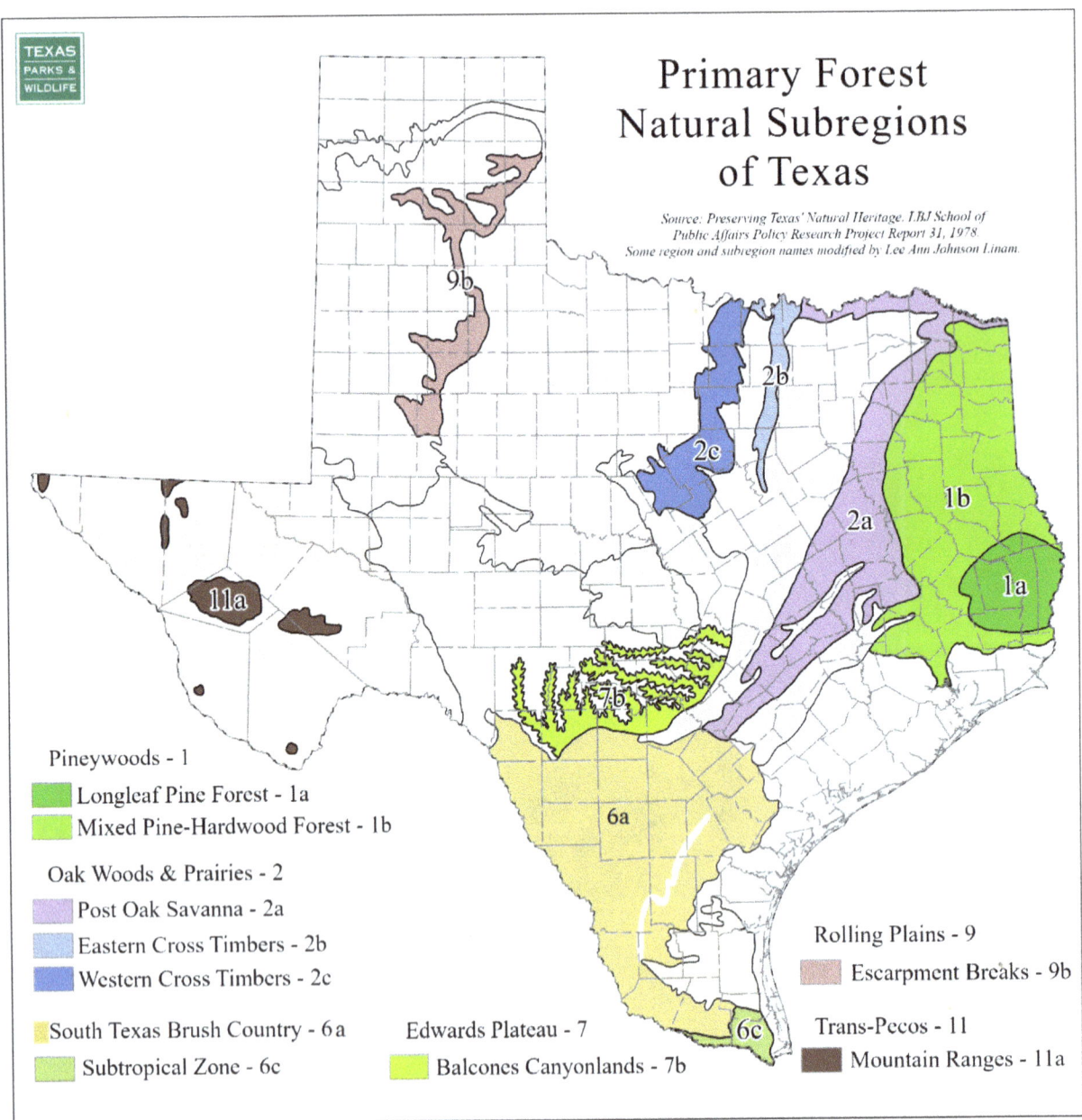

Figure 7.5. Woodland habitats are dominant in several natural regions in the state, including the Pineywoods, Post Oak Savanna, Cross Timbers, South Texas Brush Country, the Balcones Canyonlands of the Edwards Plateau, the Escarpment Breaks of the Rolling Plains, and the Mountain Ranges of the Trans-Pecos. Woodlands also occur along the rivers of the state. Map source: LBJ School of Public Affairs. Prepared by Paul Daugherty, TPWD IT-GIS.

PINE-HARDWOOD FORESTS

Figure 7.6. Loblolly pine forest, W.G. Jones State Forest in Montgomery County, Texas.

When one thinks of forests in Texas, the Pineywoods ecoregion comes to mind. East Texas is home to 12 million acres of forest. These woods are part of the Eastern Deciduous Forest that once stretched across the eastern United States; the region that lies within East Texas is part of the Southern Mixed Forest, indicating a mixture of hardwoods and evergreen pines. Pines dominate these woodlands because of the presence of sandy soils and a history of frequent fire. There were originally three distinctive pine forest types in Texas—loblolly, shortleaf, and longleaf. Loblolly historically grew in the lower slopes and bottoms; longleaf grew in the very sandy soils of southeast Texas; shortleaf grew in association with longleaf in the middle Pineywoods; and in the northeast, shortleaf dominated or grew with an oak-hickory association. Today loblolly is most widely planted because it is preferred for its fast timber growth.

Despite its label as the "pineywoods," East Texas is a mixture of forest types. As noted above, oaks, hickory, and other **hardwoods** sometimes occur in pine forests, producing a diverse mixed pine-hardwood forest. Throughout the region there is a natural mosaic of hilly uplands, mid-slopes, bottomlands, stream and river drainages and swamps that create a variety of forest types supporting hardwood species such as magnolia, beech, ash, pecan, and a variety of oaks.

Several wildlife species are closely associated with mature fire-maintained pine forests, including the red-cockaded woodpecker, Louisiana pinesnake (now thought extirpated), and Texas trailing phlox. There are many lovely spots to view various forest types in East Texas. The I.D. Fairchild State Forest in Cherokee County has a well-managed example of shortleaf pine forest. Boykin Springs Recreation Area in Angelina National Forest and several tracts in the Big Thicket National Preserve have mature longleaf pines. W.G. Jones State Forest and Huntsville SP, both just outside Houston, along with all the national forests in Texas, provide recreational access to loblolly pine forests.

OAK FORESTS

Figure 7.7. An ecotone, or edge habitat, in a post oak savanna in Lick Creek Park, Brazos County.

As one moves west from the Pineywoods, there is a pattern of forest alternating with grassland. These forests, made up mostly of hardwoods, occupy an **ecotone**, a transition zone, between the Eastern Deciduous Forest and the prairies of the Great Plains. The sandy soils of the Post Oak Savanna support mottes of post oak and blackjack oak scattered in a tallgrass prairie parkland. Once maintained by regular fire, the understory of yaupon, winged elm, and eastern red cedar have increased in recent years as fires occur less often on the landscape.

The Houston toad, large-fruit sand-verbena, and American burying beetle are uniquely associated with these sandy savanna ecosystems. Lennox Woods, a lovely property owned by the Nature Conservancy in Northeast Texas, preserves one of the few examples of old-growth post oak woodlands left in the state. The Gus Engeling WMA demonstrates active management of oak savanna habitats.

The Eastern and Western Cross Timbers lie west of the Post Oak Savanna. These two belts of woodlands are underlain by limestone and separated from each other by the Grand Prairie. Early settlers are thought to have given the Cross Timbers their name because, as they were traveling east to west, they had to cross through thick bands of timber interspersed with prairie. The forests, dominated by post oak and blackjack oak, lie at a crossroads of several biomes, so they contain few unique plant or animal species. Once, however, they hosted thousands of wintering passenger pigeons that fed on the abundant acorns. Although passenger pigeons are gone, ecologists have identified thousands of acres of old-growth Cross Timbers forests that are worthy of conservation. Fort Worth Nature Center provides an opportunity to view a mixture of the habitats characteristic of the Western Cross Timbers.

ASHE JUNIPER-OAK FORESTS

Figure 7.8. A view of Ashe juniper-red oak forest in autumn on slopes in the Edwards Plateau, Hays County, Texas.

The Edwards Plateau, also situated on limestone geology, is a diverse and unique ecoregion, hosting 2300 species of native plants, including more than 100 of Texas's 400 endemic species. This ecoregion was historically a mixture of grasslands, savannas, and forests. Forests were most abundant on steep slopes and canyons in the Balcones Canyonlands where fires were less common. These steeper, wetter areas still support a beautiful canopy of Texas live oak, Texas red oak, and Ashe juniper (commonly called cedar). Unique species such as Texas mountain laurel, Mexican buckeye, and Texas madrone can occur in the understory.

Ashe juniper forests provide habitat for endemic species such as the golden-cheeked warbler, cliff chirping frog, and understory trees such as canyon mock orange and Texas snowbell. Balcones Canyonlands NWR, Government Canyon State Natural Area, and Austin city properties that are part of the Balcones Canyonlands Habitat Conservation Plan are ideal locations to explore Ashe juniper forests.

In addition to the forests of the Balcones Canyonlands, woody vegetation has expanded on the Edwards Plateau uplands due to historic overgrazing and fire suppression, and many areas on the former savanna have developed into a brushland dominated by Ashe juniper. These thick stands (both invasive cedar growth on uplands and the original Hill Country forests have been called "cedar brakes") are low in biodiversity. Clearing and prescribed fire in these habitats can restore the savanna and improve the recharge of aquifers under the land's surface. As one goes west in the Edwards Plateau, the woodlands become shorter and more scattered, a habitat structure suitable for species such as the black-capped vireo.

MOUNTAIN FORESTS

Figure 7.9. Mixed ponderosa pine-deciduous woodland in McKittrick Canyon, Guadalupe Mountains National Park, Texas.

Although many parts of the Trans-Pecos ecoregion are desert, several mountain ranges run through the region. With heights above 7,000 feet, rainfall can be twice as high in the mountains as in desert basins below. These mountain slopes and peaks support pinyon-juniper forests with Mexican pinyon pine, alligator juniper, and several unique oak species. Ponderosa pine and Douglas-fir grow in the highest, coolest elevations. In addition, canyons in the Trans-Pecos often have colorful forests of bigtooth maple, Texas madrone, and several oak species.

Mountain forests provide habitat for black bears. The gray-footed chipmunk and Davis Mountains cottontail are found only in the montane forests of West Texas. The Colima warbler, Mexican jay, common black hawk, and acorn woodpecker are popular bird species in the woodlands. Visitors can explore mountain forests in Big Bend and Guadalupe Mountains National Parks.

SUBTROPICAL FORESTS

Figure 7.10. Subtropical sabal palm forest on Sabal Palm Sanctuary in Cameron County provides habitat for a wide variety of unique South Texas species. Photo: William L. Farr. CC 4.0.

At the far southern tip of Texas, subtropical forests dominated by sabal palm once lined the banks of the Rio Grande. These lush forests, with an understory of anacua, Texas ebony, and tepeguaje (also called leadtree), provide habitat for southern yellow bats that roost in the drooping palm fronds. Most of the palm forests were cleared to make way for vegetable and citrus crops in the warm South Texas climate. Now, only small preserves, such as the Sabal Palm Sanctuary, exist, providing habitat for subtropical species such as green jays, Altamira orioles, speckled racers, northern cat-eyed snakes, and Mexican white-lipped frogs.

Other forests in South Texas occur mostly along drainages, both further up the Rio Grande and along inland arroyos (the name given to watercourses formed in the arid regions of the Southwest). These communities are dominated by huisache, honey mesquite, and several subtropical tree species, including guayacan, Texas ebony, and even native citrus trees. Guayacan and ebony produce the hardest wood in North America, and extracts from guayacan are used in medical tests and treatments. These thick forests provide habitat for other South Texas specialties, such as the plain chachalaca and the endangered ocelot. The Lower Rio Grande Valley NWR system manages subtropical woodlands for wildlife species in South Texas.

OTHER WOODLAND AREAS

Outside the subtropical forest zone, the South Texas Brush Country hosts Taumalipan thornscrub communities that are high in diversity of woody species and wildlife. Honey mesquite, huisache, blackbrush, guajillo, cenizo, Texas mountain laurel, and Spanish dagger are found on flats and ridges; additional species such as spiny hackberry, bluewood condalia, Texas persimmon, lime prickly ash, guayacan, and kidneywood occur in arroyos and rocky draws. These brushlands are highly valuable as wildlife habitat, especially for white-tailed deer. Examples of the South Texas Brush Country can be seen at Choke Canyon SP and Chaparral WMA.

Although forests do not dominate in the Coastal Prairies and Marshes or Coastal Sand Plains ecoregions, there are scattered live oak woodlands on sandy ridges along the coast. These oak mottes, with a varied understory of yaupon, redbay, Turk's cap, mustang grape, and other species, are vital bird habitats. During the spring, millions of songbirds comprised of dozens of species figuratively "fall out" of the sky to feed and rest during their northward migrations. Picturesque wind-swept oak mottes can be explored at Aransas NWR, San Bernard NWR, the Candy Cain Abshier WMA at Smith Point, and Goose Island SP.

Woodlands also occur in the Rolling Plains in the rugged topography along the edge of the Caprock Escarpment (subregion 9b in Fig. 7.4). Erosion in that area has created impressive canyons, such as Palo Duro Canyon. The slopes and floors of these canyons support Rocky Mountain juniper (the "hard wood" for which "Palo Duro" is named) and mesquite, redberry juniper (also called Pinchot's juniper), oneseed juniper, cottonwood, willow, western soapberry, and netleaf hackberry. The strings of woodland provide habitat for species such as the Palo Duro mouse.

In addition to the upland forests described above, woodlands can be found in many low-lying areas and along many waterways in Texas. The forested wetland types listed below are explored further in Chapter 10.

In East Texas, **swamps** dominated by bald cypress and water tupelo can occur where water is permanent year-round. The ivory-billed woodpecker and Carolina parakeet once frequented these ancient habitats, but, as swamps were drained, cut or flooded by reservoirs, the birds disappeared. Caddo Lake SP, WMA, and NWR are accessible sites for visiting swamp habitats.

Bottomland hardwood forests occur in other areas of East Texas and the upper Coastal Plain where low-lying soils are frequently flooded by streams overflowing their banks. These forests are some of the most species-rich habitats in Texas, hosting willow oak, water oak, pecan, blackgum, hackberry, overcup oak, cherrybark oak, water hickory, and various elms. They may include up to five times as many species as upland forests. The Rafinesque's big-eared bat finds roosting habitat in the mature trees of these habitats. The Columbia Bottoms in the Mid-coast NWR, Village Creek SP, and the Neches River and Trinity River NWRs host lovely examples of old-growth bottomland hardwoods.

Riparian forests occur in ecoregions across the state along streams and waterways. Pecan, elm, American sycamore, cottonwood, black willow, and bald cypress are components of these communities that help stabilize streambanks and protect water quality. Nesting birds, such as bald eagles, red-shouldered hawks, and kingfishers, find habitat in riparian forests. Several Edwards Plateau state parks, such as Guadalupe River and Lost Maples, provide access to lovely riparian forest. Guadalupe Mountains National Park in the Trans-Pecos is home to a riparian forest dominated by colorful bigtooth maples.

Figure 7.11. Bottomland hardwood forests rely on intermittent flooding and host a rich diversity of species. As seen in this East Texas photo, they may be interspersed with swamps. Photo courtesy TPWD © 2025, (Chase Fountain, TPWD).

Chapter 7 - Texas Woodlands | 97

THREATS TO TEXAS FORESTS

In the Pineywoods, unsustainable timber harvest was the original threat to this ecosystem. Early settlers set up the first known sawmill for processing timber in Texas in 1825 in San Augustine. Timber harvest increased heavily in East Texas throughout the 1800s, and by late in the century, timber was being exported all across the U.S., to Central and South America, and even to Europe. In 1907, Texas lumber production peaked at 2.2 billion board feet harvested (a board-foot is 12" wide by 1" thick by 1' long). In addition, Texas was the nation's largest producer of charcoal from 1883 to 1920.

Unfortunately, early timber harvests in Texas did not plan for the future, and there was little focus on restoring cut forests. When forests were replanted, they were usually planted in **monocultures** of one species—loblolly pine, since it was the fastest-growing local pine species. By the time the Texas Forestry Association formed in 1914, forests in East Texas had changed forever. Now shortleaf pine forests, once dominant in the region, are rare due to land use changes, fire suppression, and hybridization with loblolly. Longleaf pine, which was heavily harvested due to the desirability of its timber, only occupies about 3% of its former range. More than 60% of bottomland hardwood forests have been cut or flooded. The era of the steamboat was the beginning of their demise, as steamboat operators heavily harvested the streamside timber for fuel. However, the greatest impacts to bottomlands came from dam-building and other river modifications affecting flow in the mid-20th century.

Insect pests, some native and some non-native, also threaten East Texas forests. Southern pine beetles (SPB), a native species, have traditionally been a threat to pine production throughout the Southeast. The larvae of SPB feed under the bark of native pines and then bore into the sapwood,

Figure 7.12. Clearcuts of longleaf pine and other species supported an intense and unsustainable logging industry in the 19th and early 20th centuries. Logs were transported by rail and were floated down rivers. Photo 1: USFS, WC (public domain). Photos 2 & 3: Book of Texas (1916), WC (public domain).

eventually killing the tree. Once an infestation begins, the beetles spread quickly into neighboring trees and can result in the deaths of hundreds of acres of pines and the loss of tens of thousands of dollars in timber. SPB are controlled by early detection and cutting of infected trees. Some foresters also cut an additional buffer of trees. Infestations can be discouraged by greater spacing of trees, planting longleaf pine that are more resistant to beetle infection, and planting a mixture of pines and hardwoods. Fortunately, SPB outbreaks have declined in Texas since the 1990s. Populations of several ips bark (or pine engraver) beetle species are known to surge in East Texas during times of drought, also affecting pines.

Non-native threats have also emerged. The emerald ash borer has killed tens of millions of ash trees across the United States and has been detected in Texas. Emerald ash borers have the potential to wipe out all ash species in Texas—seriously affecting rural and urban ecosystems. TFS and USFS are currently studying methods of control.

In the Post Oak Savanna and Cross Timbers, population growth and fire suppression are the primary threats. The Dallas-Fort Worth metroplex in the Cross Timbers has experienced one of the fastest growth rates in the country over the past decade. Along with habitat loss, urbanization makes habitat management more complicated. Prescribed fire, necessary to maintain savanna habitat, is more difficult in urban and suburban areas. Without fire, shrubby plants such as yaupon and eastern red cedar can overwhelm woodlands and grasslands. In addition, the lack of prescribed fire increases the likelihood of wildfires. Texas averages about 10,000 wildfires totaling two million acres per year, but in drought periods those numbers can jump. In 2011, a record drought year, Texas had 31,000 fires totaling 4 million acres. A devastating fire in the Lost Pines of Bastrop that year caused substantial property losses, impacted the endangered Houston toad, and created a primary succession management challenge (see Highlight 7.1).

The Edwards Plateau also faces the loss of forest habitat due to rapid growth of cities. Much of this growth has occurred in the eastern Hill Country, where most of the mature Ashe juniper forest occurs. Fortunately, habitat conservation efforts for the endangered golden-cheeked warbler have protected some of these forest habitats. A couple of natural threats to forests also exist in this region. Oak wilt is a fungal disease affecting millions of red oak and live oak trees in dozens of Central Texas counties. The fungus spreads through roots and through beetles that carry the fungal spores. One of the most destructive tree diseases in North America, it has killed over one million oaks in 76 Central Texas counties. Treatment is complex and expensive because it involves creating barriers between the roots of trees. To prevent oak wilt, it is essential to avoid pruning oaks during the February through June active season and transporting wood at any season. The second native threat to forests is an overabundance of white-tailed deer. Thanks to reduced numbers of natural predators, the Edwards Plateau has the highest density of white-tailed deer in the country. Deer are browsers, especially feeding on young woody growth. Their overabundance has severely impacted the ability of oak-juniper forests to regenerate hardwood species.

In South Texas, most subtropical forests have been converted to citrus groves and agricultural land. Only about 600 acres of the original 40,000 acres of palm forest remain, and other patches of mature forest are fragmented. Though several national wildlife refuges, state parks, and Texas wildlife management areas exist in the Lower Rio Grande Valley, it is challenging to manage and connect those fragments in the face of growing urban populations and issues with border security.

Threats to the mountain forests of West Texas are less direct. Habitats are afforded protection in national parks, state parks, nature preserves, and on large privately-owned properties managed for wildlife. However, forests can only persist in West

Texas because they occupy "sky islands" of cooler, wetter weather at higher elevations. If climate changes result in lower rainfall, along with the predicted rising temperatures, forests in more southern mountains, like the Chisos Mountains of Big Bend National Park, may struggle to survive.

HIGHLIGHT 7.1 - THE BASTROP COUNTY COMPLEX FIRE – A LESSON IN FIRES AND FORESTS

Figure 7.13. 2011 Bastrop County Complex Fire. Photo courtesy TPWD © 2025, (Chase Fountain, TPWD).

In 2011, Texas was experiencing the hottest summer and worst one-year drought ever recorded. In addition, it was setting records for the most wildfires ever in the state. In early September, another record was about to be set in the Lost Pines of Bastrop County.

The Lost Pines is a unique ecological community located near the southern end of the Post Oak Savanna. This 13-mile-wide patch of loblolly pines is more than 100 miles from the Pineywoods ecoregion and is quite different in composition from the oak woodlands typical of the Post Oak Savanna belt. Ecologists think pine forests once stretched from Bastrop County to the eastern border of Texas. As the climate dried over thousands of years, only one small patch of pine forest was left behind in this area. The Lost Pines are the habitat stronghold for the highly endangered Houston Toad, which has also lost habitat due to a drying climate. Much of the ecosystem is protected in parks and preserves in Bastrop County, but the towns of Bastrop and Smithville also have many residential areas in the Lost Pines.

100 | Chapter 7 - Texas Woodlands

Figure 7.14. An endangered Houston toad sitting among pine needles in the Lost Pines. Source: USFWS, CC 2.0.

Figure 7.15. The fire quickly climbed into the tops of the trees, requiring the assistance of helicopters to battle the blaze. Photo: USDA.gov, CC 2.0.

The Lost Pines were heavily harvested in the decades following the establishment of the city of Bastrop in the 1830s. However, as human settlement increased in the 20th century, forest management shifted, and over the last half of the 20th century thick stands of pine with a heavy understory developed. By the 1990s, land managers recognized that the dense foliage represented a fire hazard and began implementing carefully controlled prescribed burns. They had made slow but steady progress in Bastrop SP when Mother Nature caught up with them.

On the afternoon of September 4, 2011, high winds associated with a nearby tropical storm caused powerlines to ignite fires at three separate locations northeast of Bastrop. Under the dry, low humidity conditions, the fires spread rapidly into forests nearby, rising from the thick understory into the flammable pine canopy. Fire-fighting crews responded from all over the country, but the fires raced out of control. It was 55 days before the last of the fires were out. Ultimately, the fire consumed 1.5 million trees across more than 32,000 acres of forest, along with more than 1600 homes. About 96 percent of Bastrop SP was burned. Two people were killed. The Bastrop County Complex Fire was the most economically destructive wildfire in Texas history.

The fire was so fierce that it consumed all the organic matter in the soil, leaving it covered with ash. Knowing that the environment would be very susceptible to erosion, park managers decided to assist in the process of primary succession. The first step was to apply hydro-mulch – a slurry made of native grass seed, straw, and water. Next, a decision was made to replant pine seedlings, but where would they find the right loblolly pines? The Lost Pines are genetically unique, with adaptations that help them thrive in the lower rainfall of the Post Oak Savanna. Luckily, only three years before the fire, TFS had begun to store seeds from the Lost Pines. Several nurseries germinated seedlings, and volunteers stepped in to help plant over two million young pines.

And what about the Houston Toad? Already at perilously low numbers, biologists were worried that the absence of tree shading and the siltation of breeding ponds could result in the loss of the species. Fortunately, for several years, the Houston Zoo had been involved in a head-starting program that maintained Houston toads in captivity and then stocked eggs or young toads in the wild. As habitat recovers, the Zoo is poised to reintroduce strands of fertilized toad eggs into breeding ponds to give that species a boost as well.

It will take many years—perhaps 100 years—for the pine forests in the burn zone to reach maturity. As it does, however, land managers will not forget the prescribed fire lessons that the 2011 wildfire in the Lost Pines taught.

Figure 7.16. The hot fire killed all vegetation and burned away the organic layer of the soil. Bastrop SP decided to replant loblolly pines to speed recovery of the habitat.
Photo 1: joni, CC 2.0. Photo 2 courtesy TPWD © 2025, (Chase Fountain, TPWD).

FOREST MANAGEMENT

Forests are often managed for multiple benefits, including wildlife habitat, forage for livestock, recreation, water quality protection, wilderness, and recreation. Gifford Pinchot, the first director of the U.S. Forest Service, helped develop that agency's philosophy of multiple use: "the greatest good, for the greatest number, in the long run." As a result, forests may often be in several stages of maturity, providing different types of habitat at once.

With or without human intervention, forests are constantly changing. Although mature climax communities may persist for decades, a disturbance that disrupts the tree canopy can change the structure and composition of forests. Such disturbances can be natural or human-made and include factors such as fire, ice, wind, insects, disease, invasive species, and clearing. Evidence indicates that disturbances such hurricanes, tornadoes, and destructive landscape-level events such as wildfire were frequent occurrences in the Western Gulf Coast Plain. Fortunately, natural vegetation communities are resilient, and succession can allow forest ecosystems to recover (Fig. 7.17).

Succession in forests proceeds through several stages or **seres** dominated by different types of vegetation (Fig. 7.18). In the stages described below, primary succession begins in stage 1 with bare rock. If soil remains intact after a disturbance, then secondary succession can start in stage 3 or 4. As succession proceeds, forests tend to increase in biomass, biodiversity, and depth of soil.

Figure. 7.17. Bastrop SP is progressing through ecological succession after the Bastrop County Complex Fire (Highlight 7.1). First photo is taken seven years after the fire. (Source: Mark Gustafson, CC 4.0). Second photo is taken 13 years after the fire (Photo: Lee Ann Johnson Linam).

Stages of Succession:

1. Stage 1: Bare ground – If the disturbance is severe, such as a dramatic flood, lava flow, or landslide, then primary succession may have to start with rebuilding the soil.
2. Stage 2: Pioneer species – Lichens, mosses, and fungi arrive via spores, converting rocks into soil and building organic matter.
3. Stage 3: Grasses and forbs – Species that prefer abundant sunlight get established. Seeds in the soil or brought in by the wind germinate. The area has the look of a grassland, with open areas used by reptiles, rodents, seed-eating birds, and raptors.
4. Stage 4: Woody shrubs – The wind and animals feeding on the new growth in the pioneer stage bring in seeds of woody plants that begin to establish and shade out some of the pioneer plants. These young woody shrubs are often heavily used by white-tailed deer.
5. Stage 5: Early trees – Trees that thrive in sunlight, such as pine, cedar, and hackberry, establish first. Large mammals start to move in, bringing additional seeds from nearby plants.
6. Stage 6: Mature forest – Shade-tolerant hardwoods like oaks and hickories begin to overtake the pines and limit their ability to germinate in the new forest until another disturbance arrives.

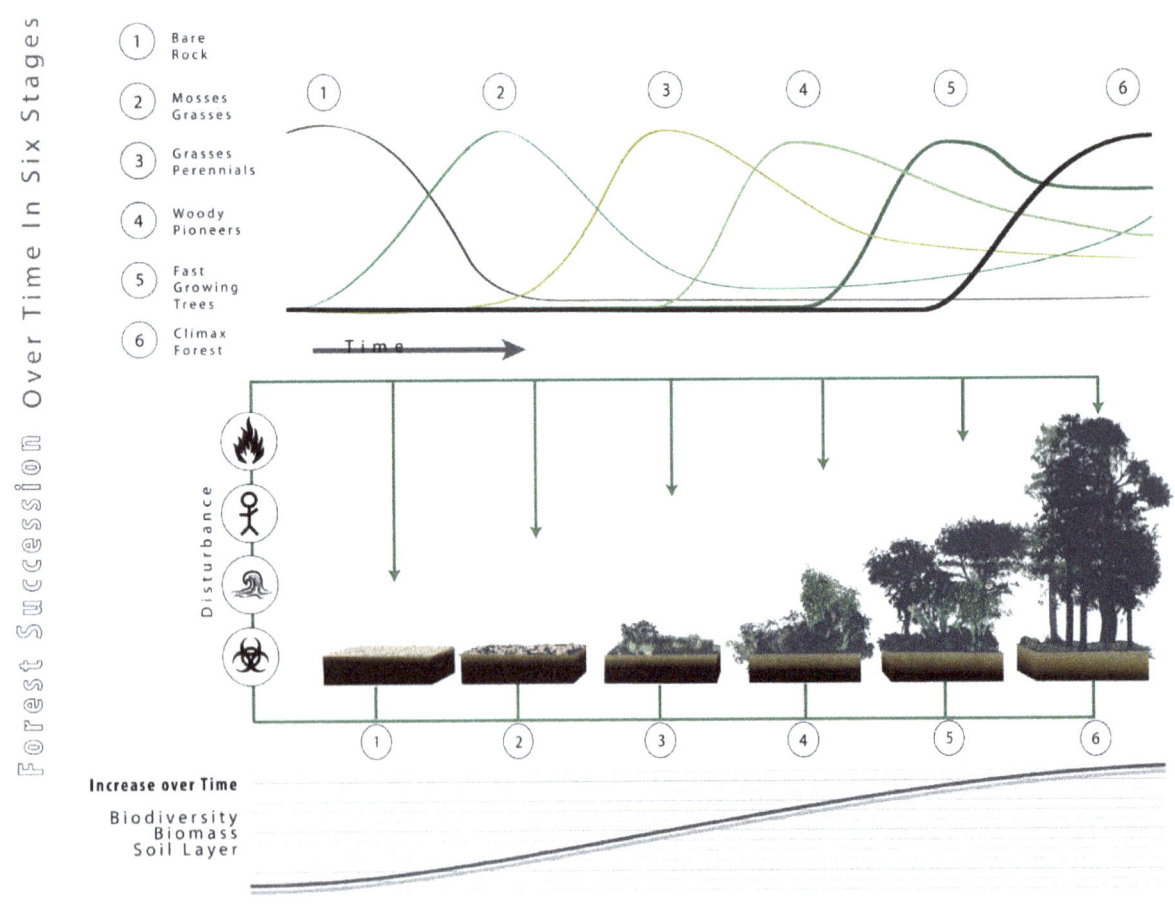

Figure 7.18. Stages of forest succession. Notice changes in the dominant plants, biodiversity, biomass, and depth of soil over time. Source: LucasMartinFrey, CC 3.0

Fire is a crucial tool in forest management. Although wildfires and fires that are too hot can damage trees, controlled fires have many benefits. Benefits of prescribed fire include removal of competing vegetation, especially invasive species; rejuvenation of wildlife habitat by increasing the abundance of young tender leaves and grasses; improving soil nutrients and providing open soil for plants to germinate; and changing forest structure to benefit recreation and species that need open forest. Many forest species are adapted to frequent low-intensity fires, such as longleaf pine. Longleaf pine seeds germinate better on bare soil uncovered by fire. Then longleaf enters a "grass stage" that recovers quickly from fire until it rapidly shoots up. Use of prescribed fire requires careful attention to weather conditions, the amount of fuel in the forest (especially grasses and mid-story vegetation), and the spacing of fire so that not all habitat is impacted at once. Even when wildfires occur, with careful management, forests can recover through the processes of succession (Fig. 7.18).

The "axe" is also used in forest management. Timber is a valuable commodity, and many landowners rely on the income they receive from cutting and selling trees. A careful consideration of tree harvesting techniques can help maintain wildlife value even in a commercial forest. Thinning trees or creating small openings can allow sunlight to reach the forest floor, stimulating the growth of forbs, shrubs, and shade-intolerant trees. Foresters can modify the sizing and location of thinning based on the species that are being managed. There are two primary timber harvest strategies.

The first strategy is uneven-aged management with natural regeneration. This **selective cutting** technique removes individual trees or small groups of trees every 5-10 years. To benefit wildlife, selective cutting should leave all the different tree species in the habitat so that natural regeneration (or regrowth) of the forest can occur. Removing trees of various sizes and species can help preserve the diversity and age structure of the habitat. This technique has the advantage of providing continuous wildlife habitat while creating openings so that shade-intolerant species can regenerate. One disadvantage is that prescribed fire may be hard to use in uneven-aged stands, as new germinating plants can be sensitive to fire. In addition, taking equipment into the forest to cut trees in many locations can impact the ground vegetation. A variation of individual tree selection is **group selection**. Group selection involves cutting trees in groups to create gaps in the forest canopy. The goal of group selection is to mimic natural forest disturbance while maintaining an uneven-aged stand.

A second alternative is even-aged management with artificial regeneration. Also called **clearcutting**, this is harvesting all trees in a tract of land, then replanting young seedlings to regenerate the forest. Ideally, clear-cuts should be small and scattered around the property so that mature trees remain in nearby tracts. Even-aged management is the most economical way to harvest trees. If long rotations are used, high-value mature fire-climax communities such as longleaf pine savannas can still be produced. The disadvantage is that clearcuts can impact forest species for many years after the timber harvest, biodiversity can only recover if a variety of species are replanted, and the land is vulnerable to erosion with the complete absence of trees.

There are intermediate strategies between selective cutting and clearcutting that involve leaving some mature trees to provide shelter and regeneration. These strategies, called **shelterwood** and **seed-tree** cuts, can make regeneration easier and lessen impacts on wildlife. It is also important to note that old, dead trees provide essential wildlife habitat, such as cavities for birds and

Figure 7.19. This snag left after timber harvest provides habitat for several woodpecker species. Photo: Mark Alan Storey.

squirrels. Whatever timber harvest or thinning practice is used, it is valuable to leave these dead **snags** in place.

Because of the potential impacts of forest management on soil and water quality, a set of guidelines called "Best Management Practices" (BMP) are recommended when harvesting timber. The Texas A&M Forest Service provides an app and a publication that lists BMPs such as identifying Streamside Management Zones, protecting riparian vegetation, setting up erosion barriers, establishing roads along contours, and avoiding work in wetlands and wet conditions.

It can take many years for a forest to reach a maturity that benefits a variety of wildlife species. In locations where mature trees are lacking as roost sites for species such as osprey, bald eagles, or wild turkey, installing artificial roost poles can draw these birds to a site. Similarly, if habitats lack sufficient cavity trees, nest boxes benefit eastern bluebirds, screech owls, wood ducks, prothonotary warblers, and other cavity nesters. Wildlife biologists have even installed artificial boxes inside living pine trees to assist in establishing new populations of red-cockaded woodpeckers. These artificial nesting structures can overcome limiting factors in the habitat in the short term while waiting for forest habitats to reach maturity. Perhaps Leopold could have added hammers to his list of tools to restore wildlife!

Like other aspects of wildlife management, forest management is both a science and an art, and it is essential to identify the management goals. Old-growth forest in Texas is rare and should be conserved for its biodiversity value. In other locations, game species such as white-tailed deer often use early successional stages, so disturbance activities that create openings and shrubby growth in the forest can benefit these species. Squirrels rely on acorns for food, so maintaining mature oak species is important

Bluebird nest box	Osprey nesting platform	Bat box for roosting

Figure 7.20. Man-made structures can benefit wildlife when mature trees are lacking.

for them. Conservation of endangered red-cockaded woodpeckers depends on management activities that use fire to maintain a mature, open pine-dominated forest. Bobwhite quail also utilize the open understory in these woodpecker habitats.

Long-range goals are key when managing long-lived habitats like forests. With planning and use of a wide variety of management tools, it is possible to benefit multiple species and maintain forests full of diversity.

Figure 7.21. Replanting after clearcutting can restore trees on the landscape, but it often results in forests that are monocultures with evenly spaced trees, such as this planted loblolly stand in Jones State Forest. Wildlife diversity in such tracts is low.

CHAPTER 8 - TEXAS GRASSLANDS

Texas was once mostly grassland—miles and miles and miles of grassland! It must have been an impressive expanse. Explorers reported that the original tallgrass prairie of North America was up to nine feet high. German immigrants heading west from Houston in the 1830s marveled at the colorful plains, perhaps in spring after a recent burn, stretching to the horizon, "With scarcely a square foot of green to be seen, the most variegated carpet of flowers I have ever beheld lay unrolled before me...red, yellow, violet, blue, every color, every tint was there." Still, not everyone was impressed with the damp coastal prairie. Another old account reported, "Hardly had we left Houston when the flat prairie loomed up as an endless swamp...darkness fell and we had not reached the end, nor did we find a dry place to lie down [nor] a stick of wood to kindle a fire" (Trust for Public Land, "A Prairie Called Katy").

Farther west in the Cross Timbers and Rolling Plains, the grasses were not as tall—only "stirrup-high" or "high as a horse's belly"—but they still dominated the landscape. In the Texas Panhandle, Francisco Vásquez de Coronado, searching for the fabled City of Gold, spent days crossing an endless flat plain full of shaggy "cows" (bison), a landscape that later explorers described as "the great Zahara [Sahara] of North America."

What has happened to these vast, beautiful prairies that once swept across a landscape occasionally broken by woods or rivers? It is a story of appreciation, abuse, rediscovery, and restoration that is still ongoing today.

(opposite) Figure 8.1. Big bluestem is an indicator species for the now-rare tallgrass prairie. Photo: USFWS (public domain).

WHAT IS A GRASSLAND? AND WHY ARE THEY IMPORTANT?

Grasslands are areas dominated by grasses, with tree or shrub canopies covering less than 25 percent of the ground's surface. Grasslands occur all over the world; in North America, grassland biomes are often called **prairies**. These prairies once covered much of the landscape from the Mississippi River to the Rocky Mountains. Another frequently used term is "rangeland." The Society for Range Management defines rangelands as "land on which the indigenous vegetation (climax or natural potential) is predominantly grasses, grasslike plants, forbs or shrubs and is managed as a natural ecosystem." In general, rangelands are usually thought of as habitats that support grazing and may include shrublands, brushlands, savannas, and grasslands.

Around the world, grassland biomes usually occur in the interior of continents where there are significant seasonal temperature variations, including hot summers and cold winters. Precipitation usually peaks in summer and affects the types of grasses that occur. Higher rainfall produces taller, more robust grasses ("tallgrass prairie"), dominated by species such as big bluestem, Indiangrass, and switchgrass. Lower precipitation produces smaller, lower-profile grasses ("shortgrass prairie") where buffalograss and blue grama are common species. Intermediate amounts of rainfall result in "mixed-grass" (or "midgrass") prairies of medium or varying height. Depending on the variability in the habitat location, little bluestem, side-oats grama, and wheatgrass can dominate mixed-grass prairies.

Grasslands are not just grasses. Forbs and small woody species make up essential parts of the prairie flora. These non-grass species are critical as food sources for wildlife and help support the invertebrate diversity of the ecosystem. Legumes, plants that have a symbiotic relationship with nitrogen-fixing bacteria in the soil, play a key role in soil fertility. Woody shrub species can provide support and shelter for nesting birds.

In general, forests occur in higher rainfall areas, and grasslands occur in more moderate rainfall areas, but as Chapter 5 describes, Texas ecosystems are complex and intermingled. As one travels west from the Pineywoods, a pattern of woodlands alternates with grassland. Ecologists think differences in fire frequency and soil explain the mixture. Sandier soils or soils over fractured bedrock accumulate moisture deeper in the soil profile and tend to support trees and shrubs that can send tap roots down to water at lower levels. Clay soils hold more moisture at the surface, but may lack moisture deeper in the soil profile, leading to more lush growth of grasses. Lush growth of grasses supports more frequent fires, especially on flatter topography. Fires kill young woody plants. Thus, there is a positive feedback loop that maintains grasslands on heavier clay soils.

In *The Natural History of Texas,* Brian Chapman and Eric Bolen note that three characteristics of grasses give them an advantage in a burned landscape.

First, when grasses burn, only the current year's growth is lost to the flames, whereas woody plants lose several years of growth that take many years or even decades to replace. Second...grasses can replace their seed-bearing abilities rather quickly after a fire, commonly within weeks. In comparison, most woody plants require many years to reach sexual maturity. Finally, the growing points (buds) of most grasses lie protected in the middle of the plant at or just below ground level, whereas the buds of woody plants develop at the tips of branches where they are fully exposed to fire damage.

~Chapman, B. and Bolen, E. 2018. *The Natural History of Texas.* Texas A&M University Press.

Although prairies may not be as impressive in stature as forests, there are many reasons for their conservation and appreciation. The first is their relationship to soil. The large amount of plant matter that grasses produce yearly has built the world's richest soils. As a result, grassland regions are very important in agricultural production. Furthermore, the roots of native grass species help to hold the soil in place. Roots of tallgrass species may extend five to ten feet or more below the ground's surface (Fig. 8.2).

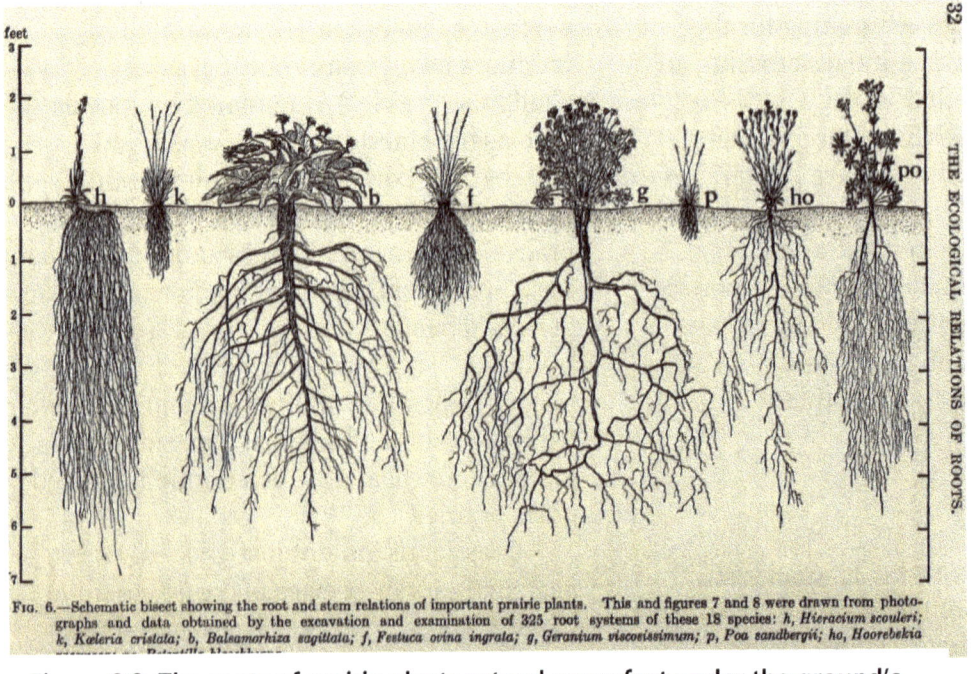

Figure 8.2. The roots of prairie plants extend many feet under the ground's surface, holding soil in place. Source: John E. Weaver, WC (public domain).

TEXAS NATIVES
BLACKLAND PRAIRIES

Figure 8.3. Prairies support a wide array of biodiversity, as this poster produced by TPWD illustrates.

Secondly, grasslands are also important in water conservation. Native prairie vegetation slows runoff, and those deep roots allow water to percolate into the soil and aquifers. Prairies also have slight depressions that can provide moist habitats for flora and fauna as infiltration occurs.

A third benefit of grasslands is carbon storage. Because of their extensive root structure, they are sometimes called "upside-down forests." With most of the biomass underground, prairies store carbon in a more stable manner than forests that readily release carbon into the atmosphere during forest fires.

Grasslands provide important habitat for animals. Seeing prairie-chickens booming, pronghorns running, or prairie dogs calling is a delight. In addition, livestock producers rely on healthy grasslands and often seek to balance the habitat use by wild and domestic animals. Prairies are also critical for pollinators. The wide variety of grassland flowering plants helps maintain a healthy pollinator community. In turn, these pollinators are valuable to nearby farms and honey producers.

Finally, prairies have always played a role in the human cultures of this continent. Before prairies came to represent the opportunities of the American frontier, indigenous people considered the prairie their homeland and relied on bison herds for food and on various prairie plants for medicine and daily needs. Many tribes saw the open expanse of the prairie as a sacred space, with deep spiritual connections to the land, sky, their ancestors, and natural elements. For modern Americans, the vast open spaces and diverse plant life of the prairie hold aesthetic value, offering a sense of peace and tranquility, open space for recreation, and spectacular sunrises and sunsets.

TEXAS GRASSLANDS

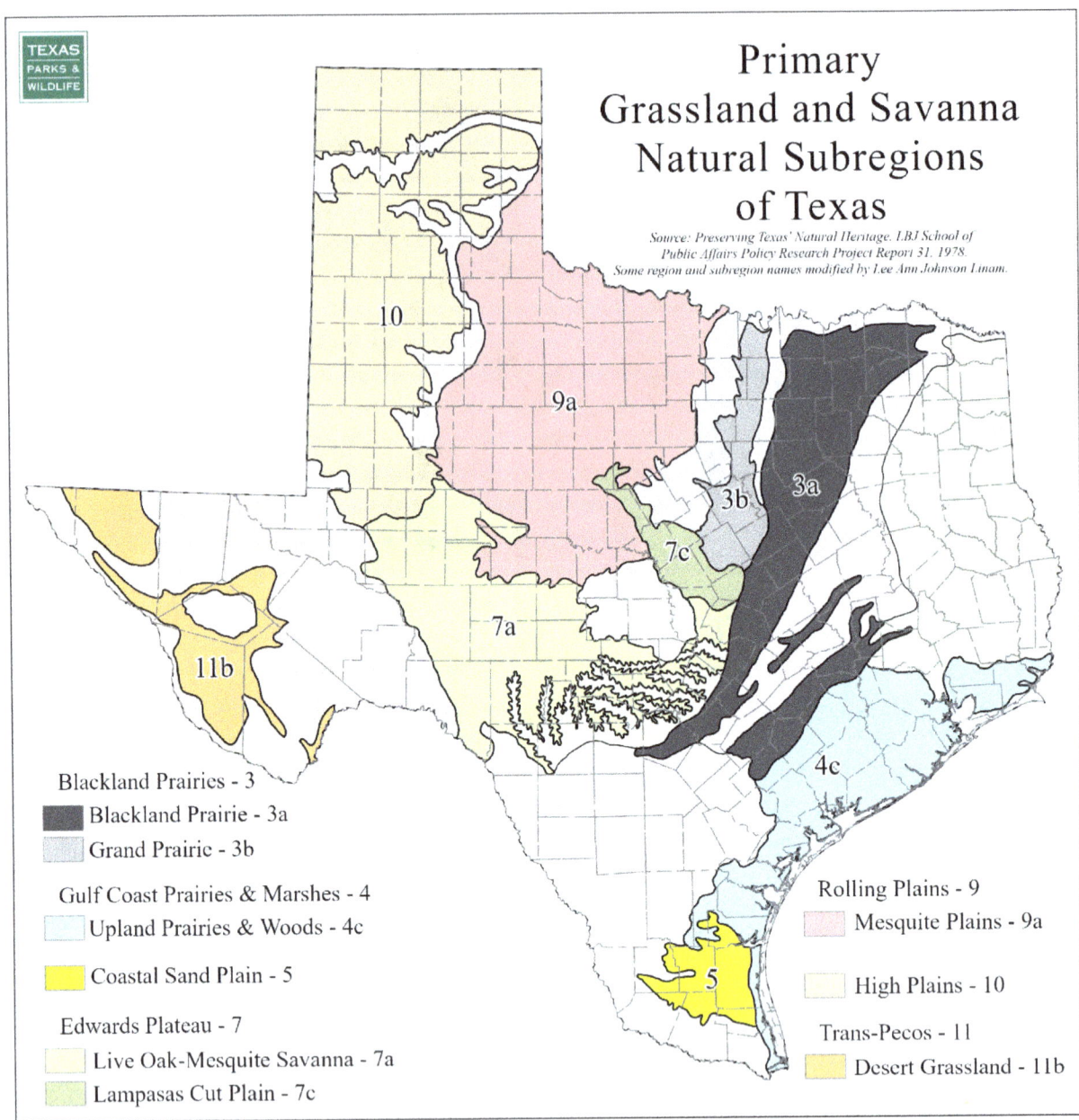

Figure 8.4. Grasslands occur in ecoregions across the state. Tallgrasses in the Blackland Prairie, Grand Prairie, Gulf Coast Prairies, and Coastal Sand Plain transition into mixed-grasses in the Rolling Plains and Edwards Plateau. Shortgrasses dominate in the High Plains and the desert grasslands of the Trans-Pecos. Map source: LBJ School of Public Affairs. Prepared by Paul Daugherty, TPWD IT-GIS.

Texas hosts at least 470 native grass species, more than any other state in the U.S. In addition, many grasses have been introduced for grazing and landscaping, bringing the total number of grass species in the state to more than 700. TPWD lists nine types of grassland communities and an additional nine shrubland types where small woody plants predominate; however, prairies can generally be placed in three or four groups. The grasslands of the Blackland Prairie, Grand Prairie, and Coastal Prairie are southern extensions of the North American tallgrass prairie, also called the True Prairie. The Rolling Plains, Edwards Plateau, and South Texas Plains are southern extensions of Great Plains mixed-grass grasslands. The High Plains support portions of the North American shortgrass prairie, and unique desert grasslands, many of short stature, can be found in the Trans-Pecos. More information about the individual species of grasses found in Texas is available online from Agrilife at their Plants of Texas Rangelands Virtual Herbarium.

TALLGRASS PRAIRIE

Figure 8.5. Bunchgrasses in restored tallgrass prairie at the Heard Natural Science Museum, Collin County, Texas.

Tallgrass prairies occur where rainfall averages 30 to 40 inches per year. Deep black clay soils, called Vertisols, dominate the tallgrass prairie regions of Texas. Vertisols are rich in nutrients, swell when wet, and shrink when dry, often leaving large, deep cracks in the soil's surface. The original climax grasslands were dominated by little bluestem, big bluestem, Indiangrass, sideoats grama, and dropseed species, with switchgrass and eastern gamagrass in wetter areas. Many of these perennial tallgrass species are considered "**bunchgrasses**" because each grass plant grows as a separate cluster of stems, providing both shelter for wildlife species and space for small animals to move around at the ground's surface (Fig. 8.5). Along the coast, saltier soils may be dominated by big clumps of Gulf cordgrass, and sandy barrier islands host grasslands with Gulf cordgrass, seaoats, and little bluestem. Alfisols are loamy rather than clayey like Vertisols but may also support tallgrass prairie in some places. Their species composition may differ slightly, with more dropseeds, sedges, and switchgrass.

The landscape within Texas's tallgrass prairies is primarily flat, but the region's soils produce some unusual features. In the Vertisols, shallow depressions called **gilgai** develop. Certain plants and animals occur only in the gilgai. On the other hand, in the loamier Alfisols, small mounds up to two meters (6-7 feet) high, called **mima mounds**, can develop. These mima mounds are like tiny mountains, with specific plants on the slopes and different species on the tops. They can provide valuable habitat for wildlife when heavy rains inundate the prairie.

Grasslands are critically important to a number of imperiled species. The National Audubon Society identifies six high-priority birds that utilize Texas tallgrass prairie, including four species that breed here—Attwater's prairie-chicken, dickcissel, lark sparrow, and eastern meadowlark. In addition, several unique pocket gopher species or subspecies occur in different segments of the coastal prairie, and one unique crayfish species—the Parkhill Prairie crayfish—lives only in gilgai of the Blackland Prairie. There are few endemic plant species, although the Grand Prairie in the Cross Timbers ecoregion hosts two yucca species—pale yucca and Glen Rose yucca—found nowhere else.

Only one percent of Texas's tallgrass prairie remains preserved in its highest quality state. It has been plowed, converted to non-native pasture, overtaken by invasive species, and lost to urban development, especially in Houston and the urban corridor from San Antonio to Dallas. Very little native prairie remains in the Blackland Prairie, and coastal grassland fragments that remain are threatened by introduced woody invasives like Chinese tallow and Macartney rose. Between 1970 and 2001, the acreage of Chinese tallow increased from 5 to 30,000 in Galveston County alone. Native species like mesquite and running live oak can also overtake poorly managed prairies.

Tallgrass prairie remnants are found on Caddo National Grasslands, Cooper WMA, Granger WMA, Hagerman NWR, Jim Chapman Lake/Cooper Dam, and Lake Tawakoni SP. Attwater's Prairie Chicken NWR manages extensively for coastal prairie. The Grand Prairie contains the largest remaining swatch of tallgrass prairie in Texas. Segments of the Grand Prairie can be visited at the Fort Worth Nature Center and Refuge, Benbrook Lake, Lake Whitney SP, and Hamm Creek Park. Conservation organizations, such as the Texas Native Prairies Association, the Coastal Prairie Conservancy, and The Nature Conservancy of Texas, are also important contributors to conserving prairie patches.

MIXED-GRASS PRAIRIE

Figure 8.6. Forbs and grasses of moderate height in a mixed-grass prairie in the Texas Rolling Plains. Photo: Mark Mitchell, TPWD.

Mixed-grass (or midgrass) prairies once stretched discontinuously from the Red River along the state's northern border to the Rio Grande along the southern border, dominating flatter deeper soils in the Rolling Plains, Edwards Plateau, and South Texas Brush Country. Mixed-grass prairies are in a transition zone where rainfall averages between 20 and 30 inches per year. Some tallgrass species, especially little bluestem, occur in this region, along with many species of grama grass.

The Rolling Plains, sometimes called the Rolling Red Plains, occur on red soils formed from shales, mudstones, and sandstones. As the ecoregion name implies, the grassland topography is generally rolling, with woodlands found in the eroded edges of the Caprock Escarpment in places such as Palo Duro Canyon and in breaks formed along gypsum deposits. The original grassland was composed of medium-height stands of sideoats grama, little bluestem, and Texas wintergrass (sometimes called speargrass). Many areas have been heavily grazed and now support more increaser grasses—species that move into disturbed habitats, such as three-awns. Mesquite and redberry juniper have also increased in heavily-grazed areas, whereas many flatter landscapes are in cultivation. The Rolling Plains provide habitat for northern bobwhites, black-tailed jackrabbits, and rodents such as packrats and the endemic Texas kangaroo rat. The rugged terrain is also home to many reptiles, including Texas horned lizards, eastern collared lizards, and western diamond-backed rattlesnakes. The Texas poppymallow, a large showy wildflower, is found only in the Rolling Plains. Caprock Canyons SP and the Matador WMA offer an opportunity to visit the mixed-grass prairie on the western edge of the Rolling Plains. Matador WMA has a healthy population of Texas horned lizards, and the state bison herd roams Caprock Canyons.

The limestone uplands of the Edwards Plateau traditionally supported a mosaic of woodlands, savannas, shrublands, and grasslands. Although the topography and soils of the Edwards Plateau are significantly different from the Rolling Plains, the mixed-grass prairie plants are similar. Little bluestem and sideoats grama dominate healthy grasslands. Sheep, goats, and cattle once grazed many areas of the Edwards Plateau heavily. Three-awns, shortgrass species, and introduced grasses dominate these overgrazed areas. As grazing reduced the abundance of native grasses, King Ranch (KR) bluestem, a grass introduced from China in the 1920s, replaced native grasses in many areas. These KR bluestem habitats are lower in bird, insect, and rodent diversity than native grass communities. Finally, as noted in Chapter 7, some former grassland areas have developed into brushlands dominated by Ashe juniper because of historic overgrazing and fire suppression. The Kerr WMA and Kickapoo Caverns SP offer a glimpse of healthy Edwards Plateau savanna and the chance to see woodland-grassland edge species such as painted buntings and black-capped vireos.

The South Texas Brush Country once supported a mixture of mesquite grassland and diverse brushland communities, with grasslands more common over loamy or sandy soils in the eastern part of the ecoregion where it abuts the Coastal Sand Plain. Soon after the Spanish arrived in South Texas, escaped cattle and horses formed large wild herds that grazed the grasslands year-round. As more settlers arrived, livestock numbers increased even more, fire became less frequent, fences were erected, and exotic grass species were introduced. Native grasslands dominated by cane bluestem, windmillgrass, and grama species now are much less common. Many overgrazed areas are dominated by mesquite, and some cleared rangelands have been chiefly planted in introduced grasses such as buffelgrass. One consolation is that large ranches remaining in South Texas generally avoid overgrazing. One example is the historic King Ranch. Private ranches also preserve large tracts of grassland in the Wild Horse Desert, the unique coastal sand sheet that lies between Kingsville and Brownsville.

SHORTGRASS PRAIRIE

Figure 8.7. Shortgrass prairie in the Llano Estacado.

In the harsher, colder, drier climate of western Texas, grasses may struggle to reach 10 inches in height. Shortgrass prairies occur in the High Plains where rainfall ranges from 10 to 20 inches annually; semi-desert grasslands of short stature occur in lower elevations of the Trans-Pecos.

The High Plains of Texas lie on a flat bedrock or "caprock" covered with multiple layers of soil deposited by wind. The Canadian River cuts the caprock into two sections—the more northern Canadian-Cimarron High Plains and the larger Llano Estacado or "staked plains." The dominant grasses in the High Plains are buffalograss and blue grama. Unlike the bunch-grasses of the tallgrass and mixed-grass prairies, these grasses form thick sods of densely packed roots. They have abundant roots in the top six inches of the soil to capture all surface soil moisture but also send other roots as deep as five feet to reach subsurface water. In addition to the shortgrass areas, the High Plains can support woody species such as sand sagebrush and Havard shin oak over deep sandy soils. Interestingly, subsurface moisture in the some deep sands of the Panhandle and in the riparian habitats in the western Rolling Plains may even support patches of tallgrass prairie!

Although millions of bison and black-tailed prairie dogs once occupied the Texas High Plains, these vast populations are gone. Bison, the American national mammal, are now found only in the official Texas State Herd at Caprock Canyons SP and in domesticated herds. The Texas State Bison Herd contains the last remnants of the Southern Plains bison populations and exists primarily due to the conservation efforts of Molly Goodnight, wife of rancher Charles Goodnight, in the 1860s in the Palo Duro Canyon area of West Texas. Prairie dogs persist in the wild, mostly in small colonies, but mammalogists estimate that 98% of their original population is gone. The list of animal associates of prairie dogs is quite extensive, and, as Chapter 3 notes, the decline of a keystone species like the prairie dog affects numerous other shortgrass species as well, including burrowing owls, swift foxes, black-footed ferrets, ferruginous hawks, mountain plovers, killdeer, horned larks, and horned lizards. Prairie dogs and shortgrass prairie habitats can be experienced at the Rita Blanca National Grassland in the Canadian-Cimarron High Plains. The Gene Howe and Pat Murphy WMAs offer a look at the transition between mixed-grass and shortgrass prairie and a chance to spot lesser prairie-chickens, an indicator species for the High Plains.

The Trans-Pecos ecoregion provides a surprisingly diverse assortment of grassland habitats. Semi-desert grasslands, composed primarily of shortgrass grama species, occur in the foothills of mountains from about 3,500 to 4,500 feet. At higher elevations, these grasslands become more similar to mixed-grass communities. At lower elevations, tobosagrass mixes with desert shrubs, whereas harsh salt flats in Hudspeth and Culberson counties support a shortgrass community dominated by alkali sacaton. At the other extreme, early settlers described "waist-high grass" along Terlingua Creek in the Trans-Pecos. Pronghorn are familiar occupants of plains communities, and grasslands on mountain slopes provide a specialty habitat for the yellow-nosed cotton rat and Montezuma quail. The National Audubon Society notes that these desert grasslands are essential habitats for 10 of its 19 highest-priority grassland bird species. Davis Mountains SP is a great location to explore Trans-Pecos grasslands.

Figure 8.8. Desert grasslands near Marfa.

THREATS TO TEXAS GRASSLANDS

Nationwide, only 54% of the shortgrass prairie, 24% of the mixed-grass prairie, and 11% of the tallgrass prairie that once covered much of the continent remain. Dramatic declines have also occurred in Texas. The native grasslands of the state have been steadily decreasing since settlement began, primarily due to urban development and conversion to row crops and tame pastures. Texas A&M AgriLife reports that only about 96 million of the original 148 million acres of native grasslands remain. Overgrazing and brush encroachment have degraded much of the prairie that continues to exist. Misuse of three of Leopold's tools—fire, cow, and plow—are responsible for most grassland degradation.

FIRE

Fire suppression may be one of the most significant impacts on Texas grasslands. As described above, the structure and nature of grasslands naturally lend themselves to burning, and naturally-open landscapes once allowed fires to burn in large swaths. In addition to wildfires, Native Americans may have purposefully used fire as a hunting and wildlife management tool. Indigenous people saw fire as part of their culture, even as "medicine" for the land. A modern ecologist of Native American descent, Frank Kanwha Lake, notes, "When you prescribe it, you're getting the right dose to maintain the abundance of productivity of all ecosystem services to support the ecology in your culture." (Roos, Dave. 2020. Native Americans Used Fire to Protect and Cultivate Land. *The History Channel*.) A study of evidence of fire intervals based on woody vegetation suggests that tallgrass prairies may have burned every one to three years, with somewhat longer intervals between fires in the mixed-grass prairies and High Plains where fuel loads would have been less.

When European Americans arrived on the scene, they often saw fire as destructive. Some Spanish officials described the indigenous burning of grasslands as "childishness" and a harmful practice that should be discontinued. Their fear of fire was understandable, but a pattern of repressing fire was doomed to lead to changes in grasslands. Brush invasion began—eastern red cedar in the east, mesquite in the mixed-grass prairies, running live oak along the coast, Ashe juniper in the Edwards Plateau, and thornscrub in the South Texas Plains.

COW

America's grasslands were always grazed. Although cattle were not present until introduced by the Spanish in the 1500s, wild bison, pronghorn, elk, and rodents fed on grasses and other leafy material. Even before these modern mammals fed on prairie grasses, species such as mammoths, camels, and ancient bison shaped their development in the Ice Age.

Wild bison herds were migratory—moving through an area and not returning until new growth had occurred. Under that scenario, even heavily-grazed grassland ecosystems had time to recover between grazing episodes. Once European settlement of Texas's rangelands began in earnest, several factors conspired to lead to the overgrazing of grasslands. First, cattle and horses escaped from Spanish missions, forming feral herds that did not migrate. Second, as the U.S. military succeeded in removing Native Americans from Texas after the Civil War, there was a mass movement of settlers into the Texas prairies. These settlers were attracted by "free grass." Based on higher rainfall patterns in the eastern United States, government policy set up homestead sizes that were too small, leading almost immediately to overstocking. Third, the invention of barbed wire in the 1870s and "wolf-proof" mesh wire in the early 1900s allowed ranchers to increase the concentration of cattle and sheep, respectively. At the same time, bounty programs for wolves, mountain lions, and bears encouraged an increase in livestock numbers.

In the 1880s, a notorious "die-up" of cattle occurred when harsh weather conditions and poor range conditions contributed to the death of 80 to 90 percent of the cattle on West Texas rangelands. Sensing that grazing practices were becoming unsustainable, by the early 1900s some ranchers and ecologists called for a more scientific approach to managing grazing. Still, grazing pressure continued to mount. Finally, in the 1930s, the range management profession began to emerge in earnest. Like the new wildlife scientists, range scientists sought research and data to guide natural resource management. But, by then, grasslands in Texas were almost universally overgrazed.

The harmful effects of overgrazing are numerous. Overgrazing increases bare ground, causing erosion, increased water runoff, and decreased infiltration into the soil. Bare ground opens the rangeland to invasion by noxious weeds, brush, invasive plants, and less-nutritious increaser grasses. Lack of grass cover raises soil temperatures and lowers soil fertility, leading to less soil replenishment. Native grass species experience reduced vigor and root mass. Finally, there is diminished cover and food production for wildlife species.

Figure 8.9. Contrast between overgrazed and moderately-grazed grassland. The pasture on the left is more subject to erosion and invasion by exotic species. Photo: USDA NRCS South Dakota, CC 2.0.

PLOW

Even as overgrazing was emerging as a threat to Texas grasslands, the plow was emerging as a more potent threat. Almost immediately, settlers in the tallgrass prairies noted the fertility of the soils, but initially cultivation was slow in coming to the region. The heavy clay soils, although fertile, were challenging to plow. In 1837 John Deere developed a sharp, smooth metal plow capable of cutting through the black soil, and farming in the true prairie increased rapidly. By the 1880s, most of the Texas Blackland Prairie had been converted to cotton. Cotton was so lucrative and widespread, it was called "King Cotton." A similar pattern occurred in the coastal prairies. About three million acres of former tallgrass prairie are now planted in row crops yearly.

On the High Plains, homesteaders also shifted to row-crop agriculture as overgrazing removed the grass cover from the plains. The acreage devoted to farming in the Llano Estacado increased by a factor of six in the first three decades of the 20th century, but the intensive use was not sustainable. A historic drought began in the 1930s that devastated agriculture and decimated the soil in one of the greatest ecological disasters in recorded history, the Dust Bowl. As tons of topsoil were lost, the country responded by creating the Soil Conservation Service and placing some of the most denuded properties into National Grasslands managed by the U.S. Forest Service.

Today, many acres of former shortgrass prairie in the Texas Panhandle have been permanently converted to intensive agriculture, now aided by irrigation from the Ogallala Aquifer. Of the 50 million acres of land in the Rolling Plains and High Plains, over 11 million are planted in row crops annually—primarily corn, cotton, sorghum, and winter wheat.

The Soil Conservation Service, now called the Natural Resources Conservation Service, was created to stabilize and prevent further soil loss in Texas and across the nation. In addition, the Conservation Reserve Program, a part of the Farm Bill administered by the Farm Services Agency, offers farmers financial incentives to remove highly erodible lands from agriculture and to convert them instead into grasslands, thus improving the outlook for wildlife on the High Plains.

ADDITIONAL THREATS

Another threat to native grasslands emerged beginning in the 1800s. In an effort to support more grazing animals or produce more hay, non-native grass species were introduced, such as bermudagrass from southern Africa, bahiagrass from South America, and Sudan grass from East Africa. According to AgriLife, 10 million acres of native grassland have been converted to non-native, or "tame," pastureland in Texas.

Although some landowners favor non-native grasses for cattle grazing and hay production, these pastures have little value as wildlife habitat. They are usually monocultures, so the plant diversity of the original ecosystem is lost. In addition, bermudagrass grows in dense mats that shade the ground. It eliminates the bare ground that small wildlife species need to move about and provides little cover or nesting habitat for grassland birds. Finally, its solid cover shades out the native forbs and grasses that provide food for wildlife.

Other introduced plant species have proven to be threats to native grasslands as well. In addition to Chinese tallow and Macartney rose in the Texas Coastal Prairie, the Texas Invasives project lists dozens of species that are potentially invasive in each of Texas's ecoregions.

Figure 8.10. Former shortgrass prairie converted to farmland in the Texas Panhandle.

THREATS TO PRAIRIE WILDLIFE

Grassland wildlife species struggle to survive in fragmented and degraded grassland landscapes. Although large grazers such as bison and pronghorn are the most visible wildlife species in prairies, pollinators and grassland birds make up the bulk of the species dependent on grasslands. The news is not good for these two species groups.

Worldwide studies have documented significant declines in grassland arthropods (insects and arachnids), including in the Great Plains. These declines are higher in areas fragmented by agriculture and in areas infested with red imported fire ants. In addition, pollinator diversity is lower when native plant species richness is lower. Because of the importance of insects as pollinators and food sources for other wildlife, such declines can affect the whole ecosystem.

Regarding birds, the North American Breeding Bird Survey, an annual sampling of birds nationwide, reveals that grassland birds are declining faster than almost any other group of birds. Their populations have fallen 40% since the counts began in 1966. In addition to habitat loss and reductions in insect prey, habitat fragmentation is a significant factor in the decline of bird species. Nest predation of grassland species is up to 24% higher on small prairie patches as compared to larger fragments. Attwater's Prairie-chicken is one Texas native that has been severely impacted by conversion and fragmentation of grasslands (Highlight 8.1).

Native coastal prairie

Native coastal prairie converted to agriculture

Native coastal prairie converted to monoculture of bermudagrass.

Coastal prairie fragment invaded by brush.

Figure 8.11. Examples of habitat changes on the Texas Coastal Prairie.

HIGHLIGHT 8.1 - LITTLE GROUSE ON THE PRAIRIE – A STORY OF AN IMPERILED BIRD AND ECOSYSTEM

Something magical used to happen on early spring mornings on the Texas Coastal Prairie. Above the drone of mosquitoes, one could hear a low drumming, booming sound occasionally interrupted by a cackle. Pushing through damp bluestems, peering through the fog, one could see an amazing spectacle on a heavily-grazed patch of dark earth. There was a half-comical, half-eerie dance taking place, as small brown birds with inflated yellow-orange cheek patches stomped and threw up their tail feathers like Native dancers celebrating the earth. When a smaller, drabber hen strolled across the barren ground, the dance intensified. It was the courtship display of the Attwater's prairie-chicken.

As many as one million Attwater's prairie-chickens once danced and nested, chased insects and plucked forbs across six million acres of Gulf Coast prairie. In good years, populations of the birds, a grouse species related to quail, turkey, and other ground-dwelling birds, would soar. In drought years or when a hurricane inundated portions of the coastal prairie, the numbers would drop. But, as long as there was habitat—prairie, vast expanses of prairie—there were prairie-chickens.

As human populations increased on the Texas coast, changes happened in the coastal tallgrass prairie. The grasses covering the rich soil were cleared for agriculture. Vibrant native grass and forb communities were replaced with monocultures of "improved" non-native grasses. Overgrazing allowed the invasion of both native and non-native brush species. Lack of fire ensured that the brush would gain a foothold.

As Texas coastal prairie began to disappear, prairie-chickens began to disappear. By the time the first Endangered Species Act was passed in 1966, Attwater's prairie-chickens were so rare (less than 1,000 birds) that they were placed on the first endangered species list. Habitat conservation efforts began, and land was acquired for the Attwater's Prairie-chicken NWR, but pressures continued. Even the habitat that remained was fragmented and infested with red imported fire ants that caused the reproduction of prairie-chickens to fail. By the beginning of the 21st century, only about 50 Attwater's prairie-chickens remained in the wild.

Today, the conservation of Attwater's prairie-chickens continues to be a challenge. Prairie-chickens are raised in captivity to supplement wild populations, large areas are treated for removal of red imported fire ants, and government and non-profit groups are working with private landowners to try to maintain larger tracts of grassland managed with fire and well-planned grazing. Prairie-chicken numbers are slowly climbing. Even though it's a long way home for the little grouse on the prairie, land managers know that the work that is done to restore native grasslands has the potential to benefit many grassland species. It's the habitat that counts.

Figure 8.12. Male Attwater's prairie-chicken performing a courtship display. Photo: USFWS, CC 2.0.

GRASSLAND MANAGEMENT

As Leopold suggested, some of the same tools that have harmed grasslands are essential to its recovery. Modern wildlife managers have added a few additional tools as well.

FIRE

Prescribed fire is the most important tool in the prairie manager's toolbox. Fire invigorates grasslands by clearing rank (old, overgrown) vegetation and opening the canopy to allow the growth of forbs and browse species. It recycles nutrients back into the soil and can increase moisture infiltration. It reduces woody shrub abundance and can help to control some invasive species.

"Prescribed" fire implies that fire is used in a well-planned dosage to treat a certain condition. Therefore, the wildlife manager needs to consider the goals of prescribed burning. Fires set in cool seasons can encourage the growth of warm-season grasses. Fires set in the summer can control brush and encourage cool-season grasses. After a burn, grazing should be deferred for three months or longer to allow for the recovery of native vegetation. Generally, grasslands should be burned every three years, but burning different tracts in different years can help meet multiple goals and can help protect ground-nesting birds. Furthermore, land managers need to understand the response of non-native species to fire. The invasive King Ranch bluestem actually increases in response to fire, so a variety of tools may need to be applied, such as following fire with heavy grazing.

Fire is a powerful but potentially dangerous tool for habitats, wildlife, and even humans. The Texas Department of Agriculture and TFS provide training in the safe application of fire as a management tool. TPWD offers these general guidelines to help make a prescribed fire a "controlled burn:"

- First, prepare a disked bare-ground fire guard around all sites before burning. Disked fire guards, which can include roads and rights-of-way, should be 15 to 20 feet wide. (These disked areas can be planted to winter supplemental food plots between burn years.)
- Second, check weather conditions and timing. Humidity should be between 25 to 40 percent. Wind speed should be between 10 to 15 miles per hour. Initiate burns in the morning after 9 a.m. Plan the burn to be a size that will extinguish itself before dark.
- Finally, notify the local fire department before initiating a burn. When ready, first light a backfire that burns against the wind. After the backfire produces a 50-yard barrier, then set the headfire with the wind behind it on the other side of the burn plot. Monitor the burn continuously to make sure it does not escape the target area.

GRAZING

When increasing numbers of settlers began arriving in what would become Texas in the early 1800s, 90% of the area was rangeland. Livestock grazing quickly became an important livelihood and range management tool. Today, grazing remains big in Texas. The state is the nation's top livestock producer for meat and for fiber, such as wool and mohair.

Although overgrazing has caused much harm to grassland habitats, grazing can also be a helpful habitat management tool. Well-managed grazing combined with rest periods creates short-term disturbances in climax grass communities that can benefit wildlife.

Figure 8.13. Fire is used at Mason Mountain Wildlife Management Area and other locations in Central Texas to maintain an open grassland-woodland savanna. Here, a road is used as a fire break. The 4x4 vehicle carries a water supply to respond to emergencies. Photo: Jeff Forman, TPWD.

Grazing opens the grass canopy and stimulates new grass growth. Cattle hooves break the soil, allowing forbs and young browse plants to get started. Grazing and browsing can also keep new woody plants from being successfully established. Grazed grasslands can have higher species diversity, more foliar cover of perennial grasses, and fewer invasive species than similar ungrazed tracts.

Grazing management is founded on four basic principles—proper use, proper season of use, proper distribution, and proper kind and class of animals.

Proper use focuses on the stocking rate or the number of grazing animals in a grassland. The stocking rate must consider the proportion of the plant tissue removed. In native grasslands, livestock should remove only 15-25% of the annual growth. This rate allows native species to maintain themselves without fertilizers or irrigation. It also enables some uneaten portions of the plants to become litter (decomposing plant material), helping to hold and build the soil. It is important to remember that stocking rates may need to be adjusted based on climatic conditions. In East Texas, less than 10 acres of grassland may be required for each cow, whereas in West Texas, over 150 acres may be needed per cow. The landowner also needs to be ready to adjust stocking rates lower during periods of drought to avoid long-term damage.

Proper season considers the composition of the grassland and how to utilize and

HIGHLIGHT 8.2 - COOL-SEASON VERSUS WARM-SEASON GRASSES

Everyone has their favorite season, including grasses! Understanding how to encourage different types of grasses can be important for range and wildlife managers.

Warm-season grasses grow predominantly during the warm months of the year, from about April through November in Texas. The tops are dormant during winter, but the roots store energy to begin growing again in the spring. Most warm-season grasses are long-lived with deep roots and nutritious forage. Many native warm-season grasses are "bunch grasses," providing a clumpy structure that allows wildlife to move around at ground level and seek shelter and concealment in the foliage. Some of the most common native warm-season grasses in Texas include big bluestem, little bluestem, Indiangrass, switchgrass, sideoats grama, dropseed, sand lovegrass, purpletop tridens, and eastern gamagrass.

Cool-season grasses thrive during cooler, wetter months. They are frequently faster-growing, with less extensive roots and less robust structure than warm-season grasses. Cool-season grasses can hold the soil as a grassland is recovering from disturbance and provide substantial winter and early spring forage. Native cool-season grasses include Texas wintergrass, Virginia wildrye, western wheatgrass, and Texas bluegrass.

A healthy wildlife habitat can include both warm-season and cool-season grasses. However, a pasture dominated by cool-season grasses often provides less diversity and a less suitable wildlife structure than one managed to provide abundant warm-season grasses.

conserve both warm-season grasses and cool-season grasses. A pasture with cool-season grasses can provide forage in cooler months. Still, it must also be rested at some time during cooler, wetter periods so that those grasses can regenerate. Similarly, warm-season pastures must be rested during a portion of the summer growing season so that those grasses can continue to thrive. In addition, wildlife goals should be considered. If a grassland is used by ground-nesting birds, then grazing pressure should be lessened during the nesting season so that the birds have cover for their nests and broods.

Proper distribution considers where livestock concentrate to graze, how to protect sensitive areas, and how to allow all areas time to recover from feeding. Some sensitive areas, such as bogs or riparian zones along streams, may need protection from any grazing; however, in other parts of the grassland, grazing can help create different habitat structure. For example, prairie-chickens prefer open areas with very little vegetation for their "booming grounds"—sites where the males put on courtship displays for the females. Heavy grazing can help open up these booming grounds. At the same time, females need heavy grass cover nearby to conceal their nests. As the young prairie-chickens mature, they often use areas where moderate grazing retains cover, while also allowing forb species to produce seeds and insects. For prairie-chickens, all three habitats are needed in the same grassland.

Livestock can be encouraged to shift their grazing pressure by providing water, salt, or minerals in different locations. Another alternative is to use fencing to develop a rotational grazing system. Figure 8.14 illustrates a high-intensity, short-duration rotational grazing system, showing how different pastures can be in various stages of grazing, rest, and recovery. The timing varies according to the time of year and rainfall, and some areas may be excluded if sensitive vegetation or habitats are present.

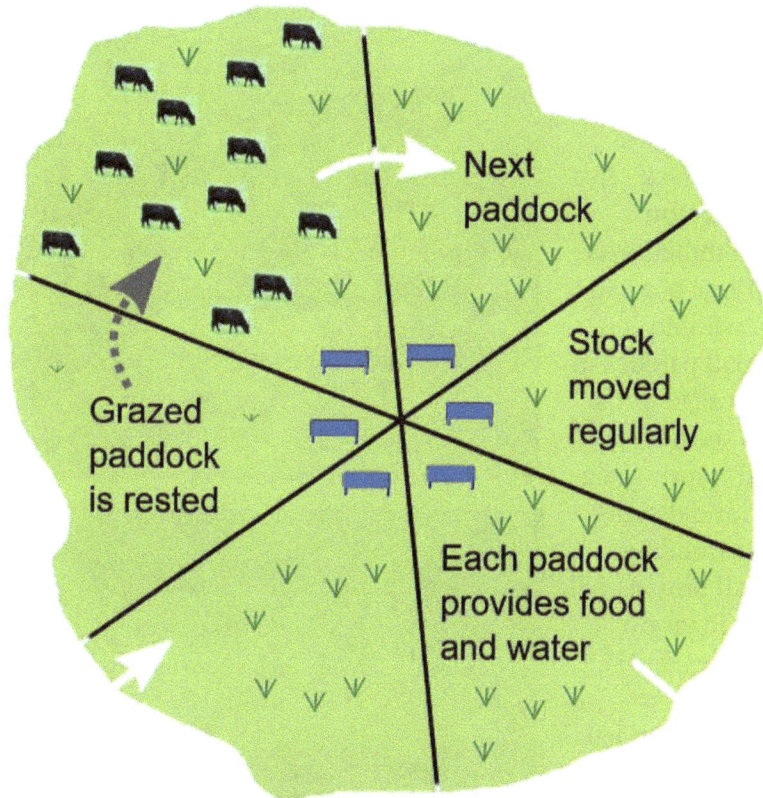

Figure 8.14. Sample design for a high-intensity, short-duration rotational grazing system. Source: Ian Alexander, CC 4.0.

Proper kind and class of animals considers that different livestock have different feeding preferences. Cattle prefer grass. Sheep prefer forbs. Goats prefer browsing woody plant species. These preferences can be used to achieve wildlife management goals. For example, goats can be added to a herd to help reduce brush problems in a grassland. Wildlife species also have different feeding preferences. Bison prefer grass. White-tailed deer prefer woody browse and forbs. Elk are grazers and browsers. Pronghorn, who live primarily in grasslands, actually mostly eat forbs. When managing livestock, it is essential to consider competition between species and the composition of the vegetation community so that wildlife management goals can be met.

OTHER MANAGEMENT STRATEGIES

Restoring prairies sometimes requires the removal of invasive brush species. Range managers may need a modern version of Leopold's axe if the brush is too mature to allow fire or goats to restore the grassland. Tractors can be used to bulldoze brush. Implements may be used to grub out the roots. Chopping devices on the back of the tractor or a hydro-axe can be used to chip woody debris so that grassland restoration can proceed more rapidly. Creating mulch this way also protects the soil from erosion after mechanical brush control. Some brush species, such as mesquite, can quickly re-sprout after being cut, and some species, such as huisache, can increase after a prescribed burn. Careful use of herbicide may be required in these circumstances.

Although well-managed native grasslands provide everything wildlife needs, using Leopold's plow to disk or plant small areas to produce high-value native plants, such as common sunflower, croton, dayflower, Illinois bundleflower, ragweed, and partridge pea can attract and benefit wildlife species. Disking small patches in the winter can encourage forbs that produce heavy seeds in the spring, disking in the spring can encourage grasses, and disking in the summer can stimulate tender forbs that will provide fall seeds. Using a plow in virgin native prairies (prairies that have never been plowed) is discouraged, however, as so few unplowed patches remain.

Even when grasslands have been converted into croplands, there are some tools and practices that can benefit wildlife. Farmers can leave a few rows of grains, such as corn or wheat, unharvested and delay plowing under crop stubble to allow birds time to forage on waste grain. Native grassland can be retained along field edges, providing habitat for grassland birds and pollinators such as monarch butterflies. Using **minimum tillage** or allowing some weeds to persist in the field can provide food sources for wildlife. Farmers can also purposefully choose to plant small food plots for wildlife. Alfalfa, corn, sorghum, millet, peanuts, sunflowers, clover, rye, oats, and wheat are often used by wildlife.

When grasslands have been converted to agricultural areas or tame pastures it can be challenging to bring back native prairie. Still, the wildlife value of these areas can be

Figure 8.15. Mechanical brush control, such as bulldozing, may be needed when grassland brush is too thick to allow a fire. Disking is sometimes used to encourage forb diversity for wildlife; however, its use is discouraged where native grasslands have never been disturbed. Photos: Mark Mitchell, TPWD.

improved over many years with purposeful management. Wildlife managers can use heavy grazing, mowing, and fire to reduce non-native grasses; however, herbicide is also often needed. Hay cut from nearby native grasslands contains native plant seeds. When that hay is spread on disturbed soil, it can help to restore native prairie. Native prairie grass and forb seeds can also be planted or sprayed on the soil in a liquid mixture called hydro-mulch. Planting is an expensive project and can have a low success rate, so it is helpful to obtain seeds produced in the same region and to plant under ideal climatic conditions. In the long run, a pasture with native species will be much cheaper to maintain than a tame pasture because it does not need the fertilizer and irrigation required for non-native grasses.

Prairie restoration and conservation can take place at a variety of scales. Many grassland species, such as bison and grassland birds, require large, interconnected landscapes in which the grassland habitat can fluctuate between disturbance and regrowth. The Great Plains Conservation Council, The Nature Conservancy, and Texas Audubon all strive to bring varied partners together to restore prairie habitats on a large scale. The Texas Wildlife Habitat Federation, Coastal Prairie Conservancy, and Native Prairies Association of Texas also partner with agencies, private landowners, and non-profits to conserve prairie landscapes.

On the other hand, some patches of prairie are small. Early explorers noted a scattering of small prairies among the forests of East Texas. Even urban or suburban areas can restore small prairie fragments or "pocket prairies" that support pollinators, reduce pesticide use, and decrease rainfall run-off (Fig. 8.16). Patches can be as small as one's yard, as explored further in Chapter 17.

Figure 8.16. A pocket prairie in a Houston yard provides a beautiful variety of native plants that support a large variety of native pollinators. Photo: Jaime González.

CHAPTER 9 – TEXAS DESERTS AND SHRUBLANDS

Early Europeans exploring the deserts of Texas had mixed feelings. The Center for Big Bend Studies shares notes from Lt. Edward L. Hartz, who explored the region as part of the U.S. Army Camel Corps (yes, the Army rode camels in Texas!). He said, "A rougher, more rocky, more mountainous, and rugged country, can scarcely be imagined." The Belgian explorer Jules Leclercq declared that "each plant in this land is a porcupine, it is nature armed to the teeth." And yet, these same explorers and others marveled over the magnificence of the landscape.

Still today, Texas's roughest ecoregion is many people's favorite. It is a vast, rugged landscape where mountain islands arise out of desert seas, columns of igneous basalt and ice-cold spring waters burst forth from broken terrains, dull tan desert vegetation explodes with color in late summer, riparian woodlands teem with birds, bears wander the mountains, and reptiles roam the nights. It is harsh; it is untamed; it is lonely and challenging. Welcome to Big Bend Country. Welcome to the Trans-Pecos!

WHAT IS A DESERT?

Deserts are vegetation communities that receive less than 10 to 12 inches of precipitation annually. In West Texas, desert habitats adjoin semi-arid regions that may receive up to 20 inches per year. Evaporation in these regions often exceeds precipitation, exaggerating the moisture deficit in the environment. The word "desert" comes from the Latin *desertus*, for abandoned or lying in waste; however, these biomes can be full of life, hosting a wide variety of plants and animals that possess adaptations to help them avoid water loss and access water through unique strategies.

Deserts cover about 20% of the earth's land area. North America is home to four deserts characterized by distinctive vegetation communities. The Great Basin, covering nearly all of Nevada and portions of nearby states, experiences very cold winters and is dominated by big sagebrush. The Mojave Desert, which encompasses Death Valley in eastern California, is characterized by the presence of Joshua trees. The Sonoran Desert, extending from southern Arizona into western Mexico, has the mildest winters of all the North American deserts and is typified by towering saguaro cacti. Finally, the Chihuahuan Desert extends from West Texas into southeastern Arizona and Central Mexico. Its indicator plant is lechuguilla, a small species of agave.

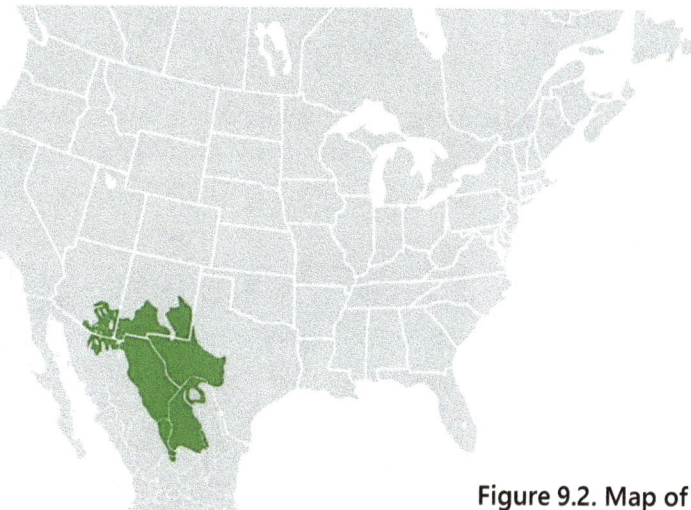

(opposite) Figure 9.1. Giant dagger yuccas blooming in Big Bend National Park.

Figure 9.2. Map of the Chihuahuan Desert, including adjacent semi-arid shrublands in Texas. Source: Cepha, CC 3.0

Chapter 9 – Texas Deserts and Shrublands | 131

DESERT AND SHRUB COMMUNITIES IN TEXAS

As one travels west in Texas, lower rainfall and higher evaporation result in more sparse vegetation communities. The western edge of the Edwards Plateau is dominated by a shrub community with mesquite, juniper, and desert scrub species. The sandy soils on the southwestern edge of the High Plains in Texas support a short, scrubby community dominated by shinnery oak and sand sagebrush. Perhaps uninspiring in appearance, these shin-oak landscapes are valuable habitats for lesser prairie-chickens, northern bobwhites, pronghorns, and several small mammals. Continuing west, the vegetation becomes shorter and sparser, until finally, crossing the Pecos River, one reaches the landscape of the Trans-Pecos.

Often considered "just desert," the mountains and basins of the Trans-Pecos ecoregion support perhaps more diversity in topography and vegetation than any other part of the state. Several processes have produced the varied topography. Long ago, the region was covered by an inland sea, laying down sediments that would form a limestone base and alkaline soils. Tectonic activity lifted sections of that limestone base, creating mountains out of former coral reefs, including striking El Capitan and Texas's other highest peaks in the Guadalupe Mountains. Later, a period of violent volcanic activity emerged, forming ranges such as the Davis, Chisos, and Chinati Mountains with acidic soils. Trans-Pecos topography is frequently described as basin and range, with low-lying basins separated by higher ridges and uplifted mountains. Several watersheds do not connect to larger rivers. Instead, they drain to low-lying areas where water evaporates, leaving behind salty soils in the dry lake beds.

The Trans-Pecos generally experiences cool, relatively dry winters. Late spring and summer temperatures are hot. Rainfall is low, from less than nine inches per year in the west near El Paso to about 12-15 inches along the Pecos River and as much as 20 inches or more on the higher mountain peaks. Nearly all the rainfall happens during the "monsoon season" between May and September. With few clouds and high elevations, insolation—or the amount of solar radiation reaching the earth's surface—is higher here than anywhere else in the U.S. The Trans-Pecos is a biologically diverse region despite the challenging climatic conditions. Over 2,000 species of plants have been recorded, with 1,000 found nowhere else in the state. It is also home to over 500 species of birds, about 170 species of reptiles and amphibians, and more than 120 species of mammals.

The Trans-Pecos hosts 117 vegetative cover types distributed across six ecological subregions (Fig. 9.4). Chapters 7 and 8 describe the mountain forest and desert grassland subregions. The remaining subregions explored below host various desert and shrubland habitats.

Figure 9.3. Chihuahuan Desert habitats in Mexico. Artwork by William Henry Holmes (public domain).

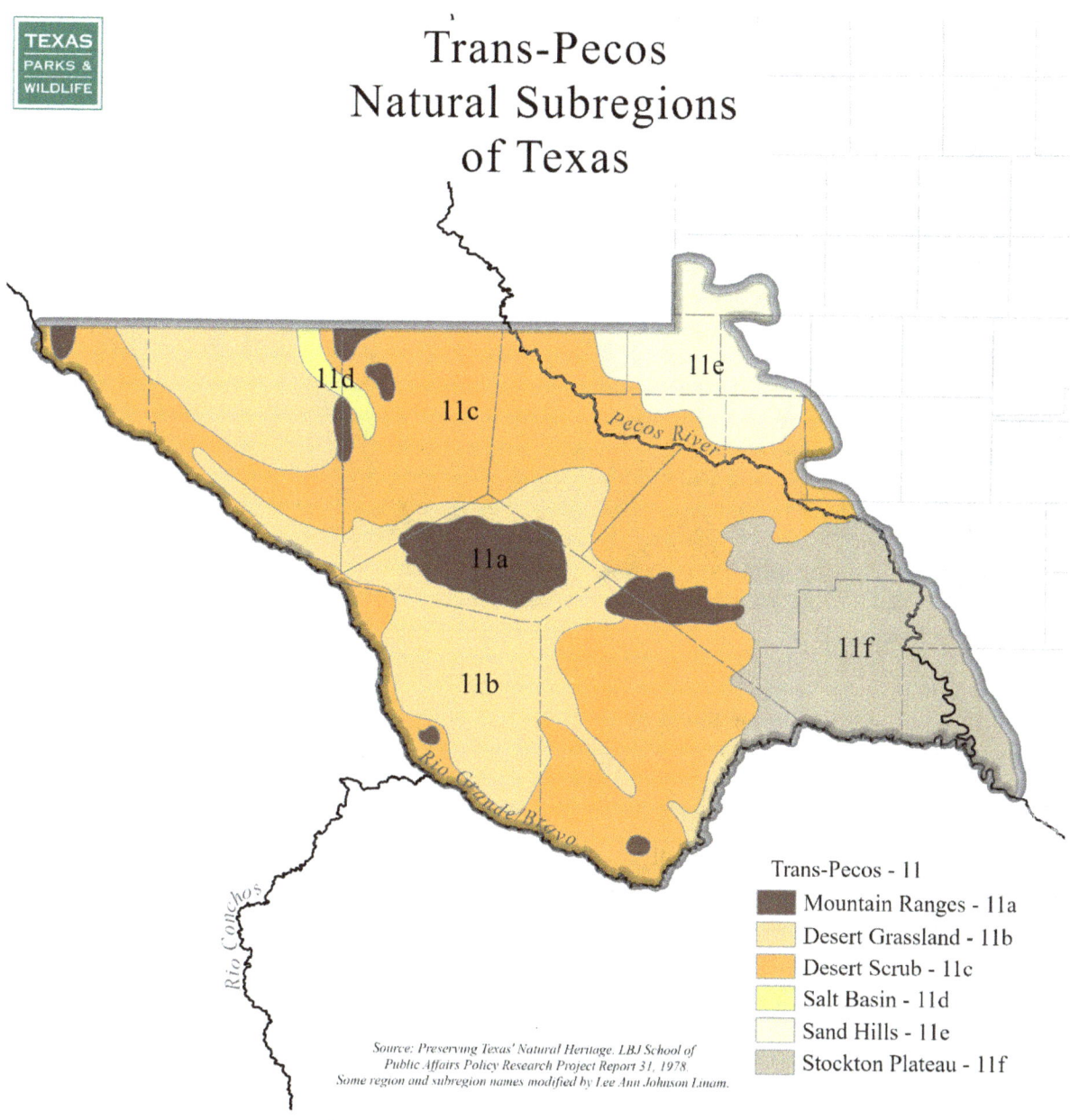

Figure 9.4. Vegetative subregions of the Trans-Pecos. Map source: LBJ School of Public Affairs. Prepared by Paul Daugherty, TPWD IT-GIS.

STOCKTON PLATEAU SUBREGION

Crossing the Pecos River gorge on Highway 90 on the highest bridge in Texas, one enters a land of flat-topped mesas dissected by steep-walled canyons and dry arroyos. Like the Edwards Plateau to the east, this subregion was once dominated by shrubby grasslands with pockets of oak, mesquite, and juniper in drainages and on the slopes. Overstocking in the late 1800s, erosion, and a reduction in fire frequency in the sparse rangelands replaced the grasses with a shrub savanna where creosote bush, broomweeds, and cacti are now dominant. Numerous caves, caverns, and sinkholes occur in the subregion's limestone geology. Springs were once abundant; however,

Figure 9.5. Mesas and breaks of the Stockton Plateau.

agriculture in the Pecos River Valley has drawn down groundwater levels so that many springs are now dry (Highlight 9.1). Springs that remain, such as Caroline Springs and Diamond Y Springs, provide critically important habitats for endemic species (Fig. 9.6) and stream flow essential to diminish salinity in the Pecos River. Fort Lancaster State Historical Site gives a glimpse into the past of the Stockton Plateau. The Nature Conservancy's Independence Creek Preserve conserves some of the most valuable aquatic habitats in the region.

Figure 9.6. TPWD educational poster illustrating the rich biodiversity of spring-fed wetlands in the Trans-Pecos.

HIGHLIGHT 9.1 - COMANCHE SPRINGS—A STORY OF PARADISE LOST

Comanche Springs, a complex of six springs in the Stockton Plateau, was once among the largest in Texas—an abundant source of clear, cool water flowing into Comanche Creek and a home for unique species, such as the Comanche Springs pupfish. The Comanche Trail passed through the area as it came off the Llano Estacado, crossed the Pecos River, and continued southward to the Chisos Mountains and into Mexico, allowing Native Americans and, later, western-bound stage routes to stop at the springs for refreshment and replenishment. In the mid-1800s the city and military post of Fort Stockton developed around the springs.

Figure 9.7. Postcard from Comanche Springs in 1940. Source: Wikimedia Commons (public domain).

Farms in the Pecos Valley first drew water from the Pecos River but later began tapping the groundwater for agriculture. At the same time, the city embraced the springs as a natural tourist attraction, developing a swimming area and hotels that drew visitors from many miles around. However, groundwater pumping for agriculture accelerated in the 1950s to a level that was unsustainable. By the 1960s, the springs no longer regularly flowed.

Fort Stockton lost its springs, but other communities and conservationists took note. When water withdrawal for agriculture threatened San Solomon Springs in nearby Reeves County about two decades later, a diverse group of partners began to work together. The community of Balmorhea, local farmers, university researchers, and TPWD conservation biologists began to collaborate to develop a plan to address water needs. The team built a **ciénega**—a desert wetland—in Balmorhea SP. The ciénega is fed by water flow from the beautiful spring-fed swimming pool at Balmorhea SP. The ciénega and the canals associated with it provide habitat for the endangered Comanche Springs pupfish and Pecos gambusia. After the water flows through the park, it is available for agricultural use in the surrounding area, thus protecting the livelihoods of farmers. This win-win project offers hope that, with careful planning, perhaps consistent spring flow can be returned to Comanche Springs.

Figure 9.8. The Pecos gambusia is a desert fish dependent on spring-fed habitats in the Trans-Pecos. Photo: Angela Palacios, USFWS (public domain).

Figure 9.9. Restored ciénega at Balmorhea SP. Photo courtesy TPWD © 2025, (Maegan Lanham, TPWD).

SAND HILLS SUBREGION

Dunes are usually associated with beaches; however, in the corner of Texas where the High Plains meet the Trans-Pecos, there are sand dunes larger than any on the Texas coast. The dunes, made up of quartz grains eroded from ancient mountains and blown by strong winds, loom 70 feet tall and cover an area about 200 miles wide. Constantly shifted by the wind, shinnery oak (or "shin-oak") communities stabilize portions of them. This unique miniature oak grows only about four feet tall. Like native grasses, most of the shinnery biomass is underground, holding soil and tapping subterranean water sources. Shin-oak communities provide habitat for other species uniquely adapted to the sand environment, such as the endemic dunes sagebrush lizard. Despite the harsh environment, semi-permanent seeps with wetland vegetation can occur in areas where the water table lies near the surface. Sandhill habitat can be visited at Monahans Sandhills SP.

Figure 9.10. Shifting dunes and shin-oak communities at Monahans Sandhills SP. Photo: Leaflet, CC 3.0.

SALT BASINS SUBREGION

As rainwater drains from Trans-Pecos mountains into basins with no riverine outlet, it collects into broad, shallow lakes called **bolsons**. When water evaporates from the bolsons, it leaves behind bright white gypsum and salt deposits. The most extensive salt basin in the Trans-Pecos occurs just west of the Guadalupe Mountains. Once a large shallow lake, in today's drier climate, it is a gleaming stretch of barren white soil visible from the heights of Guadalupe Peak.

Figure 9.11. Trans-Pecos salt basins visible from Guadalupe Peak.

Salt basins are a harsh environment for plant species, yet some survive there. Plant names such as fourwing saltbush, iodinebush, and alkali sacaton indicate their unique adaptations to exploit these challenging soils. Black-throated sparrows, who can conserve water by concentrating their excrement, are an indicator species, as are side-blotched lizards, long-nosed leopard lizards, and little striped whiptail lizards.

CHIHUAHUAN DESERT SUBREGION

The vegetation community most associated with the Trans-Pecos is desert. Specifically, the Trans-Pecos lies within the Chihuahuan Desert biome. The Chihuahuan Desert is the largest desert in North America and one of the most biologically diverse regions in the world. Its diversity of yuccas, cacti, and agaves is especially noteworthy. The impressive variety of yucca is associated with an impressive variety of pollinators. A distinctive species of yucca moth pollinates each of the seven unique yucca species in the Trans-Pecos. The Mexican long-nosed bat is the pollinator for several agaves.

The original extent of the Chihuahuan Desert in West Texas is unclear. Nevertheless, it is likely now more extensive than before human settlement. As overgrazing occurred in West Texas in the 19th century, plant species associated with desert biomes, such as creosote bush, ocotillo, sotol, and catclaw acacias, began invading former grassland areas. Desert vegetation extends from low-lying basins up the lower slopes of mountains in the region, intergrading with grassland communities. These dry mountain slopes once provided habitat for desert bighorn sheep, a native subspecies of wild sheep perfectly fit for this broken arid habitat. Its disappearance and restoration are described in Highlight 9.2.

Figure 9.12. Mexican long-nosed bats pollinate agave as they feed on nectar at night. Photos: David Cervantes Vlogs, CC 4.0.

diet of insects and seeds. Some animals have bodies that help dissipate heat, such as the long broad ears of black-tailed jackrabbits and the long tail of the Big Bend tree lizard. The tendency for warm-blooded animals in hotter climates to have elongated appendages holds for many species and is referred to as **Allen's Rule**. Birds have various strategies. Some migrate to avoid temperature extremes. Turkey vultures, vulnerable to overheating due to their black feathering, cool off by defecating on their legs, where evaporation produces cooling. Black-throated sparrows concentrate their excrement to avoid water loss, as do many lizards.

Figure 9.13. Kangaroo rats are nocturnal and need very little free water in their diet. Black-throated sparrows minimize water loss by concentrating their bodily wastes. Merriam's kangaroo rat photo: Bcexp, CC 4.0. Black-throated sparrow photo: Jeff Forman, TPWD.

Many plant and animal species have adaptations to survive the arid environments of the Chihuahuan Desert. Plants may have extensive root systems, waxy coatings on leaves and stems, modified leaf structures (the thorns on cacti are actually leaves!), and seeds that survive long periods underground until rain occurs. One species, candelilla, was once harvested to make candles from its waxy cuticle. It was one of many useful species in the harsh climate. Curanderos (local healers of indigenous and Spanish background) used other species to treat maladies—creosote bush for saddle sores, Mormon tea for jaundice, leatherplant for canker sores, and sunflower paste for heatstroke.

Animals also have adaptations to help them thrive in the desert. Many use burrows to escape extreme heat. Many species such as kit foxes or banner-tailed kangaroo rats are **crepuscular** (active in the early morning or late evening) or **nocturnal**. Kangaroo rats can draw all the water they need from their

Even amphibians have devised ways to survive in the desert. Many toads burrow. The Couch's spadefoot uses tubercles on its hind legs to dig down up to 35 inches, lining its burrow with mucus to prevent drying out. Deep in the soil, the toads can sense the vibration of raindrops associated with a summer storm. They quickly emerge after the rain shower to lay eggs; those eggs hatch within a day; and the tadpoles metamorphose within two weeks—all strategies to accomplish reproduction rapidly before scarce surface water sources dry up. Desert invertebrates are also specialized for arid environments. Desert cicadas use their piercing mouthparts to draw moisture from plants and then cool by releasing moisture through pores in their exoskeleton. The process of evaporation cools the insect. Like the spadefoot toads, two underground arthropods time their emergence perfectly with desert rains. Desert termites disperse after monsoons to begin new colonies, while giant red velvet mites (sometimes called "angelitos" or little angels) choose those same times to emerge from underground to feed on the dispersing termites. Moments like these can be a burst of life in the desert!

One can explore Chihuahuan Desert wildlife and habitat in the lowlands of Big Bend National Park, Big Bend Ranch SP, and Black Gap WMA. ("Big Bend" is the name given to this area because of the dramatic south, then north sweep of the Rio Grande in this border region.)

Figure 9.14. Typical Chihuahuan Desert landscape with indicator agave species, lechuguilla. Photo: amantedar, CC 2.0.

THREATS

Louis Harveson, founder of the Borderlands Research Institute, describes the Trans-Pecos as "the most iconic, the most diverse, and the most pristine" of Texas's ecoregions. However, it has faced its challenges. Overgrazing, invasive species, overuse of water, and collection of plants and animals are the primary threats.

In the late 1800s, just as American settlers began establishing in the Trans-Pecos, the continent was experiencing an unusually wet, cool period. Encouraged by the rainfall, by 1900 ranchers were running nine million head of livestock in the desert grasslands of West Texas. As the climate returned to normal lower rainfall years, too many livestock remained. Some areas were permanently transformed from grassland to desert scrubland in the ensuing years of overgrazing. With the settlers came other non-native species that established themselves in the wild, such as aoudad sheep and feral pigs.

Water is the scarcest resource in the Trans-Pecos. Rivers and streams in the region once supported flourishing cottonwood woodlands and beaver populations. Muskrats inhabited riverine wetlands. Many springs and streams—isolated aquatic oases in the desert—had endemic species of pupfish and other desert fishes. Now, up to 99% of the water in Trans-Pecos streams is diverted for municipal use or agriculture. Floodplain agriculture has been established in the Pecos River Valley and along the Rio Grande near El Paso and Presidio, drawing heavily from these two primary rivers for both row crops and pecans. Riparian habitats are further threatened by the aggressive expansion of saltcedar, or tamarisk, introduced in the 1800s to stabilize streambanks eroded by overgrazing.

Groundwater sources are threatened as well. Pumping is used to support agriculture in the Trans-Pecos, notably for cotton and melons. Due to heavy withdrawals, only 8 of 50 original springs remain in the Pecos River Basin. As water has diminished, so have many of the endemic aquatic species.

Many plant and animal collectors are attracted to the unique species of the Trans-Pecos. More than one-third of all cacti species in the

Figure 9.15. Trans-Pecos ratsnake crossing a road at night. Cruising to view reptiles at night is a popular activity in West Texas. Photo: Dean Stavrides, CC 4.0.

world occur in Mexico. Of the 345 types of Chihuahuan Desert cacti, more than 100 occur in the Trans-Pecos. There is great demand for these species among cactus collectors, with some experts estimating that collectors have poached hundreds of thousands of cacti throughout the Southwest. West Texas reptiles are also popular, leading the TPWD to adopt special rules regarding the collection of specimens on roads at night.

Other changes have raised concerns in the Trans-Pecos landscape. Ecosystems in desert environments that are damaged can take centuries to recover, so land use activities that are widespread in other parts of the state raise more questions in the fragile habitats of West Texas. The oil and gas industry has long been present. Recent rapid growth has occurred in this industry, as well as wind and solar. West Texas is a prime location for solar energy production, but solar arrays that cover vast acreages of desert take habitat away from desert wildlife, and some types of solar production require large amounts of water for cooling. Many large windfarms exist in West Texas. To many people, the presence of large windfarms on the open landscapes of West Texas is aesthetically undesirable. In addition, biologists are investigating the question of whether wind turbines result in high levels of bird and bat mortality.

Air quality and light pollution are similar concerns in the Trans-Pecos. West Texas has long been known for its expansive vistas and dark skies; however, visibility has declined recently, and light pollution has increased. Coal-fired energy plants in West Texas and Mexico are the primary source of hazy skies, although weather patterns can bring in pollutants from farther away. Energy projects, highways, and towns are the major sources of light pollution. The UT McDonald Observatory, which has been conducting astronomical research near Fort Davis since 1939, is spearheading an effort to create a "dark sky refuge" in the Fort Davis-Big Bend area.

Figure 9.16. The UT McDonald Observatory, located on Mount Locke and Mount Fowlkes in Jeff Davis County, depends on dark night skies. So do many wildlife species. Photo: Mark Alan Storey.

MANAGEMENT STRATEGIES

Despite the threats and changes the Trans-Pecos has experienced, it still represents some of the best wildlife habitat in the state. It has the most public land of any ecoregion, along with important habitats managed by The Nature Conservancy, other conservation organizations, and conservation-minded private landowners. Mountain lions, bears, pronghorn, and horned lizards have persisted in the Trans-Pecos even as they disappeared from many other parts of the state.

Management strategies primarily consist of using the land lightly, especially regarding grazing, and using water sparingly. Energy development projects should adopt production strategies that require little water use and include restoration strategies to help sensitive desert vegetation recover.

The outlook is good for the Trans-Pecos. Only 3% of Texans live west of the Pecos River, and 80% of those people live in El Paso. Outside of El Paso, the Trans-Pecos has the lowest population density in the state—averaging only one person per square mile (The statewide average is more than 50 people per square mile). Although some new residential developments have emerged near picturesque mountain towns like Fort Davis and Alpine, the overall population growth trend in the Trans-Pecos is flat. The average working ranch is greater than 20,000 acres in size, compared to the state average of less than 1,200 acres. There is much opportunity for conserving the species and landscapes of the region.

Although much public land exists in the Trans-Pecos, maintaining the region's unique attributes requires partnership efforts between private landowners, non-governmental organizations, local universities, and conservation agencies. Partnerships facilitated by regional entities such as the Chihuahuan Desert Research Institute and the Borderlands Research Institute have great potential to promote healthy ecosystems in the Trans-Pecos to benefit people and wildlife. The case of desert bighorn restoration is an inspiring example of partnership potential.

Figure 9.17. The Davis Mountains Preserve, owned by the Nature Conservancy of Texas, is an example of the public-private partnerships that help to conserve the Trans-Pecos. Photo: Fredlyfish4, CC 3.0.

HIGHLIGHT 9.2 - THE RETURN OF THE KING—THE KING OF THE DESERT MOUNTAINS

Desert bighorns are one of the most iconic species of the Trans-Pecos. In 1890, Vernon Bailey, reflecting on his expeditions in the Trans-Pecos, noted, "Here the sheep find ideal homes on the open slopes of terraced lime rock or jagged crests of old lava dikes." Indeed, desert bighorn sheep are a perfect fit for the desert mountains of West Texas. Their padded cloven hooves are concave, allowing them to gracefully negotiate the steepest of habitats—topography that can thwart their mountain lion predators. A complex digestive system enables them to break down fibrous vegetation. The adults can draw water from dry desert foliage, allowing them to go for as many as three days without drinking.

When Bailey made his note, there were perhaps 2,500 bighorns roaming 15 mountain ranges in Texas. Sixty years later, bighorn sheep were gone from West Texas. Domesticated sheep and goats competed with bighorns for food and brought in diseases that hit the native sheep populations hard. Aoudad sheep introduced into the wild provided more competition. Unregulated hunting diminished numbers. As water sources dried and vegetation desertified, carrying capacities decreased. The development of roads and fences in the desert basins impeded the ability of bighorns to move between different populations on desert slopes. The king of the mountains was no longer king.

Bighorns always had their supporters, and as early as the 1950s, TPWD attempted to reintroduce desert bighorns from neighboring states into the Black Gap WMA. However,

Figure 9.18. Desert bighorn being carried from helicopter to its release site. Photo courtesy TPWD © 2025, (Chase A. Fountain, TPWD).

those early reintroductions never established. Finally, in the 1970s, TPWD achieved some success in the Sierra Diablo Mountains. In the early 1980s, the newly-formed Texas Bighorn Society and the Foundation for North American Wild Sheep approached the Department about supporting reintroduction goals in earnest. Funding was secured. Additional private landowners signed up to participate. TPWD expanded captive breeding programs. Desert bighorns captured in Nevada added to the genetic diversity.

The partners used private and public funding to enhance habitat by providing watering sites at wildlife guzzlers, limiting the sheep's vulnerability to predation at natural watering sites. As populations became established, they served as a source of bighorns to translocate to other mountain ranges in the state. Although the program started slowly, by 2020, eleven herds totaling between 1,000 and 1,500 animals had been established in seven West Texas mountain ranges, allowing TPWD to permit the harvest of a few sheep in hunts drawn by lottery and auctioned off in partnership with the Bighorn Society. These hunts, limited to about 20 animals per year, generate thousands of dollars to reinvest in the restoration and management program, allowing it to be self-supporting.

Despite the success of the partnership recovery efforts, challenges remain for desert bighorns in Texas. Populations have declined dramatically in the 2020s. Biologists identified a pneumonia caused by the bacterium *Mycoplasma ovipneumonia*. This bacterium is often found in exotic aoudad sheep that are widespread in West Texas. Aoudads are carriers of the disease and are not affected by it, but bighorn sheep are highly susceptible. Bringing back bighorns continues to be a complex process of working with partners to restore the natural balance in the mountain ecosystems these magnificent sheep inhabit.

Figure 9.19. In 2024 TPWD added another site to the comeback story of the desert bighorns—Franklin Mountains SP just outside El Paso. Photo courtesy TPWD © 2025, (Chase Fountain, TPWD).

CHAPTER 10 – TEXAS WETLANDS

Texas has an arid reputation, and wetlands make up less than 3% of the state's land area. Still, among all states, Texas has the fifth-largest acreage of wetlands. Furthermore, its wetland diversity is impressive, with the state hosting swamps, bottomland hardwood forests, riparian forests, marshes, bogs, springs, **resacas**, ciénegas, playa lakes, **prairie potholes**, and saline lakes.

Though wetlands were once considered wastelands, they are some of the most valuable habitats in the state. A status report from USFWS estimates that the ecological value of wetlands is at least 11 times higher than lakes and rivers, more than 36 times higher than forests, and about 33 times higher than grasslands.

WHAT IS A WETLAND? AND WHY CARE ABOUT WETLANDS?

Wetlands are usually defined as an area that is permanently or sporadically wet with soil that is permanently or regularly saturated. The Texas Water Code defines a wetland as "an area (including a swamp, marsh, bog, prairie pothole, or similar area) having a predominance of **hydric soils**...that under normal circumstances support the growth and regeneration of **hydrophytic** vegetation." Thus, wetlands are defined by the frequent presence of water and can be recognized by their soil and plant components.

Wetland soils are frequently waterlogged and may be recognized by their "mucky" texture. Saturated soils host anaerobic bacteria that produce hydrogen sulfide gas during decomposition, giving off a rotten egg smell when the soil is disturbed. However, hydric soils may not be saturated at all times. During dry periods, soil scientists look for a characteristic thick organic matter layer that indicates the high productivity and slow decomposition rates associated with wetlands. Wetland soils also often have a green-blue-gray ("gley") colored A-horizon due to chemical changes in iron and manganese under saturated conditions. If those soils are exposed to air, then the iron may oxidize, creating a rust color.

(overleaf) Figure 10.1. Riparian wetlands provide habitat along waterways across the state. Photo courtesy TPWD © 2025, (Chase A. Fountain, TPWD).

Figure 10.2. Gley soil that is starting to oxidize. Photo: SoilScience.info, CC 2.0.

Figure 10.3. Cattails, named for the shape of their flower clusters, are a common wetland indicator plant. The interior of cattail leaves contained air-filled cells that help to support the plant in water.

Many plants in wetlands have unique characteristics. Cattails, sedges, and rushes have air spaces inside their leaves to transport air and make the leaves buoyant. Because too much water on a leaf surface can impair photosynthesis and gas exchange, many herbaceous wetland plants have a waxy coating to help them shed excess water. The roots of wetland plants also have special adaptations. **Emergent** plants (plants that stick out of the water such as cattails and arrowheads) have complex fibrous roots that help secure them in the mucky soil. Tiny floating duckweed plants have short roots that simply dangle into the nutrient-rich water. Mangroves have a complex network of aerial and submerged roots that let them draw in oxygen even when water levels are high and stabilize their growth in the soft tidal sediments. Mangroves and other coastal wetland plants also have structures that allow them to excrete salt. These plants that can survive in salty environments are called **halophytes**.

Water is not always visibly present in wetlands, but vegetation can help identify wetland habitats even during dry periods. The U.S. Army Corps of Engineers, in cooperation with the Natural Resources Conservation Service and USFWS, publishes a National Wetlands Plant List that assigns plants to one of the five categories listed below, using an abbreviation to mark the wetland status of species. Recognizing wetland indicator species can prevent accidental destruction of wetlands even when the ground is dry.

- OBL - Obligative – plants that almost always occur in wetlands. An example is cattail.

- FACW - Facultative Wetland – plants that usually occur in wetlands but may occur in non-wetlands. Example: bushy bluestem.

- FAC – Facultative – plants that occur in wetlands and non-wetlands. Example: pecan.

- FACU - Facultative Upland – plants that usually occur in non-wetlands but may occur in wetlands. Example: live oak.

- UPL - Upland – plants that almost always occur in non-wetlands. Example: juniper.

Many wetland animals also have unique characteristics. Ducks and cormorants have webbed feet for swimming. Stilts and gallinules have long toes for walking on floating vegetation or muddy ground. Pelicans have pouches for scooping fish, curlews have long bills for probing in the mud, and herons have sharp bills for fishing. River otters and beavers have fur that repels water, whereas waterfowl have oils secreted from a preen gland to stay dry. Frogs have semi-permeable skin that keeps them close to wetlands. Frogs and alligators have an extra eyelid—a see-through nictitating membrane—that protects the eye while allowing them to see underwater. Whirligig beetles have four eyes, two to see above water and two to see below.

There are many ways in which wetland ecosystems are valuable. Wetlands...

- Purify water. Wetlands are the kidneys of the water cycle. As water filters through wetlands, they remove sediments and nutrients and improve the quality of water that ultimately seeps into aquifers, streams, or bays. Riparian wetlands can remove up to 90% of the phosphorus, 50% of the nitrogen, and 80% of suspended sediment from storm and agricultural runoff. Many cities use or construct wetlands to provide the final stage of wastewater purification. In Port Aransas, the wetland associated with the water treatment facility is one of the best birding spots on Mustang Island!

- Reduce flooding. Wetlands act like sponges. Wetland vegetation slows the flow of floodwater, and wetland landscapes absorb and slowly release excess water so that surrounding uplands are not flooded. An acre of wetland can store 1 to 1.5 million gallons of floodwater. One study estimated that a landscape with wetlands can reduce flood levels by 60% to 80%.

- Prevent erosion and buffer shorelines from storms. Wetland vegetation slows water flow and absorbs wave and wind energy, protecting nearby cities. It has been estimated that coastal wetlands in the U.S. provide over $23 billion worth of storm protection each year.

- Store carbon and help to stabilize climate. The National Oceanic and Atmospheric Administration research shows that coastal wetlands store up to five times more carbon per acre than tropical forests.

- Produce commercially important species. Wetlands serve as nursery grounds for 90% of Texas commercial fish and shellfish. This supports a Texas commercial fishing industry that produces $400 million in wholesale sales and employs about 30,000 coastal residents annually. Saltwater sport fishing in Texas yields almost $2 billion in economic benefits and employs about 25,000 Texans. Waterfowl hunting in Texas wetlands produces $1 billion in spending and supports 14,000 jobs.

- Are critical to wildlife. Half of the threatened and endangered species listed in the U.S. are dependent on wetlands. About 80% of vulnerable bird species are associated with wetlands. Texas wetlands are especially important to waterfowl, sometimes hosting 90% of ducks and 75% of all geese in the Central Flyway. Texas wetlands provide habitat to four threatened reptile species and 16 threatened amphibian species.

When all these contributions are considered, the annual value of wetlands in the U.S. is an impressive $7.7 trillion each year.

Figure 10.4. This poster produced by TPWD illustrates the rich diversity of wildlife found in wetland habitats.

TEXAS WETLANDS

Texas has a wide variety of wetland types. These wetlands form along bays, rivers, lakes, springs, estuaries, or depressions in the landscape, with different types of wetlands hosting different types of plants and animals. This chapter focuses on freshwater wetlands and the salty marshlands along the coast, but Texas also offers marine wetlands like seagrass beds in its bays. Wetlands can be classified as forested, marsh, or lake-like (lacustrine), plus a few other specialty types.

FORESTED WETLANDS

Swamps are forested areas where water is permanent year-round. In Texas, swamps are restricted to the Pineywoods ecoregion and are usually dominated by bald cypress and water tupelo trees. The saturated soils and low dissolved oxygen in swamps present a challenging environment; therefore, plants are specialized, and species diversity is low. Trees in swamps often lack deep roots; instead, they develop wide, flared "buttressed" trunks at the base and elevated sections of roots called "knees" that offer the large trees more stability in the soft soils and a greater surface area for capturing oxygen. Because their seeds cannot germinate in deep water, the seeds float, enabling them to drift to shallower areas. East Texas also supports shrub swamps that are dominated by smaller woody species such as water elm, common buttonbush, and eastern swamp privet. Caddo Lake, designated a Wetland of International Significance by the Ramsar International Convention on Wetlands, provides a good example of bald cypress-tupelo swamp.

Figure 10.5. Swamps are forested wetlands and once provided habitat for the ivory-billed woodpecker and Carolina parakeet.

Riparian forests are the most widespread wetland type found in Texas, as most rivers in the state support hardwood forests along their paths. These riparian zones (riparian means "along a riverbank") rarely hold water year-round but are periodically flooded and are dependent on a shallow water table provided by the stream. In many locations, the habitat may be only a narrow string of moisture-loving tree species parallel to the river. Still, they provide important corridors of habitat for many wildlife species. Riparian tree species vary depending on the ecoregion and river system. Pecan and elm forests line streams in the Coastal Plain; bald cypress and American sycamore woodlands occur along Edwards Plateau streams; cottonwood and netleaf hackberry woodlands parallel rivers in the High Plains; and patches of willows and cottonwoods emerge along some streams in the Trans-Pecos. Forested wetlands can also be found on former rivers. **Oxbow lakes** are water bodies isolated from the ancient meanders of a river and often support forested wetlands. Resacas is the name given to oxbow lake wetlands in South Texas. These sub-tropical shallow ponds are often lined with retama and huisache trees and provide important moisture for rare amphibians such as the black-spotted newt and other wildlife in South Texas.

The bottomland hardwoods of East Texas are the most extensive of the riparian forest types. Bottomland hardwood wetlands form where rivers overflow their banks into large flat floodplains that support forests of pecan, hickory, oaks, hackberry, and blackgum. The movement of the water shapes the forest floor into ridges, swales, or flats, resulting in a diversity of moisture and plant species. Bottomland hardwoods can be incredibly diverse, supporting up to five times as many species as upland forests. More than 180 species of woody plants have been reported from bottomland hardwood communities in Texas, and a survey by USFWS recorded 273 species of birds, 45 mammals, 54 reptiles, 31 amphibians, 116 species of fish, and innumerable invertebrates in bottomland hardwood forests across the U.S. The Columbia Bottoms in the Mid-coast NWR, Village Creek SP, and Trinity River NWR all host excellent examples of old-growth bottomland hardwoods.

MARSHES

Marshes are characterized by **emergent vegetation**. Marsh vegetation can occur throughout the state along the edges of lakes, in grassland depressions, or among the outflow of springs, but the most extensive marsh habitats in the state are found in the Western Gulf Coastal Plain. The Gulf Coastal Plain is a low-lying landscape more than 360 miles long and 50 to 100 miles wide bordering the Texas coast. Its flat topography (less than one to two feet above sea level in most places) collects rainwater, and waterways such as bayous and sloughs meander slowly, holding water on the landscape. The result is a vast acreage of marshlands lining the Texas coast from Corpus Christi north to the Louisiana border.

Marshes along the Gulf Coast can be fresh, salt, or intermediate (a mixture of salt and fresh). Salt marshes lie closest to the coast and are influenced by daily tides moving in and out of the wetlands. The changing water and salinity levels are challenging for plants, and plant diversity is low. Smooth cordgrass lines the water edges, and salt-tolerant species, such as saltgrass, dwarf saltwort, and sea ox-eye daisy grow in the flats. Despite the low plant diversity, these saltwater wetlands play an important ecological role. In addition to providing nursery areas and habitat for estuarine species such as the Texas diamondback terrapin, many seafood species, and innumerable bird species, data indicate that they may store as much as ten times more carbon per acre than freshwater marshes.

Figure 10.6. A broad band of marshes borders the Texas coast from the Louisiana border down to South Texas. Federal and state properties, such as Powderhorn SP and WMA, shown here, help protect these valuable habitats. Photo courtesy TPWD © 2025 (Chase A. Fountain, TPWD).

Fresh marshes occur farther inland. They are not usually influenced by saltwater. They have a much greater diversity of plant species and have high productivity, resulting in deep organic soils. Swamp smartweed, Walter's barnyard grass, and maidencane produce seeds along shallow edges. Water lilies, cattail, and hardstem bulrush (also called tule) are found in deeper water.

In between the fresh and salt marshes lie the intermediate and brackish marshes. These habitats are somewhat tidally influenced. Plant species diversity is higher than in salt marshes and may include saltmeadow cordgrass and seashore paspalum in shallow zones. Emergents like Olney three-square bulrush provide forage for muskrats. Submerged plants like widgeongrass, southern naiad, and various pondweeds produce seeds and forage for American wigeon and other ducks and coots.

Texas coastal marshes are vital habitat for many wildlife species. They are the nursery grounds for 90% of Texas's seafood. They provide crucial wintering and migratory habitat for waterfowl and shorebirds, along with peregrine falcons, osprey, reddish egrets, and whooping cranes. Coastal marshes are the primary habitat for American alligators and also provide a home for swamp rabbits, otters, muskrats, many other furbearers, and amphibians. The freshwater wetlands of Southeast Texas are the only place one can hear the grunts of the pig frog in Texas! Fortunately, visitors can enjoy wildlife viewing and recreation in marsh habitats on many national wildlife refuges, state parks, and wildlife management areas on the Texas coast.

Figure 10.7 Freshwater marshes support a diversity of flora and fauna. Salt marshes support plant and animal species specialized for life in salty environments.

LACUSTRINE WETLANDS

Lacustrine wetlands are wetlands associated with lakes or ponds. Texas has several types. Among the most important of these are the playa lakes of the Texas Panhandle. Playa lakes are round, shallow lakes found throughout the High Plains from Nebraska south to Texas. Texas has an estimated 19,000 playa lakes—about one for every square mile of the Panhandle. These circular wetlands form in depressions where collected rainwater seeps through the clay soils to erode the caliche below. Although they are called lakes, playas are shallow water bodies often dominated by emergent wetland vegetation.

Playa lakes are not large—they average 15-16 acres in size—but they are incredibly important in the High Plains environment. They provide water filtration, flood water storage, wildlife habitat, and recreation opportunities. Recharge rates in playas are 10 to 1,000 times higher than in surrounding uplands. As a result, rainwater collected in playas contributes up to 95 percent of the recharge to the vast Ogallala Aquifer.

Figure 10.8. Playa lakes in the Texas Panhandle provide critical lacustrine wetland habitats during wet seasons. Photo courtesy TPWD © 2025.

Playas cycle through wet and dry seasons, producing a high level of plant diversity. About 350 plant species have been recorded on Texas playa lakes. Texas playas, often surrounded by extensive agricultural fields, provide winter habitat for up to two million waterfowl, along with summer nesting habitat for mallards, blue-winged teal, cinnamon teal, northern pintails, and ruddy ducks. Some 30 species of shorebirds rely on playas during migration. In total, 185 bird species, 13 amphibian species, and 37 mammal species have been documented using playa lakes. About 98% of playas occur on private lands. TPWD's Playa Lakes WMA offers a glimpse of these crucial habitats on public lands in Donley and Castro counties.

Other important lacustrine wetlands are scattered across the state. **Prairie potholes**, small wetland ponds, are found in many grasslands of the state. Gilgai, the small, depressional wetlands in the Blackland Prairie, are an example of these scattered grassland wetlands. Unfortunately, many have been lost to agricultural conversion. Sand sheet wetlands are pothole wetlands formed by wind and water in the Coastal Sand Sheet of South Texas. These clay-lined depressions can be important sources of water in the Wild Horse Desert because of the lack of rivers or creeks. In other parts of the state, low-lying portions of grasslands may drain slowly, producing a **wet meadow** habitat rich in diversity.

OTHER WETLAND TYPES

Two other wetland categories in Texas deserve special mention. Bogs are wet, acidic areas that accumulate soft, decaying vegetation, leading to the formation of peat moss. **Bogs** form in low-lying areas in the Pineywoods and Post Oak Savanna where a subsurface layer of clay causes water to pool on the forest floor. As vegetation accumulates in the depressions, the wetland acidifies and produces a spongy surface that supports acid-loving plants, such as sphagnum moss, ferns, orchids, and irises. A few woody plants, including blackgum, red maple, and wax myrtle, may be scattered in the bog, but the most unique flora are the carnivorous plants. Pitcher plants, sundews, and bladderworts all have creative adaptations for attracting, capturing, and digesting insects. The invertebrates provide essential elements to these specialized plants in an acidic environment where plants struggle for nutrients. In the Big Thicket region of Southeast Texas, large bogs called baygalls develop in old stream channels. These baygalls are characterized by almost black water and thick vegetation—producing an eerie environment and inspiration for the idea of a "bogeyman!"

Figure 10.9. Pitcher plants, such as this one found in the Big Thicket National Preserve, can colonize the challenging soils in bogs because they can draw nutrients from insects they capture. Photo courtesy TPWD © 2025 (Chase A. Fountain, TPWD).

Bogs are fragile environments, easily compromised by erosion in the surrounding landscape and by foot and vehicular traffic in the bog. Bogs are also destroyed when they are dug up to harvest peat moss for the horticulture industry (in some locations the moss in bogs can be up to 16 feet deep). Management is important, as periodic prescribed burns are required in bogs to control overgrowth by woody plants. Bogs are preserved in Gus Engeling WMA, the Boykin Springs Recreation Area of the Angelina National Forest, and the Turkey Creek Unit of the Big Thicket National Preserve.

Spring-seeps are wetlands that form where springs come to the surface and create moist pockets of wetland vegetation. Spring-fed wetlands occur in Central Texas, where springs typically flow into streams. These small seepage areas may support vegetation such as floating water primrose and marsh pennywort. Spring-seeps are also important biodiversity areas in the Trans-Pecos. Mountain springs create small wetland habitats on mountain slopes in the region, supporting "hanging gardens" of columbine, ferns, orchids, and poison ivy. On the desert floor of the Trans-Pecos, springs emerging from geologic faults can create a marsh-like oasis and moist grassland habitat called a ciénega. These ciénegas provide habitat for endemic desert fishes, amphibians, reptiles, birds, and invertebrates. Spring-fed wetlands are evident in numerous locations in Central Texas, including Honey Creek State Natural Area and San Felipe Springs in Del Rio. Balmorhea SP and the Nature Conservancy's Diamond Y Preserve offer protection for spring-fed wetlands in the Trans-Pecos.

Figure 10.10. Krause Springs is an example of a spring-seep wetland in the Edwards Plateau. Many of these Hill Country springs are home to endemic salamanders, but many also are overrun by non-native vegetation such as these elephant ears. Photo: Larry D. Moore, CC 2.0.

THREATS TO WETLANDS

Early settlers in the United States viewed wetlands as wastelands—places that bred diseases, restricted travel, impeded the production of crops and livestock, and generally got in the way. The goal was to drain and fill wetlands and put them to more "productive" use. As a result, by the 1980s, wetlands in the U.S. had declined from 221 million acres to only about 103 million. Texas similarly experienced the loss of more than half of its wetlands, with only about 7.6 million acres remaining today.

As aquatic habitats have declined, many wetland animals have become imperiled as well. More than 60% of the amphibians in the U.S. are declining. Half of the crayfish species and two-thirds of the freshwater mussel species are at risk of extinction. The populations of one-third of waterbird species are declining.

How did the U.S. lose so many wetlands?

Draining and filling caused the greatest wetland loss. Settlers discovered that wetland soils were rich in nutrients, and the U.S. Department of Agriculture originally subsidized the draining of wetlands to increase cropland acreage. Drainage accelerated, reaching a peak in the 1950s through the 1970s when about 500,000 acres of wetland were lost to agriculture each year. Losses continued until the 1985 Farm Bill passed by Congress included a "Swampbuster" provision that withheld USDA benefits from farmers who converted wetlands to croplands.

More recently, draining and filling associated with urban development have been the biggest contributors to the loss of wetland habitat. One-quarter of the Texas population lives along the Gulf Coast; Houston experienced a growth rate of 38% from 2000 to 2020. Recent forecasts predict a 69% increase in urban land cover in the state by the year 2100. As urban areas in wetlands grow, they also change the flow of water in the landscape, resulting in damage to wetlands and floods in cities.

Changes in water flow impact wetlands, especially riparian woodlands. When rivers are dammed to form reservoirs, **bottomland hardwoods** upstream of the dam die out in the deep water of the reservoir. Downstream of the dam, bottomlands are impacted because the reservoirs often hold back the flooding waters that are needed to nourish and maintain these wetlands. By 1995, 600,000 acres of bottomland hardwoods were estimated to have been lost due to reservoirs. At least 22 more reservoirs are proposed in Texas, which could bring the total wetland loss from reservoirs to 860,000 acres.

Riparian woodlands are often denied the surges of high water they need because water rights permits allow too much water to be pumped out of rivers for other uses. Those shortages are even more significant in drought years. In recent years, Texas has been working on collaborative water plans that consider the "environmental flows" needed in rivers to maintain riparian wetlands. Levees and channelized streams constructed for navigation or to prevent flooding also prevent wetlands from receiving the recharge they need. This has especially affected the resacas of South Texas.

Timber harvest also threatens bottomland hardwood wetlands. Demand for pulpwood (hardwood timber that is ground up to make paper and cardboard containers) has risen rapidly during this century and is projected to continue to grow. Texas ranks seventh in the U.S. in pulpwood production. As a result of the cumulative effect of dams, water diversion, timber harvest, and clearing for agriculture and urban development, the state has lost more than 60% of its bottomland hardwood forests.

Subsidence and saltwater intrusion have impacted wetlands on the Texas coast where freshwater marshes have declined nearly 30% since the 1950s. Many coastal areas are subsiding between two and seven millimeters annually due to groundwater pumping and mineral extraction. As subsidence takes place, saltwater reaches farther into intermediate and fresh marshes, killing vegetation that is not tolerant of higher salinities. The problem is further exacerbated by channels dug for boats and by erosion of marsh edges caused by boat traffic. The result is more open water and less marsh vegetation.

In addition, because of the large amount of water pumped from Texas rivers, less freshwater is reaching the coast to create a balance in the intermediate and brackish marshes. In the year 2000, for the first time the Rio Grande—the largest river in Texas—did not reach the Gulf. This *reduced freshwater inflow* is a threat to fish and shellfish that reproduce in those estuaries and to species like the endangered whooping crane that depend on those food sources.

Even when wetlands are not destroyed, they can be degraded by *pollution* and *invasive species*. Though wetlands are good at filtering pollutants, surrounding land use changes can overwhelm the integrity of the ecosystem. Erosion, excessive nutrient run-off, and pesticides can affect wetland species. *Invasive species* also impact wetlands. Giant salvinia, common water hyacinth, and water lettuce are examples of floating invasive plants that can clog up the surface of wetlands. Alligatorweed is a common invasive that dominates wetland edges. Eurasian watermilfoil and hydrilla are submerged plants that can displace native species that may produce better wildlife food. Island apple snails, recently established in the Houston area, can decimate wetland vegetation, as can nutria, a South American rodent introduced to the wild in the 1930s and '40s. Nutria are now well-established in wetlands across the state.

Looking to the future, *climate change and sea level rise* pose a potential threat to Texas wetlands. The Gulf is experiencing the highest relative sea level rise of any water body on the globe due to the combined effects of subsidence and sea levels rising 1.8 mm per year. As a result, many ecologists fear that the state will see big shifts in its coastal marshes. Wetlands in East Texas and Coastal Texas are also expected to experience more damaging storm events and extreme episodes of

Figure 10.11. Invasive species like water hyacinth (in the foreground) can completely fill wetland areas, closing off all open water during the summer growing season. Photo: Mark Alan Storey.

flooding. In contrast, drought and lowered rainfall is forecast to affect the few precious wetlands in West Texas.

CONSERVATION OF WETLANDS

The first efforts to conserve wetlands began with the help of duck hunters. In 1937, the Migratory Bird Hunting and Conservation Stamp Act began requiring waterfowl hunters to purchase "duck stamps" to hunt waterfowl. Proceeds from the sale of the stamp were used to purchase national wildlife refuges with a focus on conserving wetlands. In that same year, Ducks Unlimited, a conservation organization made up primarily of duck hunters, also was formed. Its initial focus was protecting wetland breeding areas for waterfowl.

However, without regulations preventing the destruction of wetlands, wetlands loss continued to increase until the environmental movement of the 1970s. In 1972 the Clean Water Act was passed. Section 404 of the Clean Water Act requires a permit to discharge dredge or fill material into the waters of the United States, including wetlands. Although the definition of "waters of the United States" has changed over the years, the Act still provides some authority for the Army Corps of Engineers to review projects that might impact wetlands.

By 1985, the values of wetlands were more widely recognized, and Swampbuster regulations in the Farm Bill discouraged the destruction of wetlands in agricultural lands. In 1986 Congress enacted the Emergency Wetlands Resources Act. The official national goal since that time has been "no net loss" of wetlands. To assess progress on the "no net loss" goal, USFWS regularly conducts a National Wetlands Inventory.

Also in 1986, USFWS collaborated with the Canadian Wildlife Service to draft the North American Waterfowl Management Plan. The plan suggested developing "Joint Ventures"—regional planning entities that bring government and private groups together to collaborate on projects to benefit wetlands and waterfowl. There are now more than 20 Joint Ventures focusing on all migratory birds, and Mexico has joined the planning effort.

States also began to get involved in more proactive efforts to conserve and restore wetlands. In 1997 TPWD developed the Texas Wetlands Conservation Plan. The plan focuses on non-regulatory volunteer strategies for private landowners and on regional conservation planning efforts.

These efforts and other initiatives, such as the National Estuary Program and several projects focused on rivers and fish, initially slowed the rate of wetland loss, but did not stop it completely. USFWS Status and Trends Report to Congress in 2019 noted, "Net wetland loss increased substantially since the last Wetlands Status and Trends study period (2004–2009), resulting in the loss of 221K ac [89K ha] of wetlands, primarily to uplands, between 2009 and 2019." Losses were especially high in forested wetlands and in the Southeast, including Texas.

WETLANDS MANAGEMENT

Thankfully, even in the face of wetland loss, many public and private landowners do their part to ensure wetland conservation and health. Wetland habitat managers recognize that water levels and vegetative cover change constantly, creating a dynamic environment. Effective land managers think about the landscape, knowing that wetlands in many stages of succession are beneficial for a wide variety of species. Several types of habitat management strategies are used.

Manage Water Levels - Most wetland habitats naturally vary in their water depths and degree of inundation. Where **hydrology** (the natural movement and distribution of water) has been altered, biologists may use levees, pumps, and water control structures such as gates or flashboard risers to release or hold water to meet management goals. Some goals might be to:

- Provide habitat at different depths for wintering waterfowl – Dabbling ducks such as pintail and teal feed in shallow water. Diving ducks such as scaup or canvasbacks feed in deeper water. Geese often concentrate in agricultural fields and shallow wet grasslands.
- Produce a good food supply for waterfowl - Summer drying can increase seed-producing plants such as barnyard grass (also called millet), that provide a food source when ducks return in winter. If drying is delayed until late summer, then those wetlands can still benefit amphibians in the spring.
- Provide habitat for summer water birds - Maintaining some wetlands 6-24" deep in summer can benefit herons, egrets, gallinules, and rails. In addition, some ducks, such as mottled ducks, wood ducks, and black-bellied whistling ducks, breed in Texas, so these summer wetlands provide them with needed habitat.
- Provide habitat for shorebirds - Texas provides key habitat for migrating shorebirds, some of whom travel thousands of miles between nesting and wintering sites. Open shallow water wetlands (<6" deep) in the fall especially benefit these species.

Figure 10.12. Water control structures can allow wildlife managers to raise or lower water levels in wetlands and greentree reservoirs, such as here at Richland Creek WMA. Photo courtesy TPWD © 2025 (Matt Symmank, TPWD).

Restore river hydrology - Several bills passed by the Texas Legislature have set up a process for agencies and stakeholders to identify "environmental flows" that are needed for the health of rivers and estuaries. These assessments are especially important in maintaining the health of riparian forests, bottomland hardwoods, and coastal marshes. The collaborative environmental flows allocation process helps guide the Texas Water Development Board and the Texas Commission on Environmental Quality as they make decisions regarding construction and operation of reservoirs and the issuance of water rights permits.

Vegetation management – Several of Aldo Leopold's tools are applicable in wetland areas. Rotational grazing can be used to create an interspersion of heavy and sparse vegetative cover. Seed-producing plants often emerge after a short but intense grazing application. When water levels are low, prescribed fire can be an effective management tool. Summer fires can be used to control woody species. Fall and winter fires can provide openings for geese to feed. Mechanical methods, such as shredding and disking, can be used to create vegetation openings during low water periods. In bottomland hardwoods, selectively cutting or deadening trees by stripping the bark can create small openings that allow more seed-producing plants in the understory. In bottomland hardwoods that have been previously cut, planting mast-producing trees such as oaks and hickory can improve the availability of wildlife food.

Controlling invasive species may also be an important vegetation management goal. Several combined strategies may be needed, including herbicides, hand-cutting, mulching, or physical removal. Even some native species, such as cattail, can become dominant and overgrow open areas. Using water level management can help to control the distribution of these aggressive species.

Figure 10.13. An aerial view of a prescribed burn in a marsh, followed by images of the marsh immediately after the burn and tender vegetation in the next growing season. Photos: Charles D. Stutzenbaker.

Manage adjacent uplands – Careful land management in uplands can help protect wetlands. Although wetlands serve an important function in filtering out pollutants and excess nutrients, it is possible to overload these systems. Fertilizers and pesticides should only be used in the watershed when absolutely needed, at specified rates, and never before rain is forecast. Swales (shallow depressions in the landscape), grassland buffers, and erosion barriers such as native grasses or brush piles can be used to slow run-off into wetlands. In addition, some wetland species need well-managed upland habitat nearby. For example, the mottled duck is a declining species found only on the Gulf Coast. Mottled ducks do not migrate but remain on the Texas coast year-round. They feed and roost in wetlands, but hens need thick coastal prairie grasses in nearby uplands to build their nests.

Create and restore wetlands – Where wetlands have been lost or where they are lacking, created or restored wetlands can provide some wildlife benefit. In some cases, landowners want to increase wetland habitat for wildlife. In other cases, wetlands may be created as **mitigation** to make up for the destruction of another wetland that is allowed under a permit. In areas where wetlands were present but have been lost, restoring the hydrology can help to recreate wetlands. Examples include:

- Flooding agricultural fields – Holding water on agricultural fields can attract waterfowl that feed on waste grain. The most common example is rice farmers that re-flood their leveed fields to attract waterfowl and other birds. These winter wetlands can also reduce soil erosion, prevent the growth of winter weeds, and help with the decomposition of rice straw. Unfortunately, although flooding rice fields has many benefits for farmers, a shortage of water for irrigation and wildlife has reduced rice production in Texas. Texas now plants only about one-fourth the amount of rice it did in 1980.

Figure 10.14. Flooded agricultural fields such as this field in Waller County, Texas, can provide valuable wildlife habitat. Photo: Beverly Moseley, NRCS, Wikimedia Commons (public domain).

- Creating moist soil management units - Levees and water control structures can be placed in natural depressions to create shallow wetland areas. At times solar-powered wells are used to provide extra water.

- Managing salinity in coastal wetlands – In some locations levees or erosion barriers can be erected to prevent saltwater intrusion and marsh erosion. Agencies have even used sediments that are dredged up when ship channels are dug out to fill in areas that have subsided or create new marsh units along the coast. At other locations, man-made levees may have been built that prevented natural tidal influences in brackish or intermediate marsh. Managers can mimic the natural balance of salt and fresh water that would have occurred by installing and opening water control gates in the levees.

- Restoring bottomland hardwoods – Sometimes removing levees can allow natural flooding of bottomland hardwoods. In other cases, permits may be obtained to construct levees to allow scheduled flooding of bottomland hardwoods for wildlife. These management areas are often called "greentree reservoirs" because the intermittent flooding does not kill trees like permanent reservoirs do.

- Building wastewater treatment wetlands – As noted earlier, many cities construct wetlands to help with the final cleansing stage for their sewage. These wastewater wetlands are often used by wildlife. One major project is the George W. Shannon Wetlands project located at the Richland Creek WMA adjacent to the Trinity River. The 1,700-acre wetland complex was constructed to be the final treatment for 90 million gallons of wastewater each day from the Dallas-Fort Worth area. The 24 wetland units are extremely effective at removing suspended sediments, nitrogen, and phosphorous and are managed by TPWD to provide excellent wildlife habitat.

- Cities can also use engineered wetlands, such as stormwater retention ponds and rain gardens to reduce flooding, improve water quality, and provide wetland habitat (Fig. 10.15).

Figure 10.15. Examples of wetland creation to manage stormwater runoff. Small rain gardens in San Marcos help to filter water running off parking lots before it makes its way into the river. A wetland pond in Wimberley helps to slow water flowing out of a detention pond and provides wildlife habitat.

Several federal and state programs assist private landowners who want to conserve or restore wetlands. The Texas A&M Forest Service offers education on voluntary "Best Management Practices" to help forestry operations avoid impacts on wetlands. The Wetland Reserve Enhancement Partnership, administered by the Natural Resources Conservation Service in the U.S. Department of Agriculture, assists landowners willing to commit to wetlands conservation. The Texas Prairie Wetlands Project, part of the Gulf Coast Joint Venture, works with landowners who want to improve wetland habitat along the Gulf Coast. Conservation organizations also contribute greatly to wetlands restoration. Ducks Unlimited has conserved and restored more than 200,000 acres of waterfowl habitat in Texas (Fig. 10.16).

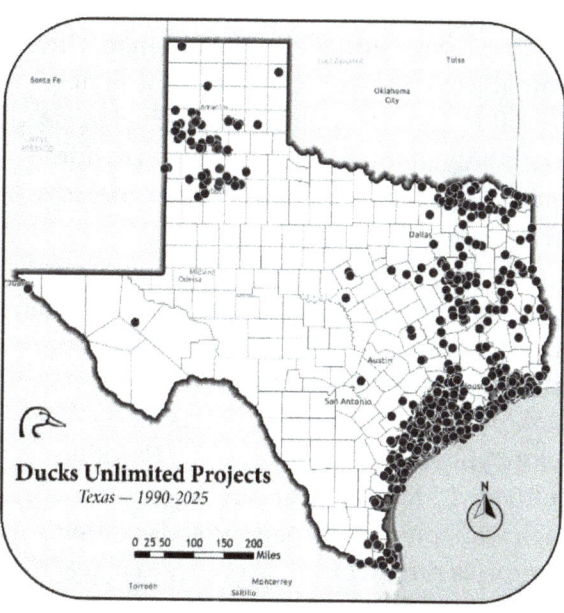

Figure 10.16. Locations of Ducks Unlimited wetlands projects in Texas. Reprinted with permission from Ducks Unlimited.

HIGHLIGHT 10.1 - CONSERVATION SUCCESS STORY – THE WHOOPING CRANE AND ITS WETLANDS

The whooping crane has often been described as a symbol of wildlife conservation success in North America. The tallest bird on the continent, its elaborate courtship displays and haunting bugling call have long inspired humans, but it is a Texas native that almost wasn't.

The whooping crane once nested in wetlands scattered across west-central Canada and the northern prairie states of the U.S. Those birds migrated several thousand miles to winter in wetlands in the mountains of Central Mexico, along the Gulf Coast of Louisiana and Texas, and in scattered locations on the East Coast. However, in the late 1800s and early 1900s, wetlands were drained for agriculture, the collection of bird eggs was widespread, and charismatic whooping cranes were hunted heavily. The Migratory Bird Treaty Act prohibited the hunting of whooping cranes in 1918, but the species was almost gone. By the winter of 1941-42 there were only 16 whooping cranes left in the last migratory flock in the world—a small group of birds that wintered on Aransas NWR on the central Texas coast and nested at a then-unknown location in the wilds of Canada.

Slowly, with protection from overharvest and with habitat conservation, whooping crane numbers climbed upward. But progress was initially slow. Whoopers lay only two eggs per year, and often only one chick survives. It then takes that chick about five years to mature and reproduce. By 1967, still only about 50 whooping cranes existed in the wild, and they were placed on the first federal endangered species list. The Texas population doubled to 100 by 1986 and surpassed 200 in 2004. From there, population growth occurred quickly, passing 300 in 2014 and 500 in 2017. (The population is surveyed by airplane during the winter in Texas.) Meanwhile, agencies developed captive breeding programs to help reintroduce whoopers back into their former range. One dramatic experiment has used ultra-light aircraft to teach whooping crane chicks raised in captivity to migrate! Another has restored a non-migratory flock of whooping cranes to the marshes of coastal Louisiana.

The key to success for whooping crane conservation has been wetland conservation. The cranes wintering in Texas require healthy coastal wetlands that produce a variety of food, including blue crabs, clams, and a halophytic succulent plant called Carolina wolfberry. And they require a lot of wetlands. Each whooping crane pair defends winter territories ranging from 200 to 500 acres in size. In Canada, the cranes defend even larger nesting territories in remote muskeg wetlands (vast peat bogs formed on glacial lakes) in Wood Buffalo National Park. Whoopers also require wetlands during migration, a 2,500-mile journey that takes several weeks. During their stops, whooping cranes roost in wetlands and feed in nearby grain fields.

As the whooping crane population has increased, it has expanded its wintering area on the Texas coast. Whoopers now uses coastal marshes on public and private lands from Port Aransas north to Palacios. Whoopers also sometimes use inland wetland habitats, including flooded rice fields. The Recovery Plan, signed by USFWS and Canadian Wildlife Service, sets a goal of 1,000 whooping cranes in the Texas-Canada population for the species to be reclassified from endangered to threatened, suggesting that whoopers will have to expand into many more Texas wetlands.

It is challenging to meet the recovery goal, primarily because threats to wetlands are challenging. Whoopers find themselves competing with condos for space in prime coastal real estate. Reduced freshwater flowing into bays has reduced blue crab populations. Black mangrove, a woody plant that grows along marsh edges, is moving north along the Texas coast in warmer winters, changing salt marshes and taking over whooping crane feeding areas. Currents

Figure 10.17. Whooping crane family feeding on wolfberries in Texas coastal marshes. Photo: Mark Alan Storey.

and boat traffic erode marsh edges, leading biologists to place miles of erosion barriers along the wetland edges. Sea level rise threatens to inundate coastal marshes, leaving biologists trying to model where new wetlands will occur.

Thankfully, whooping cranes don't face their habitat challenges alone. In an inspiring recognition of how healthy ecosystems benefit everyone, a wide diversity of players have come together in an effort called The Aransas Project. The goal of The Aransas Project is to keep coastal wetlands healthy to benefit people, wildlife, and local economies. The project is described as "an alliance of citizens, organizations, businesses, and municipalities who want responsible water management of the Guadalupe River Basin to ensure freshwater flows from the Hill Country all the way to the bays—ecosystems that support area fishing, tourism, and the winter habitat of the endangered whooping cranes (Jim Blackburn, The Aransas Project president)."

Still one of the rarest birds in the world, the ancient whooping crane still hangs on. They attract birders from all over the world who bring in millions of dollars of benefit to Texas coastal economies. For anyone who has paused on the edge of a Texas salt marsh on an early spring morning to take in the glorious courtship dance and haunting unison call of a pair of whooping cranes, meeting the challenges is worth it.

Figure 10.18. A pair of banded and radio-tagged whooping cranes calling as they fly. Photo: Mark A. Storey.

CHAPTER 11 – WILDLIFE POPULATIONS

Wildlife enthusiasts often remember individual animal experiences—the trophy **buck**, the life-list falcon, the rattler that made the heart jump, the rescued baby bird, or the playful otter glimpsed on a fishing trip. People tend to value their individual interactions with wildlife in a particular place. In fact, psychological research shows that when people encounter and interact with animals in their natural environment, their own mental health and wellbeing is enhanced.

Agencies and non-governmental organizations often focus on the big picture—the overall status of a wildlife species. TPWD and USFWS maintain lists of threatened and endangered species that reflect a picture of how the species is doing in the state or throughout its range. NatureServe is a non-profit organization that networks with other organizations and professionals to assess the status of North American species. They classify species and subspecies across different geographic areas as Secure, Apparently Secure, Vulnerable, Imperiled, or Critically Imperiled. The International Union for the Conservation of Nature (IUCN), a world-wide partnership of government agencies and non-government organizations, maintains a "Red List" for species across the globe. They classify species as Least Concern, Near Threatened, Vulnerable, Endangered, Critically Endangered, Extinct in the Wild, Extinct, Data Deficient, or Not Evaluated. These various rankings help biologists understand how species are doing overall.

Wildlife scientists tend to focus somewhere between the individual animal and the whole species. They think in terms of wildlife populations. As defined in Chapter 3, a population is a group of organisms belonging to the same species occupying a particular area simultaneously. This is the scale at which a biologist can often apply management actions to benefit a species. To manage a wildlife population, a biologist must identify their goals, measure the population, understand population dynamics (how populations tend to change over time), and apply the right tools.

MEASURING WILDLIFE POPULATIONS

How can a manager know if population goals are being met? One key is to estimate the population size. Population size rarely stays the same for long. It can fluctuate as individuals die of old age, get harvested or eaten by predators, give birth to offspring, or migrate into or out of the population.

Wildlife managers may try to estimate the total population, or it may be more appropriate to simply understand population density. Population density is the number of individuals per unit of area. Five ringtails per square kilometer, 200 post oaks per hectare, and 10 pill snails per square meter are all examples of population density. Large animals or plants generally have lower population densities because larger organisms usually need more space and resources to survive. Densities may change over time. When resources are plentiful, densities are usually higher. When resources are scarce, densities are lower.

It is also important to understand population dispersion, or spatial distribution, especially when figuring out how to measure wildlife populations. Dispersion describes the spacing of individuals in the population in relation to other individuals. Populations can be dispersed in a clumped, uniform, or random pattern. In clumped dispersions, individuals are grouped in tight clusters, such as Brazilian free-tailed bats' concentration in roost caves. In uniform dispersions, individuals are spaced even distances from each other, as when northern mockingbirds defend territories. In random dispersions,

(opposite) **Figure 11.1.** A mountain lion marked with a radio collar is "captured" eating its prey by a game camera in a research study. Photo: National Park Service, WC (public domain).

individuals can be found anywhere within an area. Wind can create random dispersion of the seeds of dandelion plants. Many lizards also have a random dispersion pattern.

Finally, to understand the health of a wildlife population, it is also helpful to assess sex ratios, age structure, and body condition. Biologists can sometimes measure the recruitment of young during wildlife surveys. Hunters can provide valuable data about sex ratio, age structure, and body condition of game species at harvest check stations. A data set that provides information on the sex, ages, reproduction, and **mortality** of a population is called a "life table." Life tables can be used to numerically model a population to predict its ability to grow or the effects of harvest.

Figure 11.2. Examples of population dispersion. Source: Projectoer, CC 4.0.

ESTIMATING POPULATIONS

How many quail live in a grassland? How many bears live in a forest? How many frogs live in a pond? To determine population size, wildlife scientists use censuses and surveys. Some organisms are easy to count. Others are more scattered. Some are difficult to see, and biologists must rely on clues or "signs" the animals leave behind.

Censuses are complete counts of all the individuals in a population. Though challenging to accomplish, censuses are sometimes used for highly-visible species with low numbers. For many years, biologists counted all whooping cranes wintering in Texas by flying in an airplane above the marshes on and near Aransas NWR. This was possible because the species was concentrated on one part of the coast, because whooper families tend to stay in defined territories, and because they are distinctively large white birds visible from a distance. These characteristics made it possible for biologists to confidently assess population size in a series of flights throughout the winter. It was even possible for biologists to determine when a death had taken place if a bird went missing from its territory. Censuses have also been used to count nesting bald eagles in Texas by watching their large, conspicuous nests, to count red-cockaded woodpeckers by monitoring nest cavities in their well-defined habitat areas in East Texas, and to estimate populations of waterbirds such as herons, gulls, and pelicans when they concentrate in nesting colonies.

However, wildlife scientists must calculate the population size for most species by conducting a survey that only counts a small portion or "sample" of the population. Just as someone might estimate the number of jellybeans in a jar, wildlife scientists count individuals in a limited sample area and then extrapolate the results to the entire population. Surveys provide an index of the population, and biologists may determine if populations are increasing or decreasing by comparing surveys from year to year and over many years.

Wildlife scientists have options for several different survey techniques. Biologists may use randomly placed quadrats to survey smaller, less mobile species such as ground-dwelling invertebrates, like the vegetation sampling methods described in Appendix B. Other survey techniques are based on the point method, where the observer stays stationary for a specified time period and estimates the distance and direction to the species being sampled. This technique is often used in songbird surveys where birds can be detected by sight and sound. Call count surveys are also used for frogs and toads. Game cameras can also be used to gather point data, a method called "camera-trapping."

The most commonly-used survey technique is the transect. In transects species are counted along a longitudinal route. In most cases, results are converted into densities. Some common examples of transects employed in wildlife surveys are:

- Walking transect – Hahn lines are a 2-mile-long walking survey used in areas with high deer populations. **Basking** surveys, best on warm sunny mornings, are transects walked to count reptiles.

- Roadside survey – A roadside survey is a count conducted while driving along public or ranch roads. They are often used for dove, quail, pheasant, and deer. They can also involve stops to conduct point counts for birds calling. Roadside surveys can be conducted at night using a spotlight to count deer and nocturnal mammals.

- Boat surveys – Similar to roadside surveys, routes are established in waterways to conduct spotlight counts for alligators or basking turtles.

- Aerial surveys – Airplanes or helicopters are flown across vast expanses with open visibility to count wintering waterfowl, pronghorn, and alligator nests along pre-determined transects.

Figure 11.3. Transects can be used in a variety of habitats. White-tailed deer and nocturnal mammals are regularly monitored through spotlight surveys conducted from vehicles. Aquatic transects can be used under water to survey organisms such as salamanders or freshwater mussels. Photo 1: Jeff Forman, TPWD. Photo 2: USFWS, CC 2.0.

Interestingly, as the whooping crane population increased and the feasibility and accuracy of the complete aerial census decreased, USFWS switched to doing an aerial transect. These transects employ "distance sampling," a method that documents the distance from the transect line to the animal that is seen. Distance sampling provides more accurate estimates by controlling for decreased detectability farther from the transect line. Any of the transects described above can include distance sampling.

In situations where individuals are difficult to see, biologists may rely on indirect measures of abundance. Observers can count animal tracks, **scat** (feces), hair, feathers, scrapings, or gnawings along a transect. The calls of songbirds, owls, and amphibians may also be recorded at listening stations. Other indirect indices include monitoring key food items, such as deer browsing on saw greenbrier—an ice cream plant for whitetails. Increasingly versatile and affordable genetic survey methods can extract DNA from hair, skin, scat, soil, water, and even the air to detect the presence of different species.

The most accurate population information can be obtained when animals are captured and marked. Animals can be captured using nets, traps, pits, and even by hand. Individuals can then be marked with leg bands, ear tags, toe-clipping, and Passive Integrative Transponder (PIT) tags that are injected. Some animals, such as bobcats, salamanders and whales, can be individually identified by photographs of their markings. Individuals can also be identified by DNA finger-printing. Capturing animals allows scientists to collect data about age and reproductive condition. When they are marked, released, and recaptured in a long-term study biologists can produce an accurate population estimate using the Lincoln-Petersen estimate (Highlight 11.1).

Finally, interesting information about movements, habitat use, and survival can be obtained if individual animals are tracked. Radio-telemetry entails placing transmitters on animals and then using a receiver to detect the signal from the transmitter. Sometimes the animal is not even seen, but the signal indicates its presence. Transmitter types are chosen based on the size, behavior, and life history of the animal being studied. Transmitters can be attached to animals using leg bands, radio collars, necklaces, glue, or even surgically implanted. Depending on the size and design of the transmitter, the signals can be received by directional antennae carried by the researcher or attached to a vehicle. Today, many large or very mobile animals are tracked using a GPS system where locations are transmitted via a satellite to a computer or other storage device. Alternatively, data can be stored in the transmitter and downloaded when the animal is recaptured. Some collars can be dropped remotely or at a specific time and then recovered to obtain the stored data. Recent technologies, such as harmonic radar, even allow tracking of insects, snails, and very small mammals over short distances.

Figure 11.4. Radio transmitters glued to the back of Texas horned lizards give off a signal that a receiver can pick up over short distances (5-15 meters) even if the lizard is hidden. Lizards and other wildlife may also be marked for long-term hand-held identification through the insertion of a Passive Internal Transponder (PIT) tag. A scanner, similar to grocery store scanners, then reads the tag inside the lizard. Photo 1: Lee Ann Linam. Photo 2: Mark Mitchell, TPWD.

HIGHLIGHT 11.1 - USING MARK-RECAPTURE TO ESTIMATE POPULATION SIZE
This section is drawn from "Nature Unbound: The Impact of Ecology on Missouri and the World," written by Matt Seek and published by the Missouri Department of Conservation.

Biologists can use mark-recapture methods to estimate population size. These techniques are especially valuable for elusive or scattered animals that may not be seen in surveys. Mark-recapture involves catching a portion of the population. These individuals are "marked" using the techniques above or by collecting unique photo or DNA attributes. After being documented, the individuals are released and allowed to mix back into the population. After time has passed, a second sample of individuals is captured. Biologists can estimate the total population size by comparing the number of individuals marked in the first sample to the number of marked individuals collected in the second sample.

The Lincoln-Petersen estimate is a simple and widely used mathematical formula that allows biologists to estimate populations over a short time period in a mark-recapture study:

$$N = n_1 n_2 / m_2$$

N = population size
n_1 = the number of individuals marked and released in the first sample
n_2 = the total number of individuals captured in the second sample
m_2 = the number of individuals with marks in the second sample

Biologists use the Lincoln-Petersen method to estimate the population size of a diversity of species, from mussels to monarchs to mockingbirds. The method does rely on the following assumptions:

- Between the two samples, the population has no births, deaths, **immigration**, or **emigration.**
- All individuals have the same probability of being caught.
- Marking does not affect survival.
- Marking does not affect the likelihood that an individual will be recaptured.
- Marked individuals do not lose their marks between captures.

If any of these assumptions prove false, the accuracy of the Lincoln-Petersen estimate decreases.

POPULATION DYNAMICS

> **population change = (births + immigration) - (deaths + emigration)**

Populations are changing or "dynamic," continually growing and shrinking. In Texas, alligators probably numbered several thousand when listed as an endangered species in the 1960s. By the late 1990s, aerial nest surveys estimated over 280,000 in Orange, Jefferson, and Chambers Counties alone. What causes this fluctuation? The answer is four basic factors—births, deaths, immigration, and emigration. Wildlife biologists often represent population change with a simple formula shown in the box above.

Births and immigration (moving into an area) increase the population. Deaths and emigration (moving out of an area) decrease the population. For alligators, unregulated hunting in the early half of the 20th century caused more deaths than births. When Texas prohibited hunting of alligators in 1969, births began to exceed deaths, and their populations climbed. Because alligators lay an average of 30-40 eggs in their nests, the populations could increase rapidly. Because adult alligators have few predators and a long lifespan (35 to 50 years in the wild) individuals could reproduce many times. Once populations increased, alligators began immigrating back into habitats from where they once disappeared. Now, numbers are stable, allowing TPWD to issue a limited number of tags for alligator hunting. The number of tags is based on aerial transect surveys of alligator nests and nocturnal boat transects along waterways.

Texas's alligator population went from thousands to perhaps half a million statewide in a few decades. What does this reveal about the ability of populations to grow? It shows that with abundant resources—food, water, shelter, and space—populations can grow, if mortality factors are not too high.

Different species exhibit different birth rates (**natality**) and death rates (mortality). Some species, such as ducks, mature quickly and produce many offspring but have short lifespans and high mortality rates. Others, such as bald eagles and whooping cranes, take longer to mature and may only fledge one young per year. These species usually have longer lifespans and lower mortality. Other species have blended characteristics. Alligators and sea turtles mature slowly and have long lifespans. They also lay many eggs, but the eggs and young experience high mortality. In stable populations, these various characteristics of birth and death balance or come into equilibrium over many years.

Mortality and **fecundity** (the reproductive rate for females in the population) can also vary with the animal's age. Table 11.1 is a life table showing survival and fecundity data from alligator populations in Florida (survival is the opposite of mortality). Hatchling alligators have low survival rates, with only 38% surviving to the following year. On the other hand, more than 80% of adult alligators (alligators that are more than eight years old and more than six feet long) survive each year. Only when alligators reach this adult stage can they reproduce. As Table 11.1 shows, there are about six young produced for each female in the population. Why only six—when an alligator lays an average of 30-40 eggs? Only some females nest each year; some nests are destroyed by predators or flooding; not all eggs are fertile; and, as the life table shows, only about half the eggs make it to hatching.

Table 11.1: Life Table for American Alligators. Data from Dunham, K., Dinkelacker, S. & Miller, Jeff. (2014). A Stage-Based Population Model for American Alligators in Northern Latitudes. *The Journal of Wildlife Management.*

Stage	Size (cm)	Duration (years)	Survival Rate	Fecundity
Egg	0	0.25	0.54	0.00
Hatchling	<30	1.00	0.38	0.00
Juvenile	30-121	3.00	0.78	0.00
Subadult	122-182	4.00	0.73	0.00
Adult	>183	>30	0.83	5.98

Figure 11.5. Female alligator guarding her nest. The female's protection of the nest and young help increase survival for this life stage. Photos: Everglades NPS, WC (public domain).

LIMITING FACTORS

What keeps wildlife populations from increasing out of control? If half of Texas's alligator population is of breeding age, then 70,000 female alligators in Southeast Texas would add about 420,000 gators to the Texas population yearly!

Yet, the Southeast Texas alligator population now appears stable at about 280,000. Competition for food, water, space, and other resources increases as populations grow. Eventually, with limited resources, some alligators die or emigrate. Fewer young may be produced due to competition for food. Offspring die of malnourishment or predation from bigger alligators. As the population becomes more crowded, it becomes easier for diseases and parasites to spread from one alligator to another. These factors that slow a population's growth or prevent it from existing in certain areas are called limiting factors.

As noted in Chapter 6, many things can act as limiting factors. Some limiting factors are abiotic, such as sunlight, precipitation, the amount of nutrients in the soil, or temperature. Alligators are limited in their distribution by moisture and temperature and only occur in 10 states in the southeastern U.S. Other limiting factors are biotic. The availability of food, a shortage of mates, an outbreak of disease, an infestation of parasites, and competition with other alligators can all be limiting factors. Predators, such as alligators, are a limiting factor for prey populations. A shortage of prey, such as turtles or nutria, is a limiting factor for predator populations.

Limiting factors can be influenced by population density. **Density-dependent**

factors affect a population more dramatically when populations are more concentrated. For example, hatchling alligators in crowded captive situations are affected by parasites and fungal diseases not seen in more dispersed wild populations. **Density-independent factors** affect a population regardless of its density. If an extreme freeze occurs in Southeast Texas, then it is likely to affect all alligators present, regardless of the population's density.

CARRYING CAPACITY

Populations tend to grow until they reach the habitat's carrying capacity. When populations are small, they may grow quickly—a period called exponential growth. However, at some point, limiting factors start to take effect. At that point, population growth slows to a level called logistic growth, until finally the population reaches carrying capacity and holds relatively steady. Resource managers must understand carrying capacity to estimate how much habitat needs to be conserved to maintain healthy wildlife populations. They must also understand safe harvest levels that will not deplete the population (maximum sustained yield). Figure 11.6 shows rapid exponential growth at first, followed by logistic growth. The exact pattern varies for different species.

Sometimes, populations reach sizes that cause habitat damage or economic loss before they settle at carrying capacity. As Figure 11.7 depicts, exponential growth may sometimes exceed carrying capacity, causing damage to habitat. Damaged habitat has a lower carrying capacity, and populations decline until they finally settle at a lower carrying capacity. This scenario often occurs when predators are absent from an ecosystem or when non-native invasive species are first introduced.

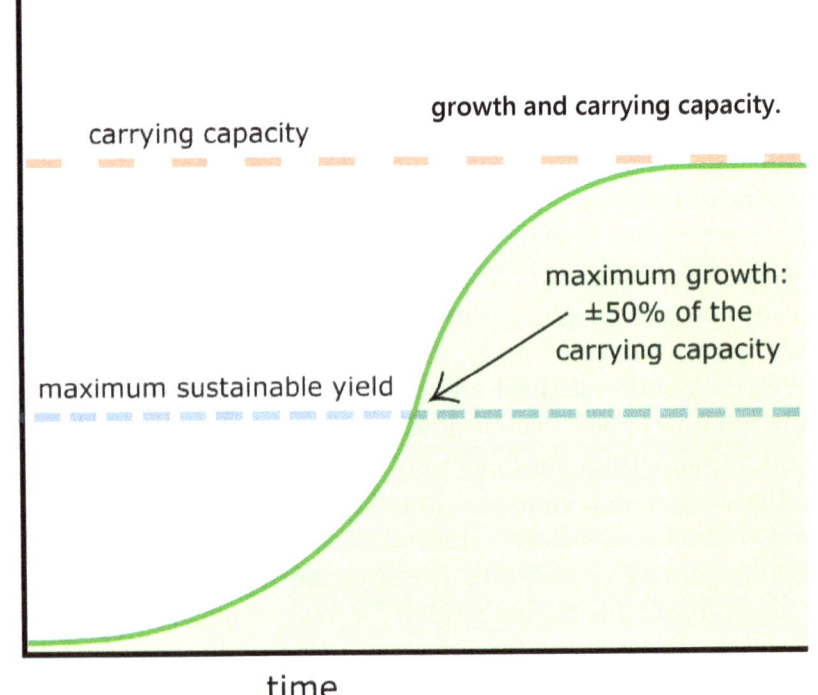

Figure 11.6. Population growth and carrying capacity. Source: Woudloper, CC 4.0.

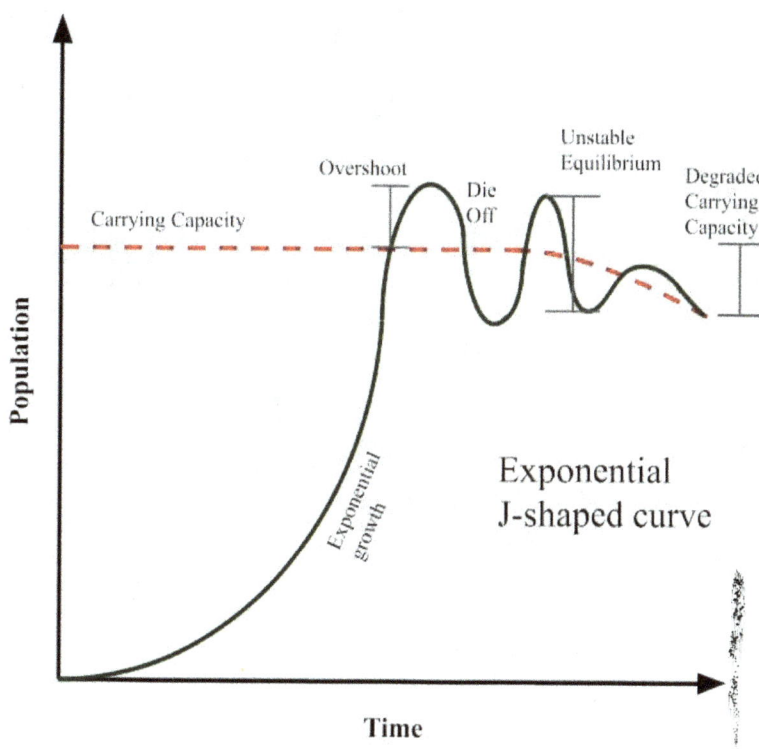

Figure 11.7. Growth of populations can sometimes exceed carrying capacity, especially if natural limiting factors are absent. Source: Nchisick, CC 4.0.

POPULATION MANAGEMENT TOOLS

Proper habitat management is by far the most effective way to manage wildlife populations and improve carrying capacity, but wildlife managers have sometimes explored other population management tools. Some are more successful than others.

WILDLIFE FEEDING

Perhaps the first idea that comes to mind to benefit wildlife populations would be to feed wildlife. Feeders are often used to bring animals close for observation, such as bird feeders or deer feeders, and deer managers sometimes use packaged supplements such as cottonseed to supplement protein levels. However, trying to increase wildlife populations through supplemental feeding carries several risks. First, if wildlife populations are artificially increased using supplemental feeding, then they are likely to exceed carrying capacity. This can cause harm to the habitat around them and degrade the carrying capacity over the long term. Second, many popular wildlife feeds, such as those based primarily on corn, are not nutritionally balanced for wildlife, leading to poor growth. Commercial feeds also have the risk of spreading a dangerous fungal toxin called *aflatoxin*. Third, the feeding of wildlife near residential areas can cause wildlife to become habituated to humans. This can increase risk of disease transmission, such as bovine tuberculosis, brucellosis, or avian influenza. It can also increase the risk of aggressive interactions of wildlife with humans. Many national parks require the use of bear-proof food containers, and the state of Florida prohibits the feeding of wild alligators for this reason. The best strategy by far is to feed wildlife by providing high quality wildlife habitat, perhaps with a few small, planted food plots.

Figure. 11.8. Bird feeders, such as this jelly and fruit feeder that is popular with orioles, may provide some supplemental nutrition for migrating birds, but they are primarily used to draw birds close for photography and birdwatching. Photo: Mark Alan Storey.

WILDLIFE STOCKING

If wildlife populations can increase through immigration, why not help the process and bring in extra animals to increase the population of desired species? Wildlife stocking, or reintroduction programs, can involve animals produced in captivity, animals rehabilitated from illness or injury, eggs or young animals captured in the wild and "headstarted" in captivity, or adult animals captured from the wild and released elsewhere.

Reintroduction programs have contributed to some inspiring recovery stories for species in the wild. One of them is the bold effort in the 1980s that brought the last 27 wild California condors into captivity. The reintroduced condors are now gaining ground in the wild, with a population of about 500. However, reintroduction programs have drawbacks, including high costs, breeding and behavioral challenges, disease concerns, and neglect of habitat issues. For some species, such as desert bighorns in Texas, reintroductions have been learning experiences with some success and some continuing challenges. For other species, such as Attwater's prairie-chickens, captive breeding and reintroduction are buying time as biologists work to restore a fragmented landscape. The conservation story at the end of this chapter (Highlight 11.2) documents how reintroduction and habitat management have both been valuable tools in conserving the red-cockaded woodpecker.

Whenever captive breeding and reintroduction are proposed, wildlife scientists must consider risk versus reward. TPWD has recently been weighing those questions regarding captive deer breeding in Texas. In 1995, the Texas legislature joined a handful of other states in allowing private citizens to obtain permits to hold white-tailed deer in captivity for breeding. The industry initially focused on importing deer with large antlers and body size from outside the state to breed more desirable trophy animals. By the 2020s, there were about 1,000 deer breeders with more than 100,000 deer in captivity. Captive-bred deer are sold to landowners for release, and some deer breeders also host hunts on their properties. Though the industry says that it generates over one billion dollars in economic benefits annually, questions arose about the risks of concentrating deer in captivity and moving them around the state. In 2015, some fears were confirmed when deer at several deer breeder facilities reported the first cases of chronic wasting disease (CWD) in whitetails in Texas. CWD, a fatal neurological infection similar to mad cow disease, has negatively impacted wild deer populations and the deer hunting industry in several states. It has now been reported from numerous deer breeding facilities in Texas. More than 1,000 deer from those infected facilities have been released in more than 40 counties in the state, raising a fear that the disease may spread throughout wild populations in those areas. TPWD has established CWD Containment and Surveillance Zones in high-risk areas where hunters must present their harvest for inspection. The agency has also implemented strict rules prohibiting the importation of deer from out-of-state and requires routine testing at captive breeding facilities.

PREDATOR MANAGEMENT

Hunters often debate the effect that predators have on game species. If there were fewer predators, would there be more game for hunters to harvest?

The answer to that question is complex. A famous 90-year data set for lynx and snowshoe hares in Canada surprised wildlife scientists when they examined it closely. Snowshoe hares are one of the primary prey items for lynx, a relative of the bobcat. One might expect that high lynx populations would cause snowshoe hare populations to decline and perhaps even disappear. Instead, the data indicate that hare populations control lynx populations, not the other way around. In the data set, after hare populations increased, then lynx populations increased. At some point, hare populations got so high that they crashed, perhaps because they exceeded carrying capacity. After the hare population crashed, then the lynx population crashed. Removing predators in this scenario might produce more prey. However, the prey would reach limiting factors in the environment and crash even more quickly.

Similar habitat-predator-prey effects have been observed in Texas. When predators

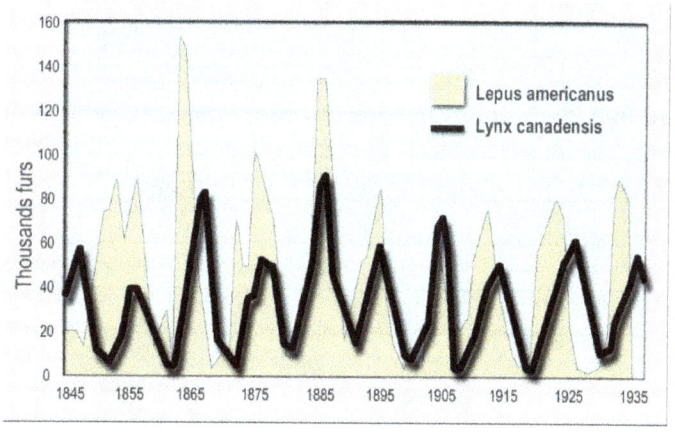

Figure 11.9. Historic data set of lynx (*Lynx canadensis*) and snowshoe hare (*Lepus americanus*) abundance. Data are based on the numbers of animals harvested by trappers from 1845 through 1935. Note how predator abundance usually only rises after prey abundance increases. Source: Lamiot, CC 4.0.

are removed, prey animals can quickly exceed carrying capacity, damaging the ecosystem. High deer populations in Central Texas are likely the result of decreased predator populations. As a result, some plants, such as the lovely bracted twistflower, are now endangered due to deer browsing. Studies in South Texas show that white-tailed deer **fawn** survival may increase when large numbers of coyotes are removed; however, deer populations then quickly reach the carrying capacity of the habitat. At that point, disease, nutritional issues, and habitat damage can cause deer populations to decline. In addition, smaller predators can increase if larger predators are removed. Data from the Rolling Plains Research Ranch indicate that coyotes have a net benefit for quail there because they eat smaller mesopredators, such as snakes, raccoons, skunks, badgers, and even feral hogs.

Similarly, alligators eat frogs, but they also eat snakes and turtles that eat frogs. Research from Florida shows that more alligators in an ecosystem can result in more frogs. It's indeed complicated. Even Aldo Leopold, the Father of Wildlife Management, struggled as a young forester in the mountains of New Mexico to understand the role of predators in the environment. In his essay, "Thinking Like a Mountain," he wrote a section called "The Green Fire" about killing a female wolf and watching her die, raising questions in his mind about the wisdom of removing predators from the environment.

"The Green Fire"

We reached the old wolf in time to watch a fierce green fire dying in her eyes. I realized then, and have known ever since, that there was something new to me in those eyes—something known only to her and to the mountain. I was young then, and full of trigger-itch; I thought that because fewer wolves meant more deer, that no wolves would mean hunters' paradise. But after seeing the green fire die, I sensed that neither the wolf nor the mountain agreed with such a view....I now suspect that just as a deer herd lives in mortal fear of its wolves, so does a mountain live in mortal fear of its deer. And perhaps with better cause, for while a buck pulled down by wolves can be replaced in two or three years, a range pulled down by too many deer may fail of replacement in as many decades. So also with cows. The cowman who cleans his range of wolves does not realize that he is taking over the wolf's job of trimming the herd to fit the range. He has not learned to think like a mountain. Hence we have dustbowls, and rivers washing the future into the sea.

-Aldo Leopold, *A Sand County Almanac, 1949.*

Predator removal may be a useful tool for short-term situations, such as when captive-bred desert bighorn sheep are being introduced into a new environment, but, in the long term, predators are an important component of healthy ecosystems and one to which prey animals are well-adapted.

HARVEST

As noted in Chapters 1 and 2, hunting has long played a role in wildlife management, has funded much wildlife conservation, and brings much economic revenue to habitat owners and the state economy. Leopold suggested that the gun can be a population management tool.

Of course, a wildlife manager would want to know whether hunting, trapping, or collecting causes declines in wildlife populations. As noted earlier, the equation for population change is:

population change = (births + immigration) - (deaths + emigration)

According to this formula, if there is a population of 5 million eastern gray squirrels in Texas, and hunters kill 350,000 during squirrel season, at the end of the hunting season, one might think there would be

Figure. 11.10. Hunters harvest about 350,000 eastern gray squirrels, also called cat squirrels, each year in Texas. However, habitat factors may affect squirrel numbers more than hunting.

350,000 fewer gray squirrels. However, it is not a matter of simple subtraction. The four variables on the right side of the equation—births, deaths, immigration, and emigration—are not independent but affect each other. A decline of 350,000 squirrels might lead to a drop in the population size next fall, but it might also cause the number of gray squirrels, sometimes called cat squirrels, to increase.

Why wouldn't hunting always cause populations to decrease? The answer has something to do with **compensatory mortality**. For gray squirrels, winter food can be a limiting factor, especially in low mast years (years when most oaks do not produce acorns). With a large squirrel population, there may not be enough food, and 350,000 squirrels might die annually of starvation. Suppose hunters kill a percentage of the population. In that case, more food is available for the surviving squirrels, and fewer animals starve.

According to compensatory mortality, the number of squirrels that die will be the same regardless of whether hunting or starvation is responsible. If hunters kill 200,000 squirrels, 150,000 will die from starvation. If hunters kill 300,000 squirrels, 50,000 will die from starvation. The different causes of death cancel or *compensate* for each other.

In other circumstances, hunting might be additive. Additive mortality occurs when different causes of death add to each other to increase overall mortality. For example, if 200 Rio Grande turkeys have been introduced into a new habitat area, 40 might naturally die because they have not yet learned to find food and shelter in the new environment. If hunters also kill 40, then a total of 80 birds would probably be lost.

Biologists debate whether hunting is additive or compensatory. For most game species, hunting seems to be at least partially compensatory, and moderate hunting does not cause population decline. Biologists try to set harvest limits within this balance, so that hunting does not add to natural mortality. In Texas, a demand exists for alligator hunting and collecting eggs for alligator farms. A recent study looked at population data for alligators and predicted that a harvest rate of 2% of sub-adult alligators, 2% of adults, and 38% of nests could maintain alligator populations at carrying capacity. However, if higher harvest levels are adopted, some portion of the harvest could exceed natural mortality, and the population might not

Figure 11.11. Severe browse line evident in Edwards Plateau habitat caused by high populations of white-tailed deer.

rebound over time. Wildlife scientists strive to avoid this scenario by carefully monitoring populations and harvest levels.

On the other hand, sometimes additive mortality is desirable, and harvest is used to address unnaturally high wildlife populations. When populations of white-tailed deer are high, especially in habitats where predators are rare, they have a severe impact on habitat, often removing all the woody growth as high as they can reach. This results in a **browse line**—a break in plant structure with growth above the line but very little vegetation below the line (Fig. 11.11). Areas with severe browse lines provide little habitat for mammals and birds that seek food and cover near the ground level, so an appropriate harvest goal might be to reduce deer populations.

Similarly, non-native animals, such as feral hogs (also called wild pigs), can severely impact habitat, ground-dwelling wildlife, and agriculture. Wild pigs have a high reproductive rate and are targeted by few native predators. In addition, they are quite intelligent, often learning to avoid traps and situations where they might be vulnerable to hunting. Now the most abundant free-ranging hooved animal in North America, there are estimated to be more than 2.6 million feral hogs in Texas. With damage estimated at over $780 million annually, TPWD continues investigating alternative control measures to protect wildlife habitat. So far, wildlife managers have not found a way to reduce these populations.

Such problems aren't limited to mammals. Fifty years ago, populations of double-crested cormorants were perilously low. Impacted by the feather trade in the early 20th century and environmental contaminants like DDT in the mid-century, these fish-eating members of the pelican order were uncommon. However, like pelicans, cormorant numbers rebounded after DDT was banned, and cormorants soon found a comfortable new niche associated with feeding at aquaculture farms and nesting in dead snags created by new reservoirs. Texas now hosts 50,000-90,000 wintering double-crested cormorants and several thousand neotropic cormorants that nest in the state. With increasing cormorant numbers, impacts on fish production have also increased. Although protected under the Migratory Bird Treaty Act, agencies now agree to allow aquaculture producers to

Figure 11.12. Habitat damage caused by feral hogs. Photo courtesy TPWD © 2025.

harass and shoot cormorants to protect fish production.

The U.S. Department of Agriculture Wildlife Services program is tasked with finding solutions to damage caused by nuisance wildlife populations statewide. However, wildlife conflicts can arise even at the household level. Private wildlife nuisance control businesses have emerged to respond to skunks, squirrels, raccoons, bats, bees, ants, etc., who make their homes with humans.

HABITAT ENHANCEMENT

As noted at the beginning of this section, habitat management is the most effective tool for benefiting wildlife populations. Prescribed burning, grazing, selective clearing, and planting can help ecosystems reach a healthy, sustainable condition. However, on small or recovering tracts, some habitat elements can be lacking. In these circumstances, habitat enhancement for specific wildlife populations may be helpful. Drilling windmills, creating ponds, and installing wildlife guzzlers to provide water for wildlife can increase carrying capacity. As noted in Chapter 7, some habitats may be lacking mature trees that provide nesting or roosting sites for wildlife. In those situations, nest boxes for species such as eastern bluebirds, wood ducks, whistling ducks, and red-cockaded woodpeckers can overcome limiting factors until trees mature. Similarly, man-made nesting platforms have been used by osprey, bald eagles, and ferruginous hawks in habitats where natural structures are lacking. In some circumstances, denning animals such as swift foxes and kit foxes may be helped by the construction of artificial dens. Selah-Bamberger Ranch Preserve in Central Texas has even constructed a "chiroptorium," an artificial cave that is used by 200,000 Brazilian free-tailed bats!

WILDLIFE MANAGEMENT PLANS

A wildlife management plan is a document that allows a biologist or landowner to apply management tools to benefit wildlife populations in a measurable manner. The first step in preparing a wildlife management plan is to set some goals.

Any one piece of land might provide several diverse vegetation communities and habitats for many types of wildlife species, depending on its management. Management that improves the carrying capacity for one species may decrease the carrying capacity for another species. For example, managing South Texas woodlands to increase the thick brush favored by endangered ocelots may reduce the carrying capacity for aplomado falcons or painted buntings because they prefer more open areas. Therefore, in preparing a wildlife management plan, it is up to the wildlife manager to set priorities appropriate for the landscape and the overall conservation goals. Examples of goals might include:

- Increasing the population of a particular species – Sometimes, managers use **focal species**, such as the ocelot in the example above, in their habitat management strategies. This goal might be appropriate when a tract of land has potential habitat for an endangered species.
- Decreasing the population of a particular species – This goal applies if invasive species harm the habitat quality or compete with native wildlife. Examples of invasive species that might be targeted are Chinese tallow, which is invasive in Texas coastal prairies, or feral hogs, which cause much habitat damage across the state.
- Creating a sustainable harvest – Because of the value of hunting on many private lands, this is a priority goal for many wildlife managers. The wildlife manager may need to consider both quality and quantity of the target population, ensuring the habitat supports an appropriate number of game animals and provides good nutrition to produce high-quality animals.
- Maintain or increase diversity – For a landowner who wants to develop nature-based tourism or simply wants to maintain a healthy ecosystem, the goal may be to manage the habitat to support a large diversity of plants and animals.

Once goals are set, TPWD offers the following steps for preparing a wildlife management plan:

1. Conduct a resource inventory – Use maps to identify what soils and vegetation communities are present. Map sensitive habitats, such as bogs or riparian areas. Conduct surveys to assess the status of wildlife populations.
2. Designate management units – Identify whether different tracts or pastures will be managed for different goals.
3. Select habitat and population management tools for each unit.
4. Create a schedule that is appropriate for the tools to be used. How often will it be grazed? How many years between burning? Will public use be allowed year-round? When will surveys be conducted?
5. Implement management for each unit.
6. Monitor results – Use vegetation surveys and animal surveys to assess success.
7. Refine management practices based on results – This adaptive management style is based on learning by doing, monitoring results, and adjusting the strategy.

In Texas, landowners can qualify for a reduced property tax assessment if they develop and implement a wildlife management plan. Following such a plan allows the landowner to receive the agricultural tax rate on their property because the property provides open space and natural resource benefits in the landscape. To obtain the wildlife tax valuation, landowners must actively manage for wildlife, and the land must have previously qualified under agricultural use. The landowner must submit an application and management plan to their county appraisal office. The plan should include at least three wildlife management practices from the following seven categories of habitat or population management:

- Habitat Control (Habitat Management)
- Erosion Control
- Providing Supplemental Water
- Providing Supplemental Food
- Providing Supplemental Shelter
- Census Counts (or surveys)
- Predator Management

Some wildlife professionals offer services to help the landowner write the wildlife management plan and conduct the management activities and surveys.

HIGHLIGHT 11.2 - CONSERVATION SUCCESS STORY – USING THE MANAGEMENT TOOLBOX TO SAVE THE RED-COCKADED WOODPECKER

When European settlers entered the longleaf pine savannas of East Texas, there were holes in the mature pines that towered over the grassy park-like understory—holes made by families of little black-and-white woodpeckers, the red-cockaded woodpecker.

Since ancient times, red-cockaded woodpeckers (RCW) have been associated with fire-maintained southern pine forests, especially longleaf pine. RCW excavate cavities in living pine, preferring the most mature trees (60 to 100 or more years old) that have begun to develop red heart fungus in the trunk's interior. As they excavate, they drill small holes near the opening so sap drips around the cavity to deter predators such as snakes. The woodpeckers live in a colony or "clan" that usually includes one breeding male and female and several related helper males. They feed together on insects found under the bark of the pines. At night, each woodpecker roosts in its own separate cavity. The open, fire-maintained forest facilitates interaction between the clan members, keeps predators from using midstory plants to reach the cavities, and ensures that hardwoods do not overtake the pines.

RCW were once the most abundant woodpecker in the Southeast but are now endangered. The story of the disappearance of mature native pine forests throughout the South is also the story of the decline of RCW. As vast acreages of old-growth pines were cut

Figure 11.13. Female red-cockaded woodpecker entering its cavity surrounded by pine sap drippings. Source: USFWS, CC 2.0.

and fire was suppressed, RCW in Texas were restricted to small remnant populations. By the 1980s, there were less than 1,000 birds in the state. USFS attempted some habitat management for known colonies on national forests; however, efforts were insufficient, and RCW numbers continued to decline throughout the 1980s. The trend was so

Figure 11.14. Fire-maintained southern pine forest allows red-cockaded woodpeckers to move easily between their roost trees.

discouraging that the Sierra Club brought a lawsuit against the Forest Service in 1988, claiming that their failure to implement more aggressive management in national forests in Texas violated federal responsibilities under the Endangered Species Act.

The judge ordered USFS to begin intensive management of RCW. The Forest Service thinned pine stands (pine forests planted for commercial harvest are often too thick for RCW). When timber was harvested, they began leaving older mature pines as shelterwood trees to provide transitional habitat. They also removed hardwood midstory and started a regular prescribed burning program in RCW areas. There was a cavity shortage because few pine trees were old enough for RCW to excavate. USFS began to place metal restrictor plates around existing cavity holes to prevent other woodpeckers, such as the larger, more abundant pileated woodpecker, from enlarging and taking over the cavities. They also built artificial cavities that were inserted into living pines

The final step was for the Forest Service biologists to play "match-maker." Because many Texas RCW populations had been fragmented for decades, there was a need to introduce new genetic material into small populations. Biologists captured unmated RCW from larger populations in Louisiana and brought them to Texas locations that needed supplementation. Introducing a male and female to a new territory happened during a single intense overnight period. The male and female to be moved were roused from their home cavities (sometimes dozens of miles apart from each other), captured in a net, transported to their new territory (sometimes hundreds of miles away), placed in their individual artificial cavities 50 feet up in the tree, and locked in with a piece of wire mesh, all under the cover of darkness. When dawn broke, the sleepy biologists pulled off the wire mesh covering the hole, and the birds met each other for the first time. The technique has proven valuable and is still in use today. Over half the translocated woodpeckers successfully form new pairs.

In addition to the national forests in Texas, RCW exist on state forest lands, timber company lands, and other private lands. In

Figure 11.15. Biologist about to climb a pine tree to insert an artificial RCW nest cavity. Photo: WC (public domain).

1998, USFWS, TPWD, and Texas A&M Forest Service signed a Habitat Conservation Plan for RCW for non-federal lands in the state. Private landowners are often leery of hosting endangered species on their property, but these "Safe Harbor" Habitat Conservation Plans provide guarantees for landowners participating in woodpecker conservation that their liability will not increase if their good habitat management increases woodpecker populations. RCW now exist in 16 of the 33 Texas counties they originally occupied; however, populations are extremely small in nine of those counties.

Why all this fuss over a little woodpecker? Why not let forests grow into mixed pines and hardwoods without active management? The primary reason is that pine savannas maintained by natural fire once dominated much of the southern pine forest. These open forests provide habitat for bobwhite quail, Louisiana pinesnakes, and more than 30 other plants and animals that are rare or declining. Pine savannas would not be the same without the keystone RCW. It

Figure 11.16. Male red-cockaded woodpecker captured for banding and reintroduction at a new site. Photo: U.S. Army Environmental Command, CC 2.0.

is the only woodpecker that excavates cavities in living pines. More than two dozen other species have been documented using RCW cavities, including lizards, frogs, snakes, squirrels, and other birds. Thanks to the Sierra Club, USFS, various state and private partners, Leopold's axe and fire, and some creative new tools, these species and their habitat have a more secure future in Texas.

CHAPTER 12 – TEXAS BIRDS

Birds are big in Texas! Some are literally big—the continent's tallest bird, the whooping crane, makes Texas its winter home. But Texas birds are big in other ways as well. According to the Texas Bird Records Committee, more than 670 species have been recorded in Texas, more than half the total reported in the United States. Texas contends with California as the state with the most bird species.

Birds are also big business in Texas. About 2.2 million birdwatchers in Texas generate roughly $1.8 billion in economic impact annually. In addition, Texas is a top destination for out-of-state birders, with a reputation for having some of the country's best birding locations. Texans watch birds, and Texans also hunt birds. Texas leads the nation in the number of dove and waterfowl hunters and in doves harvested annually. With bird hunting adding $6.3 billion annually to the nation's GDP (Gross Domestic Product), Texas reaps its share of economic benefit from bird hunters.

Texas is also big in bird diversity. The eastern forests, Great Plains grasslands, Chihuahuan Desert, and Mexican subtropics converge in Texas, providing a diversity of habitats. Texas hosts 26 of the 41 bird orders found in the world. Bird lovers here can find iridescent hummingbirds, colorful parrots, velociraptor-like roadrunners, and ocean-loving magnificent frigatebirds. It's the only state in the country where birders can glimpse golden-cheeked warblers, Attwater's prairie-chickens, green jays, plain chachalacas, and a host of other Mexican species.

WHAT IS A BIRD?

What makes a bird a bird? Birds are vertebrates (animals with a backbone) making up the Class Aves. And, just as the Latin word *aves* is synonymous with flight (forming the root of words such as "aviation"), birds are closely associated with their ability to fly. Though some species of birds do not retain the ability to fly, all Texas birds can, and most of the unique characteristics of birds seem designed for flight.

First, birds have a lightweight structure. Feathers cover their bodies. The hollow structural framework of feathers is lined by fibers linked with hooks called barbs—providing lightweight insulation and a strong, flexible vane perfect for flight movements (Fig. 12.2). Birds have several types of feathers—long, sturdy flight feathers, softer contour feathers to cover the body, very soft down feathers without barbs that provide insulation, and several types of specialty feathers, such as bristles.

Bird skeletons are also adapted for flight. Bones are hollow but strengthened by internal struts, a design that inspired aircraft wings. Numerous bird bones are fused to increase skeletal stability. The breastbone, or sternum, has a prominent keel, allowing for the attachment of bulky flight muscles.

Birds also have organ systems and a physiology that provide the energy needed for flight. They are **endothermic**, creating warmth through internal body processes (often called warm-blooded) and **homeothermic**, able to maintain a constant non-fluctuating internal body temperature. On average, their body temperatures are much higher than mammals—as high as 110°F for some thrushes! Birds have a large, strong heart with the fastest heart rate of any animal. A hummingbird's heart may beat at more than 1200 beats per minute. They have a unique respiratory system. Five pairs of air sacs connected to the lungs are scattered throughout the body, providing cooling and an efficient uptake of oxygen. The respiratory system takes up about 20% of the body, compared to 5% for humans.

(opposite) Figure 12.1. A northern mockingbird, the Texas State Bird, engages in a flight display to defend his territory.

Figure 12.2. Bird feathers demonstrate a remarkable lightweight combination of strength and flexibility, enabling agility in flight. Northern cardinal photo: Mark Alan Storey.

Even egg-laying benefits flight. Birds don't have to carry around their developing young. Instead, they lay calcium-covered eggs in a nest. An amniotic membrane protects the eggs from drying out. Because the developing young also require a constant body temperature, incubation of eggs by the parents is an essential behavior for birds.

Birds eat a high-energy diet and maintain high blood sugar levels as well. Bird bills (Fig. 12.3) and feet reveal diverse strategies for obtaining food. Roseate spoonbills and ducks have comb-like projections along their bills called lamellae for straining food from the water. Ducks have webbed feet for swimming, whereas spoonbills have long toes for walking in muddy habitats. Birds of prey have hooked beaks and sharp talons for capturing and tearing apart prey animals. Woodpeckers have chisel-like beaks for digging into the bark of trees, and feet are adapted to perch vertically on trees. Hummingbirds have long beaks and tongues for sipping nectar from tube-shaped flowers. Songbird beaks reveal that some are seed eaters, some are insectivores, and some are fruit eaters. Some songbirds, such as loggerhead shrikes, are even meat eaters!

Birds also have strategies to deal with food shortages. Many migrate to locations with better winter food supplies. Woodpeckers may stash acorns in crevices in "granary trees." Nuthatches, titmice, chickadees, and members of the crow family create seed caches that they revisit during winter. Loggerhead shrikes impale prey, such as insects and lizards, on thorns or barbed wire.

Figure 12.3. Bird beaks reveal different feeding strategies and preferences. Source: L. Shyamal, CC 2.5.

IDENTIFYING BIRDS

Biologists and birders have long relied on books called field guides to help identify birds in the field. These compact books provide illustrations, range maps, and notes on behavior and calls. Field guides are also now available as apps on phones and computers. Some can even identify birds by photo or sound. "Merlin," an app produced by the Cornell Laboratory of Ornithology is especially popular. Although automated identification methods are helpful, they are not a substitute for a careful observer. The author of some of the earliest field guides, Roger Tory Peterson, suggested that an observer of birds should use the following characteristics to aid in identification:

- *Color, Size, and Shape* – Note the size and shape of the legs, tail, wings, beak, and overall bird. Some species have brighter colors in the breeding season than in the non-breeding season. Males and females differ in color in some species. Juveniles may take a year or two to reach full color.

- *Field marks* – Look for breast markings such as streaking, tail patterns, rump patches, wing bars, eye markings such as rings or brow markings, crown markings on the top of the head, and wing color patterns in flight.

- *Behavior* – Observe the bird's behavior, including its foraging patterns. Blue-gray gnatcatchers constantly flit in trees; northern mockingbirds love to put on aerial territorial displays from high on a perch; piping plovers run, stop, and probe to feed; spotted sandpipers bob; and eastern phoebes flick their tail as they sit on a perch between quick swoops after insects.

- *Flight pattern* – Watch the motion of the birds in flight. Swifts and swallows swoop after prey. American white pelicans and geese migrate in groups in a V-formation, but pelicans often soar, and geese usually flap. Vultures soar with a V-shape to their wings. Woodpeckers flap during flight but then pause and dip. Doves and woodcocks both produce whistling sounds with their wings.

- *Sounds* – Learning to recognize different species' sounds is a valuable skill in bird identification. Birds often call in courtship and defense of territories, and each species has a unique song. Birds also may give a call note, a simple call for foraging or aggressive interactions. Some species have a distinctive flight call given when flying.

- *Location* – Examine whether the bird expected in this ecoregion or habitat. What kind of microhabitat is the bird using? Is it high in the canopy like a warbler? Or low in the underbrush like a thrasher?

- *Timing* - When is it here? Permanent residents are species that live in the state year-round. Summer residents are primarily in Texas during their breeding season. Winter residents arrive in Texas for the colder months of the year. (Interestingly, some bald eagles breed in Texas during the winter!) Birds classified as "migrants" occur in Texas only during spring and fall migration as they pass through from northern breeding grounds to southern wintering grounds.

HIGHLIGHT 12.1 - THE FATHER, AND THE MOTHER, OF THE FIELD GUIDE

It has been said that no one has done more to promote an interest in living creatures than Roger Tory Peterson. Peterson was an artist and self-taught naturalist whose work set the standard for the modern field guide.

Peterson was born in western New York in 1908 to immigrants from Europe. He immersed himself in exploring the outdoors as a child and developed a love of sketching birds, a skill that was encouraged by a high school teacher. Peterson had hoped to attend Cornell, which was emerging as the foremost university for the study of **ornithology**, but lacked the funds. Instead, he supported himself as a furniture maker while attending the Arts Student League and the National Academy of Design in New York City. In 1934, Peterson published *Guide to the Birds*. He went on to work as the National Audubon Society's educational director and art editor of Audubon Magazine. Like Rachel Carson, Peterson sounded an alarm about the effects of DDT on birds, saying, "The side effects of most [DDT] spraying programs go undocumented and more than 100 million birds are probably killed yearly. I just could not live without birds, frankly. I would hate to live in a lifeless world."

Although Peterson is often called the father of the field guide, and many types of field guides bear his name, the person who actually published the first known field guide and introduced the idea of birdwatching was a woman with a Texas connection, Florence Merriam Bailey.

Florence Merriam was born in 1863 to a family of outdoor enthusiasts (Merriam's elk and several other species were named for her brother, the naturalist C. Hart Merriam). When Florence was a young woman in the late 1800s it was highly fashionable to adorn hats with the feathers of wild birds. She reported that she once counted the feathers of 40 different species on hats as she took a walk through Manhattan. Those fashion trends resulted in the harvest of about five million birds per year and the endangerment of several wading birds. Florence thought the way to stop this fad was to help people appreciate birds. "We won't say too much about the hats," she mused. "We'll take the girls afield, and let them get acquainted with the birds. Then of inborn necessity, they will wear feathers never more."

When Florence began to study biology at Smith College in 1882 most ornithologists studied birds that had been killed, skinned, and mounted for private or museum collections. (In fact, Audubon created his famous bird paintings from dead specimens.) Few biologists observed the behavior of birds in the field. Florence suggested that naturalists and birdwatchers use an opera glass to study birds in their habitats without the necessity of killing them. Binoculars were not commonly available at the time, but opera glasses were! "The student who goes afield armed with opera-glass," she declared, "will not only add more to our knowledge than he who goes armed with a gun, but will gain for himself a fund of enthusiasm and a lasting store of pleasant memories."

Florence Merriam Bailey set several unique precedents in the field of ornithology. At the age of 26, soon after her observations in Manhattan, she published an early field guide—*Birds Through an Opera-Glass*—under her own name, an uncommon practice for women at the time. The ideas she promoted formed the basis of modern birdwatching. She went on to document populations of birds at previously unexplored locations across the country (including in Texas alongside her husband, Vernon Bailey, who focused on mammals), author over 100 journal articles, and become the first woman associate member of the American Ornithologists' Union.

Figure 12.4. Great egret in breeding plumage, sporting the beautiful feathers that made them a target for the millinery (women's hat) industry. Photo: Mark Alan Storey.

TEXAS BIRD DIVERSITY

With nearly 700 species in 26 orders, Texas birds could fill an entire book! TPWD classifies birds as Migratory Game Birds, Upland Game Birds, and Nongame Birds for management, regulatory, and conservation purposes. Rare species may be further classified as threatened or endangered. This section uses these management categories to explore the characteristics, monitoring, and management opportunities for some of the more significant orders of birds in Texas. The Cornell Laboratory of Ornithology Bird Guide is an excellent resource for exploring details about individual Texas bird species.

MIGRATORY GAME BIRDS

Migratory birds hunted in Texas include waterfowl, coots, rails, gallinules, snipe, woodcock, doves, and sandhill cranes. These species represent four different orders of birds. Not all the members of these orders are hunted, and not all of the species even migrate. However, to meet national conservation goals across state boundaries, harvest regulations for these birds are guided by a framework provided by USFWS. A federal Duck Stamp and a state Migratory Game Bird Endorsement (essentially a state duck/dove stamp) are required to hunt any migratory game bird. Fees from the purchase of stamps help fund the conservation of migratory birds. In addition, hunters who hunt migratory game birds must provide information about their harvests through the Hunter Information Program administered by TPWD.

Waterfowl – Order Anseriformes

Waterfowl refers to ducks, geese, and swans in the order Anseriformes. Texas has records of 38 species of ducks, six species of geese, and two species of swans. The variety of ducks is divided into two main groups, dabbling and diving. Dabbling ducks, such as mallards, pintails, and teal, tip their tails up to feed on seeds or vegetation in shallow water. Diving ducks, such as scaup, ring-necked ducks, and redheads, dive entirely underwater to feed on the bottom in deeper waters. Their diet is more omnivorous and includes mollusks, insects, and crustaceans.

Texas provides the wintering habitat for around three-fourths of the waterfowl in the Central Flyway. WMAs, NWRs, and private lands, many restored by conservation organizations like Ducks Unlimited and grants from Joint Ventures formed under the

Figure 12.5. Texas wetlands provide habitat for millions of waterfowl, such as these northern pintail, American wigeon, mallard, green-winged teal, and blue-winged teal. Numerous nongame waterbird species also find habitats in these wetlands. Photo: Mark Alan Storey.

North American Waterfowl Management Plan, help provide the habitat needed for waterfowl. Texas participates in mid-winter waterfowl surveys annually. These surveys consist of transects conducted by airplane in the various ecoregions of the state. Harvest limits are set for different species based on breeding and wintering grounds surveys. Therefore, duck hunters must be good at identifying ducks in flight. Biologists gain additional data by banding waterfowl. To catch ducks for banding, bait is placed in the trap area and a "rocket net"—a large net carried by dynamite charges in canisters—is fired over the feeding birds (Fig. 12.6). Biologists place a USFWS metal band with a unique nine-digit number and sometimes a colored band on the birds' legs before release. If birds are shot or re-trapped, the bands can provide information about survival and movements.

Waterfowl concentrate on their wintering grounds, sometimes in flocks of thousands of birds. Because of these concentrations, disease management can be an issue for waterfowl. Wintering snow geese in Texas have experienced outbreaks of avian cholera. More recently, highly pathogenic avian influenza (HPAI or bird flu) has emerged as a threat among migratory waterfowl populations as well as domestic poultry operations and other wild birds. Wildlife managers are developing strategies to monitor bird flu outbreaks and minimize its spread.

Doves – Order Columbiformes

Texas has recorded nine dove species and four pigeon species in the order Columbiformes. These short-necked, stout-bodied birds usually feed on the ground where they search for seeds. Doves breeding in Texas construct flimsy nests made of sticks in trees and on ledges. Species in Columbiformes have the unique ability to produce "milk" to feed their young. The parent bird produces the liquid, called crop milk, from the lining of its crop and drops it into the nestlings' mouths.

Figure 12.6. Rocket nets can be used to capture waterfowl in Texas during the winter. Individual ducks and geese are then removed from under the net and banded. Photos: Charles D. Stutzenbaker.

Although classified as Migratory Game Birds, not all individuals in this order migrate, and not all species are hunted. Only mourning doves, white-winged doves, and white-tipped doves may be harvested, along with the introduced Eurasian collared-doves and rock doves. (Rock doves, also called rock pigeons, were introduced from Europe in the 1600s and make up the large flocks of feral pigeons seen in cities.) Dove hunting is extremely popular in Texas, and Texas harvests more doves than any other state. TPWD biologists monitor dove populations through roadside surveys conducted each spring. Biologists conduct distance sampling along the roadside and stop to listen for doves at various points—a combination of a transect and point sampling method. In addition, TPWD has engaged staff and volunteers in banding doves. Birds are captured with a baited walk-in trap (Fig. 12.8).

Figure 12.7. Mourning doves, named for their mournful cooing sound, occupy a wide variety of habitats all across the state. Photo: Mark Alan Storey.

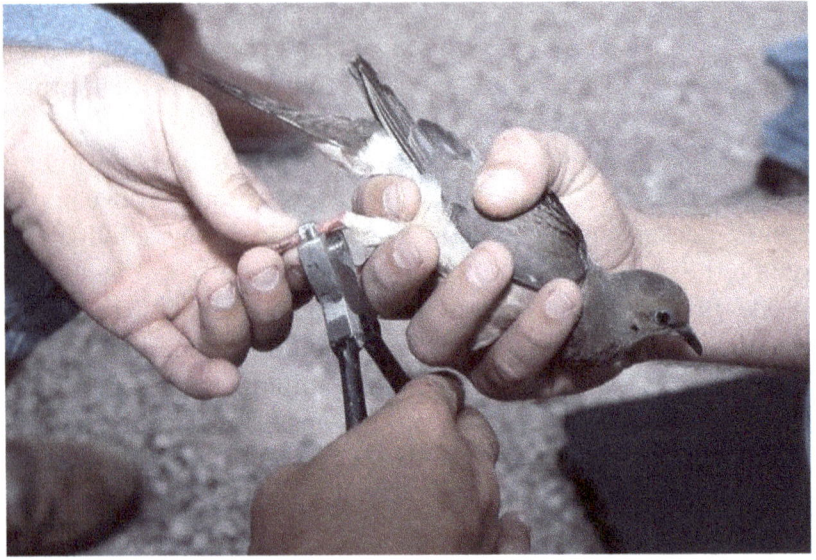

Figure 12.8. Two photos illustrating mourning dove banding. Photo of biologist removing dove from walk-in trap: Mark Mitchell, TPWD. Photo of placing metal USFWS band on dove: Joe Herrera.

Cranes and Allies – Order Gruiformes

The order Gruiformes is quite diverse. In Texas, it includes two species of cranes, six species of rails, two species of gallinules, the American coot, and a recent immigrant to Texas, the limpkin. Although all members of the order are associated with wetlands, they differ in appearance.

Cranes are tall, with long necks and legs that are stretched out in flight. About 700,000 sandhill cranes winter in Texas, primarily in the Texas Panhandle and along the coast. Sandhill cranes usually occur in large flocks, feeding in grasslands and agricultural fields. As the most populous crane in the world, the species allows a sustainable harvest. Hunters in Texas must apply for a Sandhill Crane Hunting Permit designed to track hunter numbers and to provide them with education to ensure they do not mistake an endangered whooping crane for a sandhill crane. Sandhill crane numbers are monitored as they stop at feeding areas during their spring migration. These concentrated feeding areas, such as the Platte River in Nebraska, are called "staging" areas. Limpkins and whooping cranes are types of cranes, but they are not hunted. The flagship story of the whooping crane is presented in Chapter 10. Limpkins are currently restricted to wetlands on the upper coast near Houston. These recent arrivals benefit wetlands by feeding on the invasive apple snail in those habitats.

The other members of Gruiformes are less popular as game birds but are common in wetland habitats. American coots are abundant in winter in Texas, feeding in large flocks on aquatic vegetation. They resemble diving ducks but have lobed toes instead of webbed feet and a comical, tropical-sounding call. Gallinules—purple and common—are frequently seen walking along wetland edges and even on tops of aquatic vegetation with their long toes. Rails are secretive, well-camouflaged species, taking cover in reeds and grasses near wetlands. Coots, gallinules, and four species of rail are hunted. The less common black rail and yellow rail are not hunted but are popular among birders.

Figure 12.9. Sandhill cranes use a mixture of agricultural and wetland habitats. Like other cranes, they are known for their vocalizations and courtship displays. They also "paint" their gray feathers brown with mud as part of their displays.

HIGHLIGHT 12.2 - THE PHENOMENON OF MIGRATION

Every year, as days get shorter, birds get fatter—at least some birds do. But the weight gain is not a failed New Year's resolution. Instead, it is a preparation for an ancient bird behavior—migration.

Migratory birds spend their breeding season at one latitude and then fly south to spend the winter at a different location. This behavior allows birds to minimize stress from extreme weather conditions and to access food resources when they are seasonally abundant. About 70% of North American birds migrate. Many of them are Nearctic-Neotropic migrants—birds that nest in temperate or arctic latitudes and winter in tropical parts of Central and South America.

In the fall, the process begins when shorter days and lower sun intensity stimulate the pituitary and thyroid glands to cause food consumption and fat accumulation to increase. These fat stores provide energy for long flights. The ruby-throated hummingbird may add enough fat to fly 26 hours non-stop at 25 miles per hour—enough fuel to cross the Gulf! Weather patterns, especially cold fronts, often trigger the southward flights. The process is reversed in the spring as days get longer, reproductive hormones kick in, and southerly breezes push birds north.

The length of migrations varies. Ruby-throats may travel about 8,000 miles round-trip, from the northern U.S. to southern Mexico or Central America and back. The pectoral sandpiper passes through Texas on its 18,000-mile round-trip between the Canadian North Slope and Argentina. Arctic terns are the migration champions. One bird tracked with a geolocator tag in 2015 traveled nearly 60,000 miles in one year! On the other hand, some birds, such as American robins and red-winged blackbirds, are short-distance migrants, traveling perhaps only several hundred miles as they move among food sources.

Some birds fly in small family groups, whereas others group together to migrate as flocks. Waterfowl are well known for their V-formations that increase the aerodynamics of their flights. Some raptors leave behind a solitary lifestyle to group together in migration. Groups of broad-winged hawks and Mississippi kites often form circling swirls called "kettles" as they ride air thermals upwards. Other species, such as purple martins, gather in enormous roosting concentrations before heading south. Such massive movements, popular with bird-watching crowds in Austin and other cities, may even be seen on Doppler weather radar!

How do they know where to go? Parent birds in some species teach migration paths to their young, but others seem to just "know where to go." Recently, scientists, in a blend of quantum-mechanics physics and biology, found that unpaired electrons in birds' eyes (essentially, free radicals) may be able to sense the earth's magnetic field. In addition, topographic landmarks, vegetation zones, and air masses guide birds along the migration pathway. In North America, those features drive birds into shared migration pathways. For the purposes of species management, agencies recognize four major flyways—the Atlantic Flyway, the Mississippi Flyway, the Central Flyway, and the Pacific Flyway (Fig. 12.10). Texas lies at the lower end of the Central Flyway.

Some birds migrate during the day. **Diurnal** migrants include birds of prey and other swift, strong fliers, such as cranes and pelicans. But tiny hummingbirds and insect feeders such as swifts and swallows also migrate during the day. Some migrate at night, using the stars to guide them. Nocturnal migrants include species vulnerable to predation, such as cuckoos, flycatchers, thrushes, warblers, orioles, and buntings.

Why do they do it? As creatures with a high metabolism and a need for energy-rich food, migration allows many species to "have the best of both worlds."

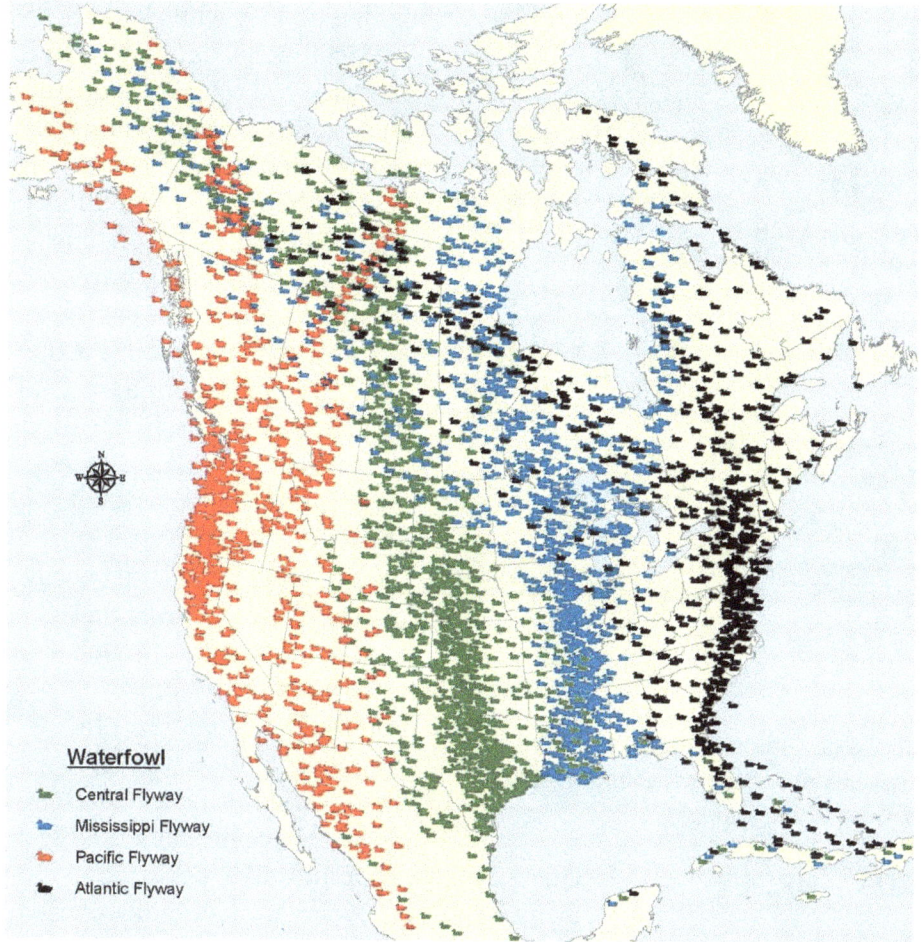

Figure 12.10. Bird migration flyways in North America. Note that there is overlap in the flyway paths. Source: USFWS, CC 1.0 (public domain).

Shorebirds – Order Charadriiformes

The final set of migratory game birds are part of the order Charadriiformes. The order is huge, with 98 species reported from Texas. It includes small, long-legged shorebirds, such as sandpipers, plovers, stilts, and avocets, commonly seen feeding in shallow wetlands and on the beach. Gulls, terns, and skimmers, familiar sights soaring on the Texas coast, are also members of this group. Some members of this order are known for their long-distance migrations (see Highlight 12.2). Texas provides critical habitat for migrating shorebirds that funnel through the state and for those that winter here. Several species also breed in Texas. Killdeer breed in many locations across the state, and high numbers of laughing gulls, black skimmers, royal terns, Forster's terns, and sandwich terns, among others, breed along the coast.

TPWD and the Texas Colonial Waterbird Society conduct an annual breeding estimate on the bare, open flats where these birds construct their simple ground nests.

The common snipe and American woodcock are the only members of the shorebird order that are hunted. Snipe occur all across Texas in winter, occupying wet grassy habitats. Woodcock are found in the state's eastern half in young forests, brushy fields, and riparian woodlands, including breeding populations in Northeast Texas. Both species are **cryptic** (hard to see), but the males are well known for the buzzy sounds produced by their tail feathers during their showy courtship flights.

Figure 12.11. People really do go "snipe hunting!" The Wilson's snipe is one of the two species of shorebirds that are hunted in Texas. Photo: USFWS, CC 2.0.

UPLAND GAME BIRDS – ORDER GALLIFORMES

TPWD classifies wild turkey, northern bobwhite quail, scaled quail, Gambel's quail, plain chachalaca, and introduced ring-necked pheasant as upland game birds. All are in the order Galliformes, also called the gallinaceous birds. Galliformes means "chicken-like," and members of this group share many characteristics with chickens, such as heavy bodies, short wings, and strong beaks and toes. They primarily feed on the ground on a variety of plants and invertebrates, and the males often display dramatically in courting the females. Other nongame galliforms in Texas include prairie-chickens.

Texas hosts four quail species—bobwhite, scaled, Gambel's, and Montezuma. A favorite among landowners and wildlife watchers, quail have experienced significant population declines over the last half-century. Quail populations are notorious for "boom and bust" cycles—rising rapidly when weather and habitat conditions are good and falling quickly when bad. Unfortunately, data collected in roadside call counts conducted since 1980 show an average long-term decline of 5% per year for bobwhite populations. Scaled quail, also called blue quail or "cottontops," have declined at about 3% yearly. Not surprisingly, habitat is the issue. Quail need large tracts of diverse native grassland with a scattering of woody cover. As with the story of prairie-chickens told in Chapter 8, habitat restoration must occur at a landscape scale so that healthy populations in one area can replenish declining populations in another. In response to these trends, TPWD and the Texas Quail Council have implemented a Quail Conservation Initiative designed to bring agencies, conservation organizations, and landowners together to work toward restoration of landscapes for quail and all the species that share their habitat.

The wild turkey is the largest gallinaceous bird in the world. It occurs nearly statewide in Texas. Three **subspecies** (Highlight 12.3) are recognized. The Rio Grande subspecies is the most widespread; the eastern subspecies occurs east of the Trinity River; and the Merriam's subspecies occurs only in a few highland areas in West Texas. Unregulated hunting once nearly wiped out the wild turkey (see Highlight 12.4). Now, with statewide populations over 500,000, sustainable hunting is possible in most of the state. Both spring and fall hunting seasons are offered for the Rio Grande subspecies. In contrast, only a spring season for gobblers (male turkeys) is allowed for the less abundant Eastern subspecies. The Merriam's subspecies is not hunted.

Bobwhite quail – Photo: Mark Mitchell

Scaled quail - Photo: Bettina Arrigoni, CC 2.0.

Gambel's quail – Photo: Bonnie McKinney

Montezuma quail – Photo: Sloalan, CC 1.0.

Figure 12.12. Quail species in Texas.

HIGHLIGHT 12.3 - WHAT'S A SUBSPECIES?

Subspecies are populations of a species that live in different areas and may vary in size, shape, or other physical characteristics from other major population segments. Subspecies are biologically able interbreed with each other, but geographic barriers may prevent them from doing so. Often, subspecies show characteristics that are well-adapted to their habitat area. Subspecies may have a three-part scientific name—the genus, species, and subspecies. In plants, subspecies are sometimes called variants.

HIGHLIGHT 12.4 - CONSERVATION SUCCESS STORY – TALKING TURKEY AND TRANSPLANTS

Wild turkeys have many admirers. They have keen hearing and the sharpest vision of all Texas game animals. They can hear noises from a mile away, and their eyes pinpoint even the slightest movement. Although Benjamin Franklin did not really propose that they should be our national bird, he was a fan, writing that the turkey was a "respectable Bird, and withal a true original Native of America...He is besides...a Bird of Courage." Although Franklin described the turkey as a bit vain, the male's courtship display—with puffed-out iridescent feathers, fanned tails, dragging wings, and colored snoods (wattles)—is truly impressive.

Figure 12.13. Release of transplanted Eastern wild turkeys. Photo: Mark Mitchell.

However, turkeys almost disappeared from North America due to overhunting and habitat loss, especially forest clearing. By the early 1900s, only about 200,000 turkeys remained in North America. In 1942, less than 100 birds remained in the Pineywoods and Oak Woods of eastern Texas.

Fortunately, the science of wildlife management had begun to emerge at that time. Hunting regulations were put in place, and habitat management began—restoring tall trees needed for roosting, the grassy layer required for raising poults (turkey hatchlings), and plant diversity to provide food sources such as mast (acorns, nuts, fruits), seeds from grasses and forbs, greens, insects, and other invertebrates such as worms, snails, spiders, arthropods.

Turkey numbers began to increase in Texas, but they were still far below historic levels in 1973 when The Wild Turkey Federation, a conservation organization, began supporting states' efforts to transplant wild-caught turkeys to depleted areas. As restocking efforts increased in Texas, the program was a success for the Rio Grande subspecies. The statewide number of turkeys climbed to 460,000.

However, recovery efforts have been a bit slower for the eastern subspecies. Early attempts using pen-reared turkeys failed, as did transplants of Rio Grande wild turkeys into eastern wild turkey habitats. In 1979, TPWD began releasing wild-trapped eastern turkeys from neighboring states in groups of 15 to 20 birds per site. Survival was a bit better, but many locations still failed to establish sustainable populations. Biologists hypothesized that the habitat was too fragmented to allow such small releases to survive. Research by Stephen F. Austin State University suggested that "super-stocking" sites with at least 80 turkeys would improve survival and establishment. TPWD began stocking with this protocol in 2014 on properties where landowners agreed to restore and manage habitat.

After almost disappearing, there is now hope that all three of Texas's turkey subspecies can once again vainly strut and gobble across the state.

Figure 12.14. Wild turkey gobbler displaying. Photo: Jeff Forman, TPWD.

NONGAME BIRDS

Less than 10% of bird species in Texas are hunted, so the vast majority of bird diversity in the state is nongame birds. Birds with varied characteristics fill every habitat niche available. Some bird orders (presented below) are especially significant in the state due to their abundance or uniqueness.

Hummingbirds – Order Apodiformes

The order Apodiformes includes the hummingbirds and the swifts. Texas is champion again when it comes to the jewel-colored hummingbirds. Observers in the state have recorded 19 species of hummingbirds, more than any other state. According to Hummingbird Roundup, a 20-year TPWD citizen science monitoring program, the most common species are the ruby-throated, black-chinned, rufous, and buff-bellied. The list of unique attributes of hummingbirds is awe-inspiring. A ruby-throated hummingbird weighs less than a nickel. Its heart beats at 1200 beats per minute, and their wings beat 60 to 200 times per second! In their courtship display dives, they can reach 60 miles per hour in less than three feet. If they were as large as humans, they would need to consume more than 150,000 calories daily. The hummingbird's forked tongue, which wraps around its skull when not feeding, can flick up to 13 times *per second* as it laps up nectar. Their glowing colors come not from pigments but from light refraction in the physical structure of the feathers. Hummers are truly impressive! Most hummingbirds are summer residents; but Texas also funnels a vast number of hummingbirds through the state during migration. Some coastal towns even sponsor hummingbird festivals during fall migration.

Figure 12.15. Many people put out hummingbird feeders with sugar water, but hummingbirds may be more attracted to native flowering plants where they collect both nectar and insects. Photo: Mark Alan Storey.

Waterbirds – Order Pelicaniformes

The order Pelicaniformes encompasses a wide variety of waterbirds, including herons, egrets, spoonbills, pelicans, and cormorants. Texas hosts two pelicans, the year-round brown pelican and winter resident American white pelican, and two cormorant species—all with fleshy pouches attached to the beak. However, the greatest diversity is found in the wading birds—tall birds with long legs and long toes designed to feed in wetlands.

Texas hosts 27 species of waders, including herons, egrets, bitterns, ibises, and the roseate spoonbill. These year-round residents have a variety of beak types reflecting their dietary preferences. The three species of ibis have long, curved beaks for probing. Roseate spoonbills have a spoon-shaped bill with laminae on the edges designed for swishing and filter-feeding in water. (They also have pink plumage—colored by the pigments in the crustaceans they eat.)

Herons, egrets, and bitterns have sharp, pointed beaks ideal for capturing fish, but different species use different strategies in fishing. Great blue herons watch and wait—remaining motionless until a fish passes. Snowy egrets have bright yellow feet ("golden slippers") on black legs that they use to lure fish in close. Reddish egrets use an active, almost spastic, feeding strategy—running to stir up prey and using their wings to cast a shadow over their head to better draw in and view the fish. In a different twist, the western cattle egret, an African species that reached Texas in 1959, uses its pointed beak to feed mainly on insects in uplands in a commensal relationship with grazing cattle.

Most members of this order are colonial breeders, forming large breeding colonies of mixed species near water. Nests are usually in trees, although brown pelicans sometimes nest on the ground. Near the coast, the

Figure 12.16. The great blue heron is one of the most widely distributed wading birds in Texas. In the spring, the species gathers in groups to form breeding colonies. Photo: Mark Alan Storey.

colonies are often on islands, where the surrounding water (sometimes replete with alligators) and large group size help protect against mammalian predators. A century ago, wader numbers were drastically reduced because of the overcollection of birds for their showy breeding feathers. TPWD and the Texas Colonial Waterbird Society now census breeding colonies from the ground, boats, or the air to monitor population changes.

Raptors

"Raptor" means to "seize or take by force"—an appropriate description of these birds of prey found in several orders. In Texas, raptors include hawks, falcons, eagles, kites, osprey, owls, and vultures. There are 29 diurnal species (species that hunt during the day) and 17 species of nocturnal owls. Raptors host a variety of characteristics to enable their predatory lifestyle. They are swift flyers, have keen eyesight for detecting prey up to two miles away, strong feet with sharp talons for grasping prey, and powerful, curved beaks for tearing off flesh. Preferred prey varies among different species. Owls and many larger hawks seek mammals. Falcons, Cooper's hawks, and sharp-shinned hawks are maneuverable flyers adept at catching birds. Bald eagles and osprey can catch fish while in flight. Gray hawks specialize in capturing reptiles. Kites often feed on insects. Although most raptors capture prey live, vultures primarily scavenge **carrion**. Other raptor species, especially eagles and the crested caracara, opportunistically feed on carrion.

Raptors declined in Texas primarily due to DDT and other persistent pesticides. Some of the first nongame surveys at TPWD focused on peregrine falcons, one of the species that experienced those declines. TPWD, USFWS, and The Peregrine Fund, a conservation organization, have collaborated in reintroducing peregrine falcons in West Texas and aplomado falcons in South Texas through a technique called "hacking." Young

Figure 12.17. Red-tailed hawks, recognized by their rusty red tail and characteristic dark belly band, can reach speeds up to 120 miles per hour in a dive. Peregrine falcons may reach 240 mph! Red-tailed hawk photo: Mark Alan Storey.

captive-reared falcons are placed in a pen on a release platform in a suitable habitat. After a holding period in the pen, biologists open the doors but continue to provide food at the hack box. Gradually, biologists decrease the feeding, as the birds learn to hunt. When feeding is ended, the birds may still use the box as roosting habitat. As DDT in the environment decreased, bald eagles in Texas made a comeback on their own. TPWD biologists monitored nests through aerial surveys and even climbed into nest trees to band some eaglets. Citizen scientists also help to monitor raptors. "Hawk Watches," conducted by volunteers at Hawk Watch stations each fall, help enumerate the many raptors that pass through Texas on migration.

Interestingly, although raptors are not game birds, TPWD and USFWS issue permits that allow individuals to capture raptors from the wild for falconry. Falconers use trained raptors to hunt other birds, such as ducks. Falconers must spend an apprenticeship period learning the skill from other falconers to ensure good sportsmanship and bird care. Falconers and researchers that study raptors capture the birds using a bal-chatri trap—a mesh trap covered with numerous monofilament nooses that uses a live bird inside as a lure.

Woodpeckers – Order Piciformes

Of the approximately 23 species of woodpecker found in the United States, Texas is home to 16. In addition, Texas once hosted the ivory-billed woodpecker now thought to be extinct. Most woodpeckers have a similar body form and lifestyle. They use their strong beaks to make homes in trees, find insects, and drum on tree trunks to communicate. The sticky tongue is extraordinarily long and wraps around the back of the skull, providing cushioning for all that hammering activity. Woodpeckers' **zygodactylus** feet (two toes pointing forward and two pointing back) and stiffened tail feathers enable them to perch on vertical tree trunks.

Despite their similarities, woodpeckers are specialized in several ways. As noted in Chapter 11, red-cockaded woodpeckers are the only species that excavates cavities in living pines. Sapsuckers are named for their habit of drilling parallel rows of holes designed to make trees exude sap and trap insects. Acorn woodpeckers are known to be prolific "cachers"—storing tens of thousands of acorns in holes that they drill. Red-headed and red-bellied woodpeckers also cache acorns. Northern flickers dig ants from the ground. The pileated woodpecker, the largest in Texas, creates massive excavations that sometimes cause trees to break in half. TPWD lists pileated, red-headed, and red-cockaded woodpeckers as species of concern.

Figure 12.18. Red-bellied woodpecker caching an acorn.

Perching Birds – Order Passeriformes

Members of the order Passeriformes are called "perching birds" because their small feet are designed to grasp and perch on branches or grass blades. Also called songbirds or passerines, this is the largest order of birds in the world and in Texas. More than 300 songbird species have been recorded in Texas. Our Texas state bird, the northern mockingbird, is a perching bird. Common backyard feeder birds, such as robins, chickadees, sparrows, and cardinals, are passerines, as are more diverse and specialized birds, such as flycatchers, ravens, swallows, shrikes, and blackbirds. Various species use a wide variety of foods, including seeds, fruits, insects, nectar, and even occasionally small vertebrates. Many people feed songbirds from bird feeders, but homeowners also have great success landscaping yards with a variety of native plants at different canopy heights to provide a variety of habitats and food sources.

Passeriforms are extremely popular among birdwatchers, with many birders traveling to Texas to observe the variety of species present. Birding is especially intense during spring migration, when nearly two billion songbirds may pass through Texas, some even flying across the Gulf! As described in Chapter 7, the oak mottes found on the Texas coast provide essential resting and feeding places for these Nearctic-Neotropic migrants. Texans throughout the state are encouraged to help these nocturnal migrants by promoting dark skies during peak migration. The "Lights Out, Texas!" campaign encourages businesses, cities, and households to reduce outdoor lighting from April 22 to May 12 and September 5 to October 29.

In addition to migrants, some passerines are permanent residents, some are winter residents, and some breed in Texas. Texas is central to the breeding success of a few species, such as the black-capped vireo and endangered golden-cheeked warbler. Ironically, these species face a challenge from another native songbird, the brown-headed cowbird (Highlight 12.5).

Figure 12.19. Texas is home to many beautiful songbirds and may be the best state in the country to see the amazingly-brilliant painted bunting. Photo: Mark Alan Storey.

HIGHLIGHT 12.5 - BROWN-HEADED COWBIRD – CREATIVE MOTHER OR FIENDISH FREELOADER?

Habitat management can sometimes affect wildlife species in indirect ways. Such is the case with cowbirds.

Cowbirds are a passerine species closely tied to grazing animals, as they feed on seeds and insects stirred up by livestock hooves. They are also drawn to ecotones, or edge habitats, where they can feed in open areas and still have access to nesting sites. However, there's a twist when it comes to cowbird nesting. Cowbirds don't build nests of their own. Instead, they lay eggs in other birds' nests to be raised by an "adoptive" mother, sometimes even removing eggs belonging to the nesting bird. This behavior is called brood parasitism. It is sometimes practiced by other species, including cuckoos, and it gives the parasite species an advantage because they don't have the energetic costs or risks of predation associated with building a nest and incubating eggs.

When a songbird nest is parasitized by a cowbird, sometimes the mother songbird recognizes the cowbird egg and pushes it out. In some cases, the mother songbird abandons the nest or builds a new nest on top of the old nest (one yellow warbler nest was discovered to have six layers of eggs due to cowbird parasitism!). But sometimes, the mother bird incubates the cowbird egg with her own. The cowbird egg hatches quickly, giving the bigger and more aggressive cowbird chick a competitive advantage over the mother bird's chicks. It is a setback for the parasitized species; however, if cowbirds are not too prolific on the landscape, both species can still thrive.

How many cowbirds are too many? That is a question that wildlife biologists and bird lovers debate. The widespread presence of cattle on rangeland in Texas provides our three species of cowbirds (brown-headed, bronzed, and shiny) with plenty of feeding sites. Bird feeders also attract and support cowbirds. Cowbirds have been documented parasitizing more than 225 species of native songbirds nationwide, and there is fear that they may be especially problematic for Texas species of concern that nest in habitat edges, such as the black-capped vireo and southwest willow flycatcher. One management strategy is to move cattle at least five miles away from sensitive songbird nesting areas, making it difficult for cowbirds to access feeding and nesting areas. If many cowbirds are present or that kind of separation is not possible, landowners may seek permits to trap and euthanize cowbirds during the nesting season.

Figure 12.20. Eastern phoebe nest with brown-headed cowbird egg. Photo: Galawebdesign, CC 3.0.

Figure 12.21. Brown-headed cowbirds, the most common of the three Texas cowbird species, often sit on high perches, observing other birds building nests. A male and two females are shown here. Photo: Mark Alan Storey.

MONITORING

Biologists monitor many species of birds as described above, but volunteers are vital to monitoring songbirds and other bird species in various ways. The most robust dataset for bird populations comes from the North American Breeding Bird Survey (BBS). The program began in 1966 and consists of randomly-located roadside routes throughout the U.S., Canada and Mexico. BBS volunteers drive along a preset route, stopping at regular intervals to listen and watch for birds in a combined transect and point count. There are 195 routes in Texas.

An even older dataset comes from Christmas Bird Counts. In the late 1800s, it was a common tradition for groups of friends to go out on Christmas Day and see who could kill the most animals. Whichever group killed the most was declared the winner. In 1900, Frank Chapman, an ornithologist who helped form the National Audubon Society, proposed that instead of hunting teams should go out and see who could count the most birds, and the Christmas Bird Count was born. Ever since, the National Audubon Society has recruited groups of volunteers that conduct counts in locations all across the U.S. and its territories. The volunteers conduct the 24-hour count within a 15-mile-radius circle on one day between December 14 and January 5, counting all birds seen. For several years, Christmas Bird Count locations on the Texas coast have produced the highest species counts in North America!

Other volunteer opportunities include the Great Backyard Bird Count, Project Feeder Watch, and eBird, offered by the Laboratory of Ornithology at Cornell University. The Big Sit is an October event sponsored by Bird Watcher's Digest. Finally, TPWD offers the Great Texas Birding Classic during the peak of spring migration, allowing teams around the state to compete to find the most species in 24 hours. The event, which includes youth categories, draws participants and sponsors from all over the country.

Because Texas is a concentration point for many nongame birds, including songbirds, shorebirds, and hummingbirds, biologists and permitted volunteers set up many bird-banding stations in Texas. Most commonly, banders use thin monofilament nets called "mist nets" to intercept and entangle birds as they fly through. Bird banders watch the nets and then quickly move in to untangle the birds and place bands on the birds' legs. Hummingbirds may also be captured in specialized traps, and larger shorebirds may be captured with

Figure 12.22. The Texas Birding Classic is a competitive birding event that offers categories for youth participants and raises funds for habitat improvement.

rocket nets. In addition to leg bands, more advanced research uses various tags that track birds by radio or satellite, helping scientists better understand migration.

THREATS

Unfortunately, the long-term data from bird monitoring programs indicate trouble for Aves. Within Texas, TPWD identifies 111 birds as Species of Greatest Conservation Need. That's one of every six species in Texas. A report published in the journal *Science* in 2019 revealed that North America's overall number of birds declined by almost 30%, or 3 billion birds, over the past 50 years. These declines have occurred across a wide variety of species and habitats. Shorebirds, sparrows, wood warblers, blackbirds, larks, finches, and aerial insectivores (flycatchers, swallows, and swifts) are the native species experiencing the most significant losses, with the most startling losses in shorebirds and grassland birds. Concern exists that about 70 species could become threatened or endangered if trends continue. Only raptors, turkeys, wetland birds, and a few other groups increased, primarily due to effective management strategies and the prohibition on the use of DDT.

In addition to habitat loss and modification, new threats continue to emerge for birds. In recent years, Highly Pathogenic Avian Influenza (HPAI) has come to the forefront of disease concerns. HPAI, also called bird flu, has been detected in at least six bird orders across Texas, with especially high occurrences in waterfowl and scavengers such as vultures. Its impact on wild bird populations is yet unknown, but it poses significant threats to domestic poultry operations. Additional worries arise because the virus appears to be able to mutate to infect other organisms. It has already infected dairy cow herds and caused the death of a person working on a dairy farm. HPAI has also been detected in wild mammals in Texas, showing up so far in foxes, raccoons, and striped skunks, along with domestic cats.

Figure 12.23. Mist nets are used to capture songbirds and shorebirds for banding. Photo 1 shows a white-eyed vireo caught in a mist net (Photo: Mark Mitchell). Photo 2 shows a researcher removing a bird from a mist net (Photo: Lorie Shaull, CC 4.0).

CONSERVATION AND MANAGEMENT

The best answer for bird conservation and management is, of course, "habitat," but looking at the bird groups that are holding steady offers some management guidance as well. Raptor numbers increased after DDT was banned. Turkeys increased when hunting was regulated, habitat was improved on public and private lands, and transplants helped build depleted populations. But perhaps the best example for the future is "Joint Ventures," a strategy that has provided benefits for a wide variety of wetland birds.

Like other habitat types, the acreage of wetlands in the U.S. has decreased, falling to about half of their former abundance across the country. Yet, the 2019 bird population trend analysis published in *Science* showed that waterbirds increased by 18% and ducks increased by 34% over the 50-year period. In 1986, the North American Waterfowl Management Plan (described in Chapter 10) called for forming collaborative, regional partnerships between government agencies, non-profit organizations, corporations, tribes, sportsmen, and other individuals to conserve habitat for the benefit of priority bird species, other wildlife, and people. These Migratory Bird Joint Ventures are coordinated by USFWS and pull together partners to restore habitat and species at a landscape scale. Joint Ventures use a method called Strategic Habitat Conservation to set goals, identify appropriate methods, establish collaborative approaches, monitor outcomes, and design relevant research.

By 2023, 24 Joint Ventures had been implemented in North America, including five that include Texas. The earliest Joint Ventures focused on waterfowl and wetland habitats, such as the Gulf Coast, Playa Lakes, and Lower Mississippi Valley Joint Ventures in Texas. Newer Joint Ventures have expanded to other habitats and species groups, including the Oaks and Prairies Joint Venture and Rio Grande Joint Venture in Texas. As waterbirds have been able to thrive, it is hoped that a Joint Ventures collaborative model can hold promise for reversing declines in other bird species and habitats.

Within the state, TPWD and other partners continue to seek additional ways to raise awareness for bird conservation through citizen science, private landowner technical guidance, wildlife cooperatives, landscaping guidance for urban areas, and the Bird City Program. Bird City certification, implemented jointly by TPWD and Audubon Texas, is designed to recognize and promote communities that increase and maintain bird habitat, minimize threats for birds, and engage residents to learn about and protect their local species. It's going to take everyone joining in "for the birds."

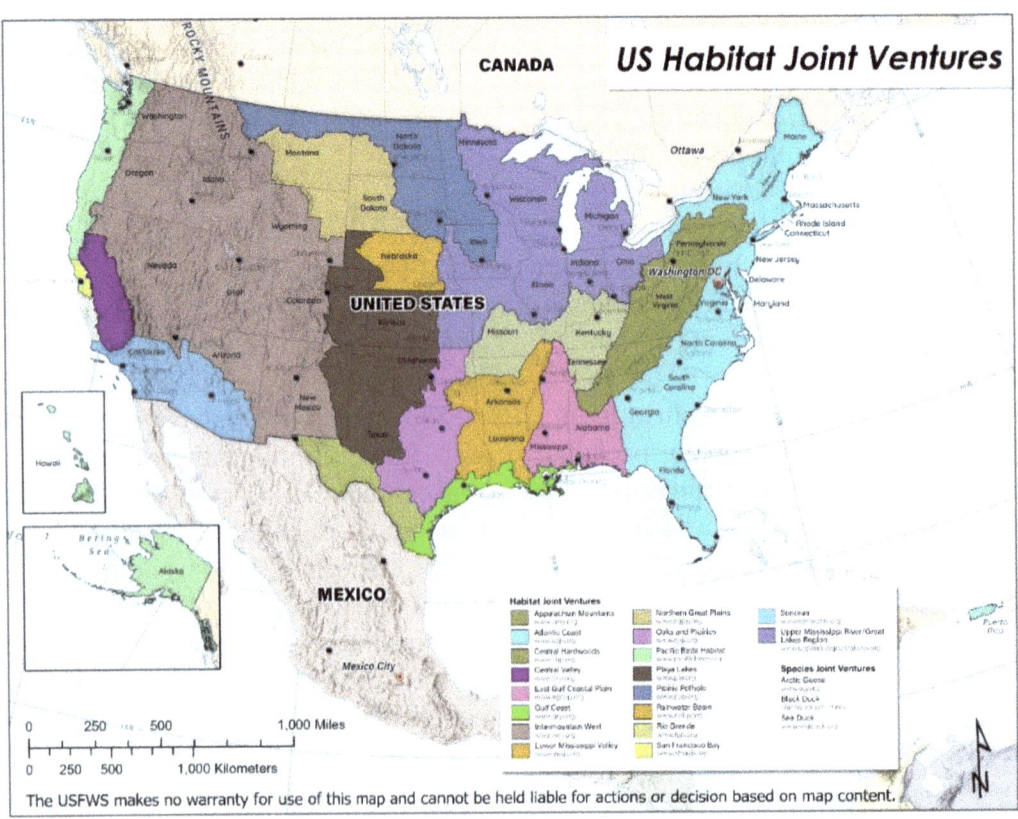

Figure 12.24. Location of Joint Venture projects in the U.S. Five projects are currently underway in Texas. Source: USFWS (public domain).

CHAPTER 13 – TEXAS MAMMALS

Figure 13.2. Texas State Flying Mammal, the Brazilian free-tailed bat. Photo: USFWS, WC, CC 2.0.

Texas mammals are diverse. The smallest, the least shrew, weighs about 5 grams, or as much as a AA battery. The largest, the bison, weighs as much as a pickup truck. There are volant (flying or gliding) mammals, cursorial (running) mammals, saltatorial (hopping) mammals, fossorial (burrowing) mammals, natant (swimming) mammals, and arboreal (climbing) mammals. Texas is home to over 20% of the nation's total deer population, over 75% of the carnivore species, and more than three-quarters of the bat species in the United States.

WHAT IS A MAMMAL?

When one thinks of mammals, one thinks of hair. Nearly all terrestrial mammals possess hair that helps regulate body temperature. (Marine mammals rely instead on a layer of blubber.) In addition to hair, the skin of mammals can have several other diagnostic characteristics, such as sweat,

(overleaf) Figure 13.1. The State Small Mammal of Texas, the nine-banded armadillo, was recently renamed the Mexican long-nosed armadillo. Photo: Shutterstock.

scent, and oil glands. Like birds, mammals are endothermic and homeothermic and have an efficient four-chambered heart. Most mammals have several types of specialized teeth for different aspects of feeding, such as cutting, tearing, crushing, and grinding.

All mammals have internal fertilization, and all female mammals feed their young with milk secreted from mammary glands. However, the birth and development of mammals varies. The five species of monotremes (egg-laying mammals found in Australia and Papua New Guinea) lay eggs that are incubated by the mother outside the body. All other mammals are **viviparous**, with young developing inside the female's uterus where a placenta connected to the mother's blood supply provides nourishment. Marsupials, such as kangaroos and opossums, give birth to very underdeveloped young that crawl into a pouch or "marsupium" on the mother's belly and complete development while feeding from mammary glands located in the pouch. The vast majority of mammals are called eutherian mammals and give birth to more developed young. Still, the degree of development can vary. Rabbits, shrews, most carnivores, and some rodents have a short **gestation period** and give birth to **altricial** young that may initially be blind, immobile, and even hairless. The mother must provide a sheltering environment and weeks of care until the young can move independently. On the other hand, species such as **ungulates** (hooved animals) and marine mammals give birth to **precocial** young that can follow the mother immediately after birth.

IDENTIFYING MAMMALS – THE IMPORTANCE OF "SIGNS"

As with birds, field guides can help with the identification of mammals. Even if the animal is not seen, animal signs, such as skulls, scat, tracks, and hair can help identify species.

DENTAL FORMULA

Specialized teeth characterize the class Mammalia. This attribute can aid in identification. In general, mammals have four types of teeth. Incisors in the front cut or bite. Canine teeth grab, pierce, and tear. Premolars and molars usually grind. The types of teeth found in a species give a clue to the animal's diet. Rodents have long incisors for gnawing. Carnivores have prominent canines. Herbivores tend to have broad, flat premolars and molars. In some carnivores, premolars and molars are modified into **carnassials**—designed for shearing meat.

Each species has a "dental formula"—a characteristic number of incisors, canines, premolars, and molars that can be used to identify mammal skulls. Dental formulas are counted on one side of the skull and are expressed as the number of each type of tooth on the top over the number of each tooth type on the bottom.

For example, adult humans have four incisors (I), two canines (C), four premolars (P), and six molars (M) on the top and on the bottom (a total of 32 teeth) (Fig. 13.3). Therefore, looking at one side, the human dental formula is:

I 2/2 , C 1/1 , P 2/2 , M 3/3

There is much variation among species. For example:

- The dental formula for a bighorn sheep is I 0/3, C 0/1, P 3/3, M 3/3. They don't need well-developed canines, but they require many grinding teeth. Like many ungulates, they lack upper incisors.
- The dental formula for a bobcat is I 3/3, C 1/1, P 2/2, M 1/1. They have well-developed canines with carnassial teeth in the back.
- The dental formula for an opossum is I 5/4, C 1/1, P 3/3, M 4/4. They have many teeth!

Teeth can also be used to age animals by looking at tooth replacement and wear. This is a valuable skill for a biologist trying to manage the age and maturity of a deer population.

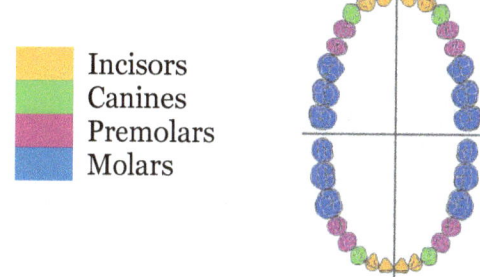

Figure 13.3. Human incisors, canines, premolars, and molars. Wild mammals also have four different tooth types that can aid in their identification. Source: Strepulah, CC 4.0.

SCAT

Scat is a term given to the fecal droppings of wild animals. Scats can be identified based on size, shape, texture, content, and location. (Observers should use gloves and a mask when examining scats closely.) Deer, sheep, pronghorn, and rabbits produce dark, rounded pellets. Because they have multi-chambered stomachs, the contents of the droppings are usually well-digested and are not identifiable by the naked eye. Size and location can be a clue to the exact animal; for example, pronghorns often scrape an area before defecating. Canines, felines, and raccoons all produce tubular droppings. For canines, the scat is usually twisted with a tapered end. Hair may be visible in the scat. Coyotes and foxes often mark their territory with scat. Their scat may be perched on a rock or centered on a road. Feline scat is more segmented and compressed (harder to break apart). For raccoons, there is usually more evidence of seeds and plant parts in the scat. Rodents produce small, elongated pellets, often scattered over a wide area. Finally, bison scat looks much like a cow's—a layered deposit that appears to have been dropped down wet. Modern techniques allow the DNA in scat to be analyzed to reveal information about genetics or diet. Many field guides also provide illustrations of scat.

TRACKS

Tracks in mud, snow, or soft dirt can also reveal an animal's presence. Field guides and state wildlife agency brochures often provide helpful illustrations for common mammals (Fig. 13.5).

Several features are helpful to note as one learns to recognize tracks. First, notice the overall size and shape of the print. Rabbits' long feet leave long tracks. A similarly-sized ringtail leaves a round print. A larger round print may be from a bobcat. Next, note how many toes are evident. Many carnivores leave imprints from four toes and the central pad, but raccoons, bears, beavers, and opossums leave prints from all five digits. The front footprint of raccoons is effortless to recognize, as it is similar to a human handprint. Deer usually leave two only impressions from their cloven hoof, formed from the 3rd and 4th toes. Third, notice whether claw marks have been left. Members of the dog family Canidae leave impressions from their claws. In contrast, species in the cat family Felidae do not usually leave claw marks. Finally, notice the spacing of the prints. Pairs of prints lined up can indicate rabbits or rodents that hop. Deer that are running leave prints spaced far apart with two small marks from dewclaws (two smaller hoof segments formed from the 2nd and 5th toes).

Figure 13.4. Deer jaw bones and droppings (scat) provide a clue about wildlife presence.

Figure 13.5. Tracks of common Texas mammals. Source: TPWD PWD LF K0700_0001.

OTHER SIGNS

Several other signs can indicate the presence and abundance of different mammals. Even a partial skull can be used to identify species, either through DNA analysis or diagnostic characteristics, such as the grooved sutures on javelina skulls, the prominent ridge on the cranium of an opossum, or the hourglass-shaped sutures on the forehead of a gray fox. Occasionally, animal hairs may be found in a fence, vegetation, or scat. In addition to observing the color, length, and coarseness of the hair, mammalogists place hair samples under a microscope and use the pattern of the keratin scales, the hair layers, and the pigmentation to identify species. Other species leave a sign of their presence through their impacts on the habitat. Deer rub the bark of trees as they shed velvet from their antlers and mark their territory. They may also scrape the ground with their hooves and antlers. Beavers leave behind pointed sticks as they cut wood for dams and lodges. Armadillos dig shallow holes searching for grubs, and feral hogs and bison may leave behind big wallows of disturbed earth.

MONITORING MAMMALS

Camera traps that use a motion-sensitive remote camera to photograph the movement of animals are probably the most commonly used method for detecting and monitoring mammal species. A variety of game cameras are available, some of which capture motion or even shoot video.

For more quantitative monitoring, transects are a common method. Often, transects are conducted by use of a roadside survey. This method can be used during the day or night and is well-suited for deer, carnivores, and other species such as rabbits, armadillos, and porcupines. Before conducting the count, biologists conduct a practice run on the route to determine how much habitat they can actually see. This average visibility can help calculate the area of habitat being sampled. For even greater accuracy, distance sampling uses computer-based models to mathematically calculate the probability of detection of species at different distances.

Typically, during a roadside transect count, two observers ride in the back of a truck for good visibility on each side of the road, using spotlights at night. The total area sampled (converted to acres or hectares) is divided by the number of animals seen. Results are expressed as the number of acres per animal. Because roadside transects assume that the road provides a representative habitat sample, the accuracy of the population estimate may be limited. However, if conducted for many years, roadside transects can provide valuable information about trends in wildlife populations.

Other variations are frequently used. One (unfortunate) version of the roadside transect uses counts of wildlife run over by cars as an indicator of species abundance. Aerial transects conducted by airplane or helicopter are well-suited for animals in rough or widely dispersed habitats, such as bighorn sheep and pronghorns. A walking variation of transects called a "Hahn line" is suitable for deer in high-density areas. The observer walks a two-mile line from west to east in the evening. Again, average visibility is determined before the counts begin. Ideally, Hahn lines should be walked on two or three consecutive evenings to calculate average deer density.

HIGHLIGHT 13.1 – TEACHING TECHNIQUES - ROADSIDE TRANSECT EXAMPLE

Some biologists decide to set up a roadside spotlight survey for deer on a wildlife management area in the Rolling Plains. The route is ten miles long. In August, they travel the route at night, measuring how far they can see perpendicular to the line every 0.1 mile. They determine that 250 yards is the maximum they can see and calculate that the average visibility is 160 yards.

In September, the biologists select a night without rain or heavy wind and start their count one hour after sunset. They count all the deer along the route and record them as buck, doe, fawn, or undetermined. They repeat the count two more nights in September. They see 60 deer on one night, 70 deer on one night, and 64 deer on one night. The average number of deer seen is 65 (10 bucks, 25 does, 15 fawns, and 15 undetermined).

The length of the transect is 10 miles or 17,800 yards. Because the average visibility from the road is 160 yards to the right and 160 yards to the left, the total area sampled is:

320 yards x 17,800 yards = 5,696,000 square yards.

There are 4,840 square yards in an acre, so the total acres sampled is:

5,696,000 ÷ 4,840 = 1,177 acres.

Average deer density on the transect is:

1,177 acres ÷ 65 deer = 18 acres/deer.

The size of the wildlife management area is 10,000 acres, so the biologists estimate that their deer population is:

10,000 acres ÷ 18 acres/deer = 556 deer.

The doe-to-buck ratio is:

25 does ÷ 10 bucks = 2.5 does for every buck.

The fecundity (number of young per female) is:

15 fawns ÷ 25 does = 0.6.

With numbers like these, the biologists conclude that their deer population seems to be in good shape.

Density and population size for mammals can also be estimated using mark-recapture methods. There are many methods of capturing mammals. Baited live traps can be set for animals ranging in size from rodents to bears. Deer and pronghorn can be captured using drop nets or rocket-fired nets over baited areas. Sometimes, a helicopter drives the animals into a long net (for deer) or a corral (for pronghorns). Snares and leg-hold traps, often associated with the lethal removal of mammals, can also be used with modifications for safe capture and release. Snares can be placed in travel pathways to capture mountain lions and bears. Leg-hold traps are often used with beaver, muskrat, nutria, coyote, and raccoons. Finally, for some animals, tranquilizing darts are used to capture and mark them. Bighorn sheep may be darted from helicopters. Dogs are sometimes used to tree bears or mountain lions for darting. All trapping devices must be checked every day to minimize harm to the animals captured.

Once mammals are captured, a variety of marking methods are available. Plastic numbered tags can be placed in ears, webbed feet, or flippers (for marine mammals). Neck collars may be placed on ungulates, bears, cats, or canines (Fig. 1.1). Freeze branding uses liquid nitrogen to mark an animal's fur so that it grows back white. Hairless skin areas can be tattooed. Sometimes, selected toes of rodents are clipped off. However, injecting passive internal responder (PIT) tags under the skin is frequently preferred. Finally, animals may also be marked temporarily using dye, paint, or fluorescent powder that allows tracking animals, even at night. Advances in genetic techniques now make it cost-effective for biologists to even monitor wildlife populations through genetic fingerprinting without the need to mark individuals.

TEXAS MAMMAL DIVERSITY AND CLASSIFICATION

Texas is home to 12 orders of mammals, including 148 native terrestrial species, 30 native marine mammals, and at least 23 introduced or feral species. The total mammal count is second only to California and New Mexico. Six species are endemic to Texas—four species of pocket gopher and two species of kangaroo rat. *The Mammals of Texas* Online Edition provides information about the different Texas mammal species.

In general, species richness is associated with diversity of vegetation, elevation, and soils. Interestingly, West Texas has the highest diversity of mammals in the state. More than 70 mammals species are found in the Trans-Pecos. Thirty of those occur only in that region.

Table 13.1. Taxonomic diversity of mammals in Texas, including recently extirpated, introduced, and feral species. Note that numbers may change as species expand ranges, are discovered, are extirpated, or experience taxonomic changes.

Order	Native Species	Extirpated Species	Non-native Species	Total Species
Didelphimorphia (Opossums)	1			1
Cingulata (Armadillos)	1			1
Lagomorpha (Hares and Rabbits)	5			5
Eulipotyphla (Shrews and Moles)	5			5
Chiroptera (Bats)	35			35
Carnivora (Carnivores)	21	7	4	32
Artiodactyla (Even-toed Ungulates)	7		11	18
Rodentia (Rodents)	68		5	73
Primates (Primates)			1	1
Perissodactyla (Odd-toed Ungulates)			2	2
Sirenia (Manatees)	1			1
Cetacea (Whales and Dolphins)	27			27
TOTALS	**171**	**7**	**23**	**201**

REGULATORY CLASSIFICATION OF TEXAS MAMMALS

Apart from federally-listed endangered species, the responsibility for conservation of mammals rests primarily with the state wildlife agency. TPWD classifies mammals as game animals, furbearers, or nongame and has the authority to set hunting seasons and bag limits for mammals that are harvested. Rare species may be classified as threatened or endangered.

Most game animals in Texas are members of the order Artiodactyla—hooved animals with an even number of toes (or hoof segments). All Texas ungulates were hunted at one time; however, Merriam's elk is now extirpated, and bison are restricted to domesticated populations and the Texas State Bison Herd at Caprock Canyons SP. Five native ungulates remain as game animals in Texas. In addition to ungulates, eastern fox squirrels and eastern gray squirrels are game mammals with set seasons and bag limits. TPWD sets harvest levels based on surveys of game animals.

Harvest of fur-bearing mammals spurred European settlement of much of North America. Motivated by the high value of furs for clothing and felt hats, trapping led to the decline of many species, especially beaver. Over time, many species have recovered. Although the harvest of furs in Texas has declined over time, TPWD still offers a trapper's license and a designated open season for those individuals who want to sell pelts commercially. In addition, anyone with a hunting license may kill a furbearer for personal purposes at any time. In some states, trappers work with wildlife officials to share population data and improve techniques for humanely capturing animals. In Texas, TPWD gathers some data on trends in furbearer numbers in the state during nocturnal roadside surveys and in questionnaires occasionally sent to trappers but does not have a systematic furbearer harvest monitoring system.

The remaining terrestrial mammals in Texas are classified as nongame. Unless a nongame animal is listed as threatened or endangered, they can be harvested or collected live by anyone with a hunting license. TPWD sets some collection limits for nongame mammals and prohibits collection from the wild for commercial sale for some species.

Some of the more significant or unique orders of Texas mammals are explored below, including a discussion of whether they are classified as game animals or furbearers.

UNGULATES - ORDER ARTIODACTYLA

The *white-tailed deer* is the most abundant and popular big game animal in Texas; however, in the early 1900s they were almost gone from many parts of the state due to overhunting. Now, successful conservation efforts, including regulated harvests, transplantation of deer to depleted areas, and elimination of the New World screw-worm—a parasite of livestock and deer that has, unfortunately, recently threatened a reappearance—have produced healthy populations statewide. The state's whitetail population is estimated to be about five million. In 2023, 750,000 deer hunters harvested about 680,000 white-tailed deer, providing an estimated $1.2 billion in benefits to the Texas economy.

Whitetails are named for their fluffy white tails that they lift as they bound away from danger. They are habitat generalists that prefer edge habitats that provide both woodlands and open areas. They feed primarily on tender woody browse and forb species. Their lifespan is six to eight years. Deer behavior (and hunter behavior) are guided by the deer breeding cycle. In the fall breeding season, called the "rut," bucks (male deer) begin to compete with each other, using

Figure 13.6. White-tailed deer does give birth to one or two fawns in late spring/early summer in Texas. Three or more fawns rarely occur. Photo: Mark Alan Storey.

their antlers to drive away competitors from herds of **does** (female deer). As the breeding season ends in mid-winter, bucks lose their antlers and begin to form bachelor herds As pregnant does approach the end of their seven month gestation period in late spring/early summer, they become more solitary, leaving the doe herd to give birth to one or two fawns. The female hides the fawns in thick vegetation for about the first two weeks of life, returning to nurse them. As the fawns begin to follow the does, they again start associating with other females and their young. Deer may breed in their first autumn, but most do not until their second year.

Mule deer are named for their large ears. They are larger than whitetails, and their tails have a black tip. Mule deer often occupy arid, rough habitats and are noted for their bouncing way of moving through the broken terrain. They are primarily browsers but may also frequently feed on succulents such as cacti. Mule deer are most abundant in the Trans-Pecos. More recently, their numbers have increased in the Texas Panhandle, perhaps due to cropland conversion to native habitat through the Conservation Reserve Program. Interestingly, some hybridization between mule deer and white-tailed deer occurs where their ranges overlap, but hybrids make up a low percentage of the population. Mule deer population numbers often fluctuate with the effects of drought and habitat management. Today, biologists are worried about an emerging threat that could affect long-term population numbers—Chronic Wasting Disease (Highlight 13.2).

HIGHLIGHT 13.2 – CONSERVATION CHALLENGE - CHRONIC WASTING DISEASE

Chronic wasting disease (CWD), a neurological disease similar to mad cow disease, was first detected in wild deer in Colorado in 1981. It is now present in more than 30 states. It was first detected in Texas in 2012 in mule deer on the New Mexico border. Though not yet widespread in the wild in Texas, free-ranging mule deer, white-tailed deer, and introduced Rocky Mountain elk populations have tested positive. It has also shown up numerous times in captive deer breeding facilities. The risk to humans is yet to be determined, but CWD has a 100% fatality rate in infected deer.

Traditionally, CWD has been confirmed by sampling brain tissue or lymph nodes, although ongoing work is examining other tissue sample options. TPWD requires hunters who hunt in specified containment zones to report to a TPWD check station within 48 hours of harvest. In addition, TPWD urges voluntary sampling of hunter-harvested deer outside of the CWD zones and can provide test kits to hunters. Hunters and outdoor enthusiasts are encouraged to be observant of sick deer in the wild. Symptoms of infection include emaciation, excessive salivation, lack of muscle coordination, difficulty in swallowing, and excessive thirst and urination. In the field, sick animals may have an exaggerated wide stance, stagger, carry their head and ears lowered, and often consume large amounts of water.

Figure 13.7. Mule deer, found in the western part of Texas, can be recognized by their large ears, black tail tip, and antlers that can have secondary forks. Photo: Bala, CC 2.0.

Figure 13.8. Bighorn sheep are a charismatic species that occupy some of the most rugged habitats in the state. Photo courtesy TPWD © 2025, (Mike Pittman, TPWD).

Desert bighorn sheep are a restoration story still in progress, as described in Chapter 9. Named for the impressive curved horns found on adult males (females have smaller horns), they live in small, scattered groups on the rocky slopes of West Texas, feeding on desert scrub plants such as sotol, ocotillo, and the fruits of yucca and prickly pear. Although listed as a game animal, very few permits are issued each year as the recovery process in the Trans-Pecos continues and biologists seek to address emerging disease issues.

The pronghorn, another big game animal, is the fastest land animal in North America. They can reach running speeds over 40 miles per hour and maintain that speed for several miles. In addition to speed, they are known for their keen eyesight. They feed primarily on forbs, including a preference for flowers and fruits. Once found across prairie habitats of the Great Plains, they are now restricted in Texas to the Trans-Pecos, Panhandle, and southern Rolling Plains ecoregions. Even where they occur, pronghorn have experienced population declines, with populations vulnerable to drought and degraded habitat conditions. In addition, pronghorn do not jump fences, so net wire fences installed for sheep and goats inhibit their ability to move between feeding areas. Because of their vulnerable population numbers, pronghorn may only be harvested through permits issued to landowners.

Figure 13.9. Pronghorn are the fastest land animals in North America. Their large eyes set wide on the head are an adaptation to help them survive on the open plans. Photo 1: Dana Wright, TPWD. Photo 2: Abigail Linam Bradbury.

HIGHLIGHT 13.3 - HORNS VERSUS ANTLERS – WHAT'S THE DIFFERENCE?

Male deer have antlers, a bony structure that is shed every year. In the spring, the lengthening daylight stimulates antlers to begin growing from two bony bumps on the skull called pedicels. The antlers are initially covered with a soft tissue called velvet that provides nutrients and minerals to the growing antlers. Antlers complete their hardening and mineralization by late summer, and the velvet dries and sloughs off. Bucks use their mature antlers to drive away competitors during the breeding season, and the size of the antlers may help does assess the health of the male. After the breeding season is over in late winter, the antlers separate from the pedicel and fall off. Soon afterward, the growth of the following year's antlers begins. The size of antlers is affected by several factors, including the buck's age and genetics and the quality of the habitat.

Antlers are characterized by a branched growth form, producing the numerous tines or "points" that hunters count. Mule deer may have secondary forks in the tines; white-tailed deer do not.

Horns are found on bison, desert bighorn sheep, and many domestic livestock. Horns are unbranched and are not shed. Composed of a bony core covered with keratin (the same substance that makes up fingernails and hair), the horns grow throughout the animal's life. Both male and female bison and sheep have horns, though they are smaller on the female.

Pronghorn have modified horns. They have a bony core covered by a keratin sheath with the characteristic prong shape. Each year, the keratin sheath is shed and replaced by a new one.

The *collared peccary* (or *javelina*) is the final ungulate classified as a game animal. Peccaries have a pig-like nose, spiky coat, and prominent triangular tusks. Still, despite their appearance, they are not closely related to feral hogs. They can be distinguished from pigs by their smaller ears and tails. They are named for their "collar" of white and black hairs around the neck and shoulder that stands out from the rest of the darker coat.

Figure 13.10. Javelina, thought to be named for their javelin-like tusks, occur only in Texas, New Mexico, and Arizona in the United States.

Javelina are found in the Trans-Pecos, South Texas, Edwards Plateau, and portions of the Rolling Plains in brushy habitats where they feed on cacti and other plant materials. They often travel in herds of related family members called "squadrons." Although not usually aggressive, when injured or threatened, javelinas sometimes defensively clack their tusks in a threat display and occasionally bite.

ORDER RODENTIA

The order Rodentia is the largest group of mammals in Texas and the world. Rodents make up about 40% of the mammal species in the world and about 40% of terrestrial mammals in Texas. They are known for gnawing and are recognized by their incisors. Their single pair of incisors never stops growing. The front of the incisors is coated with a hard enamel that is often orange. In contrast, the backs of the incisors lack enamel. As a result, the incisors are worn into a chisel shape. Texas has 73 species of rodents, including five introduced species, spanning many sizes and niches. The tiny grasshopper mouse is a fierce predator, sometimes feeding on small vertebrates in addition to grasshoppers. On the other hand, the largest Texas rodent, the American beaver, which can weigh up to 50 pounds, is a strict herbivore, feeding on leaves, twigs, and the inner bark of riparian trees. TPWD classifies some rodents (squirrels) as game animals, some as furbearers, and some as nongame.

Figure 13.11. The fox squirrel is the most widespread of Texas squirrels and is well-adapted to human habitats. Photo: Mark Alan Storey.

There are ten species in the squirrel family in Texas. Sciuridae is an interesting and diverse family that includes tree squirrels, ground squirrels, prairie dogs, chipmunks, and flying squirrels. Hunting seasons are established only for the eastern gray squirrel and the eastern fox squirrel. Eastern gray squirrels, sometimes called cat squirrels for a catlike call they make, are found in hardwood forest habitats in the eastern third of the state, where they use cavities in dead and mature trees for shelter. Eastern fox squirrels occur throughout Texas, except for the Trans-Pecos and the extreme western edge of the Panhandle. They are common in cities as well as in a variety of woodland habitats. Both species sometimes construct sleeping "nests" made of leaves in trees.

Squirrel numbers vary over time as they are vulnerable to changes in food supplies and weather. Spring frosts, summer droughts, and fall rains combine to affect the production of food each year, producing some good and bad mast years. In addition, oaks are cyclical in producing acorns, resulting in some years with an abundance of acorns and some years with nearly none. Although squirrels stash acorns when abundant, these food fluctuations and major ice storms cause fluctuations in squirrel populations. Scientists monitor squirrel populations through transect counts at selected locations. One technique involves walking one-mile lines, sitting and observing squirrels for 15 minutes at six stations along each line. The counts are usually conducted at least twice on consecutive mornings.

The American beaver and common muskrat are two aquatic rodent species that are classified as furbearers. Beavers and muskrats have a soft, thick underfur covered by long guard hairs to keep the skin dry. Although beavers were once heavily harvested across North America, they have made a comeback, including in Texas. They can now be found in most parts of the state that have perennial water. Beavers use their sharp incisors to cut trees, often mounding the cut logs to dam streams, creating habitat for themselves and other aquatic organisms. For shelter, they may burrow or build a lodge, a mounded structure built from logs for eating, sleeping, and raising young. In addition to dams and lodges, beavers can be detected by the gunshot-like sound of the slap of their paddle-shaped tails on the water, given as a warning of danger.

Muskrats also shape the aquatic ecosystems around them. They once occurred in large colonies in coastal and riparian wetlands, building houses of cattails and reeds in marsh habitats. As muskrat populations increase, they create "eat-outs"—large openings in marsh vegetation that allow feeding and movement by other aquatic species, such as waterfowl or alligators. When the vegetation becomes too scarce due to their grazing, muskrats move to another location, allowing plant succession to occur in the eat-out. Although trapping of muskrat was once intensive, high reproductive rates ensured that the harvest was compensatory and had little effect on muskrat numbers. Now, the species is much less widespread, primarily due to the loss of wetland habitat.

Gray and fox squirrels are game animals, and muskrats and beaver are furbearers, but the vast majority of rodents are nongame. Despite some people's aversion, these diverse species, including rats, mice, gophers, ground squirrels, flying squirrels, prairie dogs, and a host of others, are important components of ecosystems. They form the prey base in many food chains, help to disseminate plant seeds, create habitat, and aerate the soil. Although some rodents are pest species, especially the introduced Norway rat, roof rat, and house mouse, native rodent species are critical to Texas ecosystems.

Figure 13.12. The value of beaver pelts spurred exploration of North America. They are also valuable ecosystem engineers. Photo 1: Gilbertfilion, CC 4.0. Beaver dam: Mark Mitchell.

Figure 13.13. The North American porcupine is another species of rodent. Found across much of the western half of Texas, they feed on bark, roots, and other plant material. Photo: Mark Mitchell, TPWD

Figure 13.14. The northern river otter is famous for being playful, but it is also a fierce aquatic carnivore. Photo: Mark Alan Storey.

CARNIVORES - ORDER CARNIVORA

Texas is home to 20 native species in the order Carnivora. An additional eight species once occurred here but are now extirpated. Carnivores are a diverse group, and despite their classification, many members of the order are omnivorous, including fruits and invertebrates in their diet. Even with this variation in diet, members of this group can usually be recognized by their prominent canine teeth.

The badger, mink, otter, raccoon, ringtail, three species of foxes, and five species of skunks are carnivores classified as furbearers in Texas. Trends for these species vary. The northern raccoon, common gray fox, and striped skunk are widespread and abundant, thanks in part to the generalist habits of these species. American mink, kit fox, swift fox, and the other species of skunk are less stable due to more restricted distribution and habitat loss. On the other hand, the playful northern river otter and quarrelsome American badger seem to be increasing in number in Texas, with otters recently moving back into some portions of the state where they had disappeared.

Most carnivores are classified as furbearers, but several are classified as nongame. These include mountain lions, bobcats, and coyotes. The mountain lion, also called cougar or puma, has the widest distribution of any wild cat, ranging from Canada to South America, and once had a statewide distribution in Texas. Mountain lions can weigh up to 150 pounds and feed primarily on white-tailed and mule deer. Cougars were hunted intensely as settlers brought livestock, especially sheep and goats, into Texas. As a result, today mountain lions are primarily restricted to the Trans-Pecos and the brush country of South Texas, where large ranches and tracts of public land provide adequate space for their large home range movements. In recent years, mountain lions seem to be expanding their range in the Edwards Plateau, where an abundant prey base of white-tailed deer exists. Other scattered sightings occur in the Panhandle and North Texas. TPWD recruits reports of mountain lions by citizen scientists to better track their distribution and displays data from confirmed sightings on a public Geographic Information System webpage.

Bobcats are smaller felines, weighing up to 35 pounds. They are found throughout Texas. They may range up to five miles each night as they seek their primary prey: rabbits and rodents. They are sometimes harvested for their pelt, but because their coat is similar to endangered cats, hunters must obtain a special tag issued by the Convention on International Trade in Endangered Species (CITES) to sell the pelt. Both bobcats and mountain lions sometimes cache their prey. They may drag the carcass long distances and cover it with leaves, branches, or dirt for feeding at a later time.

Coyotes are another widespread carnivore. They are very adaptable in diet and habitat, sometimes even feeding on fruits and living in urban areas. This adaptability has enabled them to extend their range throughout the state, even outcompeting the larger red wolf that once occurred in East Texas. Coyotes often hunt alone, but they are social animals, forming packs that communicate by vocalization and scent marking. Like foxes, coyotes are also sometimes harvested for their pelts. Even when the fur is not harvested, they are frequently killed because of their perceived threat to livestock.

The largest extant carnivore in Texas is listed by TPWD as threatened, but it is slowly making a comeback. Black bears once occurred across Texas, but by the 1950s they were considered extirpated in the state. In 1988, a visitor to Big Bend National Park sighted a mother bear and two cubs, suggesting that a breeding population might be re-establishing in the park, coming from mountains in Mexico. Since that time, the population in the park has increased to 50 to 100 bears, and sightings have increased in the Trans-Pecos, South Texas, and western Edwards Plateau. In addition, sightings of a

Figure 13.15. Coyotes are adaptable predators. Though several counties and organizations in Texas offer bounties to try to reduce their numbers, they play an important role in the balance of ecosystems.

different subspecies, the Louisiana black bear, are increasing in forest habitats in Northeast Texas. As the good news about black bear recovery builds, TPWD is engaging communities and livestock producers about ways to live comfortably with the presence of this large omnivorous carnivore through collaborations such as the Texas Black Bear Association.

Figure 13.16. A game camera documents the presence of a mother black bear with three cubs in West Texas. Photo: Bonnie McKinney

RABBITS – ORDER LAGOMORPHA

The order Lagomorpha includes five Texas species. Most easily recognized by their saltation, or hopping movements, the skulls of lagomorphs are also distinctive. They have one primary set of incisors that are ever-growing like rodents and a second small peg-like set behind the first. There are two major groups of lagomorphs—the rabbits and hares. Rabbits, which include the eastern cottontail, Davis Mountains cottontail, desert cottontail, and swamp rabbit in Texas, give birth to altricial young that the mother keeps in a shallow "nest" on the ground that she lines with her fur. Hares, including the black-tailed jackrabbit in Texas, give birth to precocial young, ready to hop and follow the mother soon after birth. Both rabbits and hares may produce several litters per year (leading to the phrase "breeding like rabbits"), but mortality rates for the young are high.

Rabbits are not officially listed as game animals, but regulations allow them to be widely hunted. Licensed hunters may hunt them year-round by any legal method on properties that allow it. Lagomorphs can be counted during roadside surveys to monitor their population status.

Figure 13.17. The black-tailed jackrabbit and eastern cottontail occur across the state, but they have different reproductive and habitat characteristics. Jackrabbit photo: Jeff Forman, TPWD. Cottontail photo: Mark Alan Storey.

Figure 13.18. The Virginia opossum is the only marsupial that occurs in North America. Photo: Monica R, CC 2.0.

OPOSSUMS - ORDER DIDELPHIMORPHIA

The Virginia opossum is the only marsupial that occurs in the United States. Famous for evading predators by feigning death ("playing possum"), its distribution is statewide in Texas except in the most arid portions of the Trans-Pecos and High Plains. As a marsupial, females give birth to up to 25 tiny young that climb into the mother's pouch to complete development. However, with only 13 teats in the pouch, mother opossums can raise only that many babies. After emerging from the pouch in about five weeks, the young frequently ride on the mother's back, clinging to her fur. Opossums are omnivores and scavengers. They are easily recognized by their grizzled coat, scaly tail, and skull with many sharp teeth.

ARMADILLOS – ORDER CINGULATA

Cingulata, the armor-plated mammals, includes 22 species of appropriately-named armadillos. The Mexican long-nosed armadillo (formerly called the nine-banded armadillo) is this group's only North American representative. Over the last century, armadillos have greatly expanded their range. In 1900, Vernon Bailey reported few armadillos north of the Colorado River in Texas. Since then, they have continued their spread across the entire state, northward as far as Kansas, and eastward to the Carolinas. Armadillos usually occur in sandy habitats, where they use their sharp claws to dig burrows for shelter and small holes in search of grubs and other invertebrates.

The unique characteristics of armadillos are many. They have poor eyesight and hearing; instead, they use their sense of smell to locate food. When startled, they may surprisingly jump more than three feet into the air (unfortunately, leading to many being killed on roadways). They have little fur, and their body covering of keratin, bone, and collagen does not provide much protection from heat or cold. In response, they are nocturnal during warm weather and diurnal during cold weather. They sometimes walk across water bodies on the bottom. However, if the water body is large, they may ingest air and float on the top! Armadillos also have a unique reproductive strategy. Mating occurs in the summer, but the young do not develop until winter, a system called "delayed implantation." Four young identical quadruplets are born in a burrow in March. Because armadillos are susceptible to leprosy, they are used in laboratory research on that disease.

Armadillos are popular in legend and myth in Texas. Unfortunately, some concern has been raised about population trends in the state. In addition to the frequent road mortality, mammalogists have suggested that Texas's explosive feral hog populations may feed on young armadillos.

BATS – ORDER CHIROPTERA

Although several mammal species, such as southern flying squirrels, can glide, bats are the only flying mammals. Chiroptera means "hand-wing," and members of this group have wings formed from membranes stretched across their finger bones. Texas is home to 33 species of bats, more than any other state. `Bats are found in every county in the state and use a variety of structures for roosting and reproduction, including caves, tunnels, mine shafts, bridges, hollow trees, palm fronds, and old buildings.

Bats play an essential ecological role in Texas. Most are insectivorous, catching

Figure 13.19. Bats exiting Bracken Cave near San Antonio can travel over 100 miles in a single night eating insect pests. Photo: USFWS, CC 2.0

insects in flight using an echolocation system, sending out ultrasonic clicks and receiving the reflected sound with their complex ear and nose structures. According to a 2011 study published in *Science*, bats save farmers in Texas $1.4 billion annually by eating crop pests. Other bats have different diets. Mexican long-nosed bats and Mexican long-tongued bats serve as pollinators while feeding on nectar, pollen, and fruit. There is also one record of the hairy-legged vampire bat in Texas, a species that bites its victims (primarily roosting birds) and then laps up the blood.

Bats often have an unappealing reputation, but Texans have grown to love bats. The state is home to over 100 million Brazilian free-tailed bats, including the world's largest bat colony—Bracken Cave near San Antonio, which is a summer maternity colony for about 20 million free-tailed bats. The Bracken Cave bats are so numerous that when they emerge at night, they can be seen on the weather radar. At dusk on summer nights, people gather at numerous locations, including Bracken Cave, Old Tunnel WMA, and the Congress Avenue Bridge over the Colorado River in Austin, to watch bats begin their feeding flights. In addition, many individuals around the state choose to create bat habitat on their property by building boxes, towers, or even caves designed to provide roosting habitat for these insect-eaters.

As temperatures have warmed, the diversity of bat species in Texas has increased, with many species expanding their range northward from Mexico. In addition, mammologists have documented several Texas species that have extended their range northward in the state.

Despite these expansions, bats face challenges in Texas and elsewhere. White-nosed syndrome, a fungal disease that spreads rapidly in bat roost sites, first appeared in Texas in 2020. The deadly disease has now appeared in at least 20 counties. There is also concern about bat fatalities associated with wind turbines. Some bat species seem drawn to wind turbines, where the blades may strike and kill them. Bat Conservation International, a nonprofit organization based in Texas, is researching operating methods and deterrents to reduce bat collisions.

HIGHLIGHT 13.4 - WHAT ABOUT RABIES?

Mention bats, and many people think of rabies. However, bats are just one type of wild mammal in the state that can carry the feared disease. Any warm-blooded animal can carry rabies. Outbreaks, called epizootics, have occurred in wild populations of foxes, coyotes, raccoons, and skunks, as well as bats. Because the canine forms of rabies are especially likely to be transmitted to dogs, the Texas Department of State Health Services (TDSHS) has carried out a vaccine program for foxes and coyotes by dropping oral vaccine baits in a South Texas buffer zone to control spread of that variant.

However, the presence of rabies in wild animals does not imply an immediate threat to humans. Only a tiny fraction of wild animals are likely to be infected with the rabies virus at any one time. People can lessen their risk by avoiding interaction with any wild animal that is behaving erratically. If a dead bat or other suspected rabies carrier is found, it should be handled only with gloves and submitted to TDSHS. Since the greatest risk to humans occurs when wild animals bite unvaccinated pets, owners should make certain their pets' vaccines are current. The TDSHS website provides guidance if a bite does occur.

EXOTICS

Many mammals have been introduced to Texas. One example is the nutria, a South American fur-bearing rodent now abundant in many aquatic habitats in the state. Red foxes are another successful non-native species. Introduced to the U.S. in 1895 for sport, they now range across much of Texas except South Texas and far West Texas. There was even a small population of Japanese macaques that escaped from captivity and established a wild population in South Texas! (These monkeys have now all been captured and placed in a wildlife sanctuary.)

Most introduced mammals are ungulate species brought in from Africa and Asia for hunting. The Exotic Wildlife Association estimates that over one million exotics, comprised of at least 135 species, reside in Texas. *The Mammals of Texas* assesses that at least seven ungulate species have established stable free-ranging populations. Examples include nilgai antelope, axis deer, aoudad sheep, sika deer, fallow deer, and blackbuck antelope. Because they are non-native, they may be hunted at any time of year with the landowner's permission. There are concerns that these introduced animals may compete and hybridize with native species, spread disease, and impact native plants. In addition, many landowners choose to erect high fences for management of exotic ungulates or even for white-tailed deer. These fences are barriers to the movement of many wildlife species and limit genetic exchange.

Figure 13.20. Asian and African ungulates, such as greater kudu, are widely introduced in Texas on hunting ranches. Photo: Mark Mitchell, TPWD.

Figure 13.21. Feral hogs have a high reproductive rate, leading to high populations and habitat damage in many parts of the state. This photo of wild pigs marks some of their differences with the native collared peccary. Photo: Mark Mitchell.

Feral hogs are perhaps the most notorious exotic animal in Texas. Also called wild pigs, they are the most abundant free-ranging hooved animal in North America, with over 2.6 million in Texas alone. Feral hogs are descended from escaped domesticated pigs and introduced Russian wild boar. They have increased rapidly in Texas and throughout the United States in the past half-century. According to biologists at TPWD, the expansion of feral hogs in Texas is due to intentional releases, improved habitat, wildlife feeding, disease eradication in domestic pigs, limited natural predators, and high reproductive potential.

Although popular among some hunters, feral hogs can cause severe damage to habitats, wildlife, and agriculture. They are omnivores. As they wallow and root for food, they erode the soil, muddy streams, disrupt native vegetation, and make it easier for invasive plants to take hold. They also feed on any animal life they encounter on the ground, such as reptiles, bird eggs, and sea turtle eggs, and even occasionally prey on fawns, young lambs, and kid goats. They root up crops and tree saplings in timber plantations. Their destructive activities cost more than $780 million every year.

Feral hogs are challenging to control. They breed year-round, sometimes producing two litters of 4 to 12 piglets yearly. Although they may be hunted and trapped all year, day or night, by anyone with a hunting license, they are intelligent animals, often learning to avoid hunters and traps. Some land managers use aerial gunning from a helicopter. All these techniques only help keep numbers down, and, despite the development of new strategies such as sterilization and targeted poisoning, feral hogs have yet to be controlled successfully.

In addition to feral hogs, several other domesticated mammals roam the wild in Texas. Feral cats pose a threat because of their impacts on songbird, lizard, and rodent populations. Packs of feral dogs can also harass and kill native wildlife. In the Big Bend region, high numbers of wild burros cause damage in riparian areas. Control of feral animal populations is often difficult because of public sympathy for these familiar species.

THREATS

A recent United Nations Environment Program report estimated that almost a quarter of the world's mammals could face extinction by the end of this century. Texas has already lost some fascinating mammals. The gray wolf, red wolf, jaguar, margay, black-footed ferret, grizzly bear, and perhaps others are all extirpated in the state. (Interestingly, some red wolf DNA remains in coyote populations on Galveston Island.) Once found in Texas, the Caribbean monk seal and Merriam's elk are completely extinct. Bison have been reduced from millions of free-ranging animals to a few small managed populations. The major contributing factors to these declines were overharvest, predator control, habitat loss, and habitat fragmentation.

Today, TPWD lists 65 terrestrial and 17 marine mammals as Species of Greatest Conservation Need. Of the terrestrial mammals, five species are listed by TPWD or FWS as endangered (black-footed ferret, ocelot, jaguarundi, jaguar, and Mexican long-nosed bat). TPWD lists an additional eight species as threatened in the state. This list includes the black bear, several bat species, white-nosed coatimundi, and some endemics, such as the Texas kangaroo rat and Palo Duro mouse.

According to *The Mammals of Texas*, many species in jeopardy share characteristics that make them vulnerable to extinction. Species such as bats have low reproductive rates, so they are slow to recover from population declines caused by habitat destruction or disease. Large species with large home ranges require large tracts of habitat that may no longer exist. Other species have a limited geographic range that can make them vulnerable to extinction. Finally, specialists, such as black-footed ferrets that rely on prairie dogs for food, are vulnerable to changes in those resources.

CONSERVATION AND MANAGEMENT

Providing habitat, avoiding the introduction of non-native species, and sustainable harvest are the keys to conserving most mammal species. Research shows that larger habitat patches and more diversity of habitats result in more mammalian diversity. Microhabitat diversity (variations in habitat at a small scale) is also important for small mammals. For example, rodent diversity increases with greater levels of tree and shrub density, leaf litter, and canopy cover. These more complex habitats provide food, structure for raising young, protection from predators, and moderation of cold and heat. Aiming for habitat at a variety of successional stages can thus help small mammal populations.

Conservation of large carnivores offers both biological and sociological challenges for the wildlife scientist. Species such as black bear and mountain lions require large habitat areas with adequate food and shelter sources. At the same time, many ranchers express concern about the potential impact of large carnivores on their livestock operations. Wildlife biologists can help landowners identify potential predators and investigate livestock protection strategies such as fences, corrals, guard animals, lights, noises, and scents that can help livestock co-exist with wild carnivores.

Understanding sustainable harvest is also important in the management of mammals. To create a healthy, sustainable game population and still provide healthy habitat for other wildlife species, wildlife managers should understand the density, sex ratio, and harvest goals for focal species. Thinking of white-tailed deer as a focal species, the ideal deer density depends on the habitat's carrying capacity. It is affected by rainfall, soil and vegetation types, grazing, and other land use practices. Ideal densities may range from 10 acres per deer in productive habitats up to 50 or more acres required per deer in other areas. When deer densities are too low, they may not support a sustainable harvest. When deer densities are too high, plant diversity decreases, wildlife diversity suffers, and deer health suffers.

Hunters are often interested in harvesting the largest and most impressive males, but a healthy deer herd requires a balanced sex and age ratio. Usually, 2.0 to 2.5 does for every buck is an ideal ratio. If populations are low, a better target might be three to four does per buck. On a large property managed explicitly for trophy whitetails, wildlife managers might aim for a 1:1 sex ratio; however, managers must provide quality habitat and carefully monitor harvests to ensure the population maintains itself. Sex ratios and fawn counts can be estimated during transect surveys conducted in August or September.

Harvest goals can affect the composition of a deer herd. In general, harvesting about one-third of the bucks in a population allows a balance between harvest numbers and deer quality. One important consideration is to leave a portion of the quality bucks each year. Deer antler size is primarily affected by the nutritional value of native plants in the habitat and genetics (see Highlight 13.5). The number of points does NOT reflect the age of the deer. Removing "spike" bucks (bucks that do not have branched antlers) helps improve the overall genetics of the herd. Harvest levels for does should be based on the goal for the population size. Harvesting more does can help reduce overpopulation; harvesting fewer does can help the population size increase.

The successful comeback of the white-tailed deer in Texas is proof that, with careful management, harvest can be part of a sustainable conservation program.

HIGHLIGHT 13.5 – BERGMANN'S RULE

Interestingly, everything is not "bigger in Texas." In general, whitetails, like most mammals, are larger the farther north the population is. (Whitetail bucks may weigh more than 400 pounds in Minnesota, Manitoba, and Ontario!). This ecological phenomenon, based on adaptations to minimize heat loss in northern climates, is called **Bergmann's Rule.** Though wildlife breeders may try to import larger deer or select for larger body size, in good habitat Texas deer tend to reach an ideal body size for the vegetation type and climate of the area.

CHAPTER 14 – TEXAS AMPHIBIANS

Herpetology, the study of amphibians and reptiles, is translated as the study of creepy, crawly things. Despite some people's aversions, amphibians are more significant than one might think—no, not as the sources of warts and princes but as a barometer of the health of the environments everyone shares. At an international conference in 1989, scientists from all over the world became alarmed at what appeared to be dramatic declines in some amphibian populations. Then, in 1995, a group of schoolchildren in Minnesota was the first to notice an alarming rate of malformed limbs in a frog population (see Highlight 14.2). Because amphibians are associated with wetland habitats and because they have permeable skin, ecologists believe that declines in amphibian populations and malformations may serve as early warning indicators of broader changes in ecosystems.

Texas has a diverse array of about 30 types of salamanders and over 40 frogs and toads, ranging in variety from the bizarre cave-dwelling Texas blind salamander in Central Texas to the bleating sheep frog of South Texas to the sausage-shaped amphiuma of East Texas streams. The greatest abundance and diversity occur in the relatively wet habitats of the eastern third of the state, but a look at the various families of amphibians gives a glimpse into the diverse life histories they have developed to live even in the most arid parts of the state.

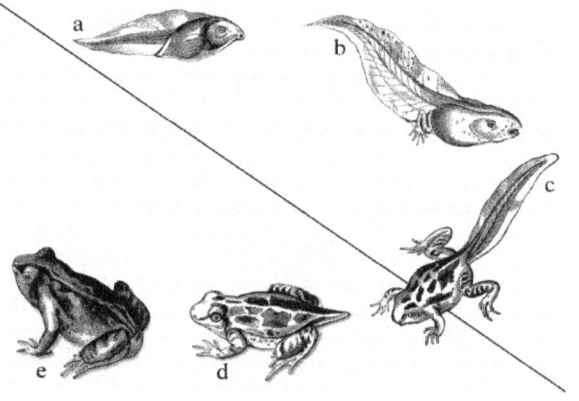

Figure 14.2. The process of metamorphosis in frogs. Source: Wikimedia Commons (public domain).

WHAT IS AN AMPHIBIAN? AND WHY CARE ABOUT AMPHIBIANS?

The word "amphibian" comes from a combination of two Greek words that translate to "life on two sides." It references the distinctive two-stage life history pattern that most members of this group exhibit. Most amphibians reproduce by laying eggs. These delicate eggs lack a shell or membranes to protect them from drying out. As a result, they must be laid in water, or at least moist situations. Most amphibian eggs hatch into free-swimming, gill-breathing larvae. Frog larvae (tadpoles) have gills inside a branchial chamber. Most have mouth parts hardened with keratin that allow them to scrape algae. In most species, these strictly aquatic larvae undergo a radical alteration of their body structure (metamorphosis), including the development of lungs, to live on land (Fig. 14.2). The general pattern applies only to "most" amphibians. Different groups of amphibians have developed different variations on this life cycle.

Aside from reproduction, amphibians are also tied to moisture because, unlike other terrestrial vertebrates, their skin is permeable to water. Amphibians can absorb water through the skin, and many frogs have a patch on their lower abdomen specifically designed to take in water. Adult amphibians also breathe through their skin (in fact, some terrestrial salamanders lack lungs completely!). Moist skin helps the gas exchange happen more efficiently.

Amphibians are organized into three orders—Anura, Caudata, and Apoda. Anura, which translates as "no tail," includes the frogs and toads. Caudata, which translates to "having a tail," includes salamanders, newts, and aquatic forms such as sirens. Apoda, which translates as "no feet," includes the

(opposite) Figure 14.1. The Texas toad is the official State Amphibian. Here, a male is using his vocal sac to produce courtship calls. Photo: Toby Hibbitts.

caecilians. This group of tropical species, not found in Texas, are fossorial or aquatic vertebrates with a worm-like appearance.

All three orders share the characteristics of a three-chambered heart and a cloaca—a single bodily opening for the digestive, reproductive, and urinary tracts. The skin of many species produces toxins that protect them from predation. All amphibians are **ectothermic**, regulating their body temperature using the external environment. They can also be described as **poikilothermic** – meaning their body temperatures are not held at one steady temperature but can vary greatly. Some species can function at very cool temperatures just above freezing, whereas the ideal body temperature for others needs to be above 85 degrees Fahrenheit. During cold weather, amphibians **brumate,** entering an inactive state similar to hibernation. During very hot or arid weather, they can **aestivate,** often burrowing to protect their bodies from drying out. They are primarily active at night for the same reason. These adaptations allow these cold-blooded animals to flourish on every continent except Antarctica, from the tropics to the Himalayas.

There are many reasons to care about amphibians. Their permeable skin allows them to serve as barometers of environmental quality in aquatic environments. Like canaries in the coal mine, amphibian decline may be an indicator of poor water quality. Amphibians are also a vital component of the ecosystems they inhabit. They are the dominant vertebrates in some wetlands—in biomass, the number of species, and/or the number of individuals present. Most larval amphibians eat algae—helping to clean aquatic systems. Others feed on mosquito larvae. Adult amphibians are carnivorous. One study determined that the cricket frogs in one pond ate more than five million arthropods (insects, spiders, and crustaceans) in one year. Amphibians are also crucial as prey in the food web. When amphibian populations decrease, the health and diversity of their predators, such as birds, fish, and reptiles, may decrease.

In addition, amphibians are directly valuable to humans. Medical researchers have used chemicals found in their skin to address drug-resistant bacteria, cancer, heart problems, and HIV. Because many amphibians can regrow limbs, doctors have studied their physiology for insight into how to regenerate tissue in humans. Frogs are eaten as a staple in countries like Burkina Faso (and states like Louisiana) and as a delicacy in others. Finally, frogs have inspired ancient artwork, occurring in drawings in ancient Peru and Egypt and in popular culture—think of Kermit the Frog!

MONITORING AMPHIBIANS

The most effective strategy for monitoring frogs and toads is based on their breeding activities. Male anurans have a vocal sac that is used to hold air and then push the air over vocal cords, producing a unique sound for each species (Fig. 14.3). When moisture is available and temperatures are appropriate, males call from wetland breeding habitats. (Most females are silent, other than distress calls when captured by predators.) In response, the female frogs approach the breeding habitat, seeking out males. The males may adjust the complexity and timing of the call when females or competitors are nearby. When a female encounters a male, the male grasps her in **amplexus** (a Greek word meaning "embrace"), a posture that has the appearance of the male holding the female and riding on her back. Eggs are laid, and fertilization takes place in the water.

Biologists or volunteers monitor frogs by listening for the calls of various species. "Call counts" are conducted at wetlands at least ½ hour after sunset, preferably within a day or two of rain. The observer listens for five minutes, noting all the different species heard. A call index value reflecting relative

Figure 14.3. Spring peepers call and breed during cooler winter temperatures in Texas. Here, a male is producing their typical clear whistle using his vocal pouch. Photo: jmeissen, CC BY-SA 2.0.

abundance is assigned for each species. A call index value of 1 represents individuals calling intermittently with space between calls. A call index value of 3 represents a full chorus of that species, with calls constant and overlapping. Under the right conditions at a springtime frog pond, a cacophony of grunts, trills, chirps, and whistles can be almost deafening. Call counts should be conducted at least three times, spanning late winter, spring, and summer, as different species are more active at different temperatures. TPWD provides a "Herps of Texas: Frogs and Toads Found in Texas" webpage that links to the calls of nearly all the anurans in the state.

Although it is possible to encounter salamanders traveling to breeding ponds in late winter or spring, some species do not lay eggs in water, and breeding activity for salamanders is less predictable. Time-constrained searches can provide a better index of abundance. In time-constrained searches, researchers explore habitat for amphibians, turning over logs on land or rocks in the water. The logs provide moist microhabitats for salamanders, and some aquatic species hide under rocks. All objects moved should be carefully replaced in their original position. The number of searchers is multiplied by the amount of time spent searching and then divided by the number of animals found to provide an abundance index for each species.

Biologists can also inventory amphibians by providing habitat structures. A grid of coverboards—12" x 12" squares of untreated plywood placed in contact with the ground in a grid pattern—can attract both amphibians and reptiles. After one to two weeks, observers revisit the coverboards, lifting them carefully to check for animals underneath. Treefrogs can be attracted to PVC pipe habitats. A length of 1" to 2" diameter PVC is cut and then stuck into the ground or

Chapter 14 – Texas Amphibians | 239

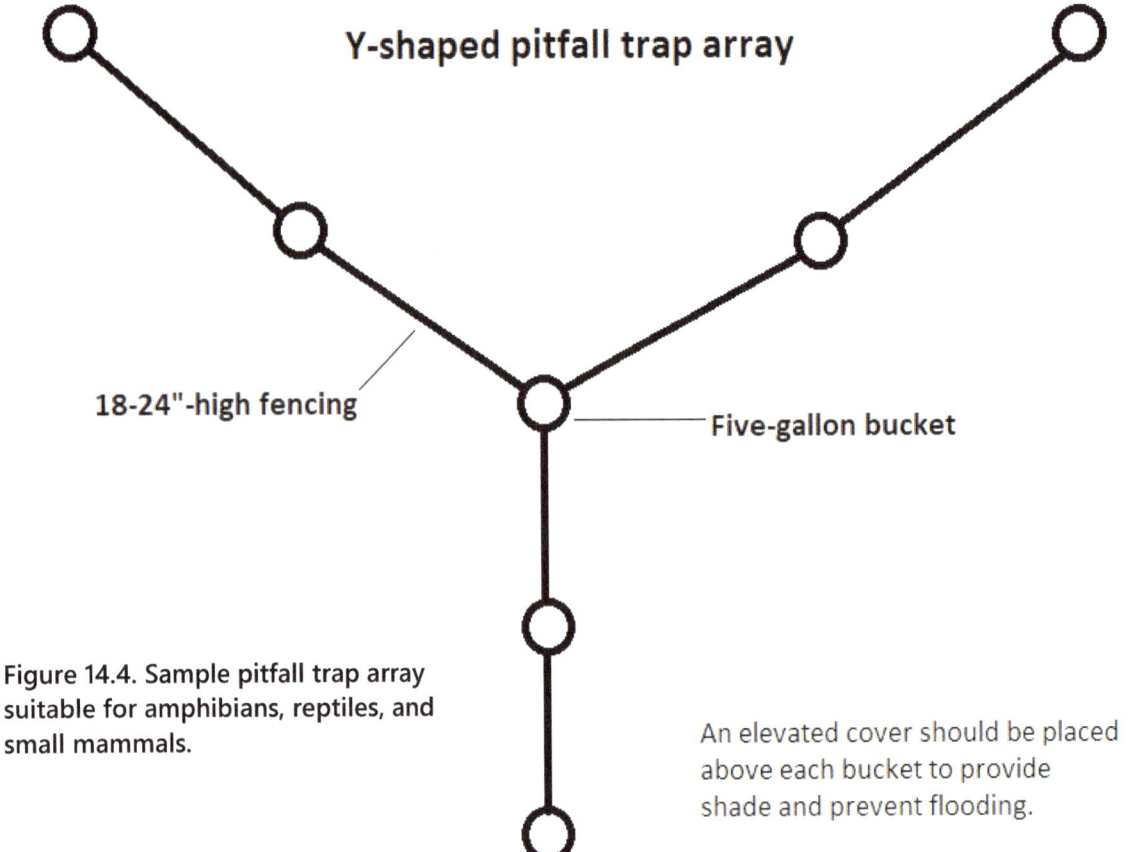

Figure 14.4. Sample pitfall trap array suitable for amphibians, reptiles, and small mammals.

attached to a tree trunk in moist habitats where treefrogs are expected. The treefrogs may crawl into the pipes to take cover.

A mark-recapture project can provide a more accurate population estimate. Baited minnow traps and seines can be used to catch aquatic salamanders. For toads, pitfall traps can be constructed. Five-gallon buckets are buried in the ground along a transect or in an X or Y pattern (Fig. 14.4). Fencing made of metal flashing or mesh is installed between the buckets to direct toads toward the "pits." Any trap must be checked regularly and is most effective if left in place for one to two weeks. A variety of marking techniques exist. The injection of a PIT tag or a colored polymer is the most common, but non-invasive methods also exist. Remarkably, many individual amphibians can be identified upon recapture using photographs because of their color patterns.

DIVERSITY OF TEXAS AMPHIBIANS

FROGS AND TOADS

Texas's frogs and toads comprise seven to eight families. (Taxonomists, scientists who study relationships among organisms, frequently revise species classification and family groups based on new insights.) Although no strict definition exists, species called "frogs" are generally associated with moist and humid habitats. On the other hand, species adapted to withstand more arid environments are often called toads. A glimpse at the different families gives a sense of how frogs and toads use their habitats differently.

Table 14.1. Frogs and Toads in Texas (Order Anura).

Family	Common name	# Species
Ranidae	True Frogs	8
Hylidae	Treefrogs	12
Bufonidae	True Toads	9
Microhylidae	Narrowmouth Toads	3
Scaphiopodidae	Spadefoot Toads	4
Rhinophrynidae	Burrowing Toad	1
Eleutherodactlyidae or Leptodactylidae	Tropical or RainFrogs	6

Family Ranidae – The True Frogs

Ranidae provides competitors for frog-jumping contests and frog legs for eating. Ranids are often called the "true frogs" because they have the stereotypical frog shape and behavior—long legs, smooth skin, and incredible leaping ability. In addition, members of this group follow the typical frog life cycle. They lay eggs in permanent water bodies. The young tadpoles hatched from the eggs are cryptically colored and tend to hide in vegetation near the bottom of the water body. After a lengthy larval period (up to one year for bullfrogs!), metamorphosis follows the characteristic pattern. As the tail is resorbed, hind limbs develop, followed by front limbs. Lungs replace internal gills. The adults can then fully emerge on land, though they tend to stay close to water.

Texas has eight species in the family Ranidae, plus one species, the northern leopard frog, that was extirpated from its limited range in El Paso County. Familiar ranids include the American bullfrog, the largest frog in North America with a body length of up to eight inches, and three species of leopard frogs—the familiar medium-sized spotted frogs of ponds and streams. Bullfrogs emit a deep, booming sound. The southern, Rio Grande, and plains leopard frogs all put their specific twist on a chuckle-like call. Interestingly, with a long tadpole stage, leopard frogs call and lay their large round clumps of eggs all year round. The crawfish frog is a true frog that is a Species of Greatest Conservation Need. It lives in crawfish burrows in the Pineywoods and Gulf Coast Prairies but is rarely seen or heard.

Figure 14.5. Ranids, like this southern leopard frog, have long legs and smooth skin. They are usually found near permanent water.

Figure 14.6. Gray treefrogs have the expanded toe pads typical of treefrogs, but the widespread, smaller cricket frog does not.

Family Hylidae – The Treefrogs

Members of Hylidae also stick close to moisture. Treefrogs in the genera *Hyla* and *Dryophytes* are known for their enlarged toe pads that allow them to climb and hang on to slick vertical surfaces, such as cattails, tree trunks, and even glass windows. Texas is home to five treefrog species, ranging from the canyon treefrog found in pockets of moist riparian habitats in the Trans-Pecos to the green treefrog, whose quonk-quonk-quonk calls fill summer nights in Gulf Coast marshes. Treefrogs are known for their ability to shift pigments in their cells to change colors in response to light intensity, background color, and temperature. Color-shifting is especially evident in the gray treefrog and Cope's gray treefrog. These two species are identical in appearance but differ in the number of sets of chromosomes they carry.

In addition to the typical treefrogs, seven other members of Hylidae occur in Texas, including the chorus frogs and cricket frogs. The diminutive chorus frogs inhabit a wide variety of habitats and breed across a wide range of temperatures, including the Strecker's chorus frog that breeds primarily in the winter. Blanchard's cricket frogs are some of Texas's most widely distributed anurans. Their range covers most of the state. Though small, with warty, variably-colored skin, their cricket-like calls can be heard many yards away. The call has also been described as sounding like two marbles clicking together.

Family Scaphiopodidae – The Spadefoot Toads

The spadefoot toads might be considered the anuran opposites of the true frogs. Named for hardened tubercles on the hind legs that allow them to burrow backward in the soil, these toads are specialized to survive in arid conditions. There are four species of spadefoot toads in Texas, all adapted to burrowing underground during dry conditions. They may remain underground for months, even up to a year, with some species able to form a cocoon made of their shed skin cells to protect them from drying out. Their skins have toxic secretions that some people describe as smelling like garlic or peanut butter, but licking is not recommended!

The sound of raindrops during heavy seasonal rains prompts spadefoot toads to emerge from their burrows. Males head to a puddle of water to call to the females (the Couch's spadefoot and Hurter's spadefoot emit a short, strained moan-like grunt—quite eerie when a large group is gathered calling). This "explosive" breeding event may bring hundreds of toads together over a one- or two-night period. The eggs hatch within a few days. Because predators are absent in the temporary water bodies, the tadpoles do not spend time hiding but are active and grow rapidly. Metamorphosis may happen in as few as 10-14 days. Most spadefoot tadpoles are omnivores, but if the habitat begins drying up, Mexican spadefoot tadpoles show an ability to switch to a carnivorous diet, even changing body form and becoming cannibalistic. After a few days to weeks of feeding above ground, the adults are ready to burrow again as conditions dry. The whole life cycle is an impressive adaptation to surviving as an amphibian in an environment with intermittent rain and short-lived wetlands.

Figure 14.7. The best time to find Couch's (and other) spadefoot toads is in the first few nights after the first heavy rain of the spring.

Family Bufonidae – The True Toads

Another widespread group that can survive in arid conditions is the family Bufonidae. Commonly called the "true toads," these are the epitome of the storybook toad—anurans with warty skin, short legs, and bulging eyes. The warts of toads do not cause human warts, but they do protect toads from predators. Bumps on the skin and a larger gland behind the eye called the parotoid gland secrete distasteful substances called bufotoxins. (A dog that foams at the mouth after capturing a toad has just gotten a good taste of bufotoxin.) Like spadefoot toads, bufonids readily retreat underground or into sheltered crevices when conditions are dry. Likewise, they often use temporary water bodies to breed after rain, and the tadpoles metamorphose relatively quickly. Unlike spadefoots, bufonids have horizontal pupils.

Texas has nine species of true toads. They can be identified by their distribution, color, parotoid gland shape, and bony crests on the head (cranial crests). Males may be distinguished by the color of the vocal sac under the chin. The Texas toad, the Texas State Amphibian, is widely distributed across the western two-thirds of the state. The Gulf Coast toad, a species adapted to both urban and rural environments, is one of the most abundant anurans in the state. Its long, low, rattling trill is frequently heard after rains on spring and summer nights. On the other hand, the endangered Houston toad, found only in woodlands on deep sands in the Post Oak Savanna, is probably the rarest amphibian in the state, surviving only with the help of captive breeding efforts described in Chapter 7.

Figure 14.8. The Gulf Coast toad, one of the most abundant species in the state, is recognized by its strong cranial crests and triangular parotoid glands.

Family Microhylidae – The Narrow-mouthed Toads

The narrow-mouthed toads live up to their names as toads. These small, bullet-shaped anurans spend much time underground, emerging only after heavy rains in the warmth of late spring and summer fill new breeding areas. When they begin to breed, however, everyone knows it! All three species produce memorable high-volume calls that seem like they would not come from such a small, inconspicuous toad. The eastern narrow-mouthed toad sounds like a gymnasium buzzer, the western narrow-mouthed toad sounds like an angry bee, and the sheep frog sounds like a very angry sheep. The western narrow-mouthed toad sometimes exists in a mutualistic relationship with tarantulas, often sharing burrows with the large spider. Since narrow-mouthed toads specialize in eating ants, it is hypothesized that the toads keep ant pests out of the burrow. At the same time, the larger tarantula protects its anuran "friend."

Figure 14.9. The western narrow-mouthed toad has a fold of skin that allows it to cover its eyes and protect them from ants. Photo: Fernando Mateos-González, CC BY-SA 2.0.

Figure 14.10. Burrowing toads spend much of their life underground. There is only one species in this unusual family of toads. Photo: Greg Schechter, CC BY 2.0.

Family Rhinophrynidae – The Burrowing Toad

Perhaps the most bizarre frog call belongs to the burrowing toad, the only species in the family Rhinophrynidae. This species, which looks like a squashed version of a large narrow-mouthed toad, also spends nearly all of its life underground. In Texas, it is found only in two counties located in far South Texas on the Mexican border, where it waits for extreme rainfall, usually associated with a tropical storm or hurricane, to emerge from its burrow. When it does, then things get really weird. Puddles, stock tanks, and resacas are filled with floating frog pancakes with a red racing stripe broadcasting sounds like a spaceship come to earth. A few days later, powder blue tadpoles fill the wetlands. And then, a few weeks later, they all go back underground.

Family Eleutherodactylidae/Leptodactylidae – The Rain Frogs/Tropical Frogs

Taxonomists are currently reviewing how this group's frogs are organized. Six species in this collection are found in Texas, including one introduced species—the glass frog on Galveston Island. The story of these species is presented last because many of them break all the rules of the frog life cycle. One example is the barking frog, a bulky, grayish frog found in Texas only in the Edwards Plateau ecoregion. Barking frogs inhabit caves and crevices and do not come to ponds to breed. Instead, during damp spring or summer weather, the males call from their hiding places, sounding like barking dogs. The female comes to the sound of the male and lays 50-75 eggs in a limestone crack or depression. The oversized eggs provide the moisture that the larvae need even without water present. In addition, the female may stay with the eggs, providing extra moisture. The tadpoles never hatch. Instead, they go through metamorphosis within the egg and emerge as tiny barking frogs. Several other species in the state follow this strategy, including the Rio Grande Chirping Frog, which has managed to migrate all over the state (Highlight 14.1).

HIGHLIGHT 14.1 - HITCHHIKING HERP – THE CASE OF THE RIO GRANDE CHIRPING FROG

An unusual sound occurs in residential landscapes across Texas on warm, humid nights. It sounds like a high-pitched "chirp"—maybe from some tiny bird? But, no, there are no birds that make that call at night. Yet, it's coming from a bush or tree, like a bird. But, when one listens closely, chirps emanate from bushes all around. What could it be?

The culprit is likely the Rio Grande chirping frog, a half-inch-long, darkly mottled frog that's *supposed* to be found only in the most southerly counties in Texas.

The Rio Grande chirping frog breaks all the rules. This tiny Texas amphibian doesn't go to water to breed, and it's been hitchhiking all over the state in recent years. Though the frogs are native to the Lower Rio Grande Valley, folks throughout the eastern half of Texas have heard their tiny nocturnal chirps, described as sounding like a bird or the squeak of tennis shoes on a gym floor. Urbanization, the bane of many wildlife species, somehow seems to be beneficial for this one.

Chirping frogs are a part of the free-toed frog group, genus *Eleutherodactylus*. With more than 700 members found primarily in tropical areas, it is the largest genus of vertebrates in the world. Two other chirping frogs occur in Texas: the cliff chirping frog of the Texas Hill Country and the spotted chirping frog found only in the Trans-Pecos. All share the unique characteristic of direct development of the young. Most frogs go to ponds to lay hundreds or thousands of eggs and the young larvae develop as tadpoles in the water before metamorphosing, but chirping frogs lay a dozen or so eggs in pockets of moist soil. The young frogs go through the larval development stage inside the egg and emerge from the egg as tiny froglets.

That unique attribute explains how Rio Grande chirping frogs have shown up all over the wetter portions of Texas, from San Antonio to Dallas to Houston and in smaller towns in between. As tropical plants have been shipped out of the Lower Rio Grande Valley, biologists suspect they have carried hitchhiking cargos of chirping frogs and their eggs. Encountering irrigated landscapes and the warmth retained in concrete structures in cities, the Rio Grande chirping frogs have flourished, expanding their range into nearby natural areas.

Little is known about how the chirping frogs cope in these new habitats — whether they compete with other species, whether predators exist in these environs, or whether they are simply a welcome addition as a predator of small insects. Volunteers who report hearing and seeing them can help biologists learn more about the little frog that breaks all the rules.

Figure 14.11. Rio Grande Chirping Frog photographed east of Houston. Photo: William L. Farr, CC BY-SA 4.0

SALAMANDERS

Table 14.2. Salamanders in Texas (Order Caudata).

Family	Common name	# Species
Ambystomatidae	Mole Salamanders	6
Salamandridae	Newts	2
Plethodontidae	Lungless Salamanders	18+
Proteidae	Waterdogs	1
Sirenidae	Sirens	1
Amphiumidae	Amphiuma	1

The other major group of amphibians in Texas are the tailed ones—the salamanders. Texas salamanders are grouped into six families (Table 14.2), with much variation in their life cycles and adult forms. Because Texas is subject to drought, many salamander species in the state spend much of their time hiding in moist microhabitats, making them hard to find. Other species never leave the water at all. Salamanders can live for decades, emphasizing the importance of conservation planning for this group.

Family Ambystomatidae – The Mole Salamanders

Ambystomatidae may be the most "typical" group of salamanders in Texas, following a pattern of eggs laid in water, aquatic larvae with external gills, metamorphosis, and terrestrial adults (Fig. 14.12). This group includes the tiger salamanders, marbled salamander, spotted salamander, mole salamander, and small-mouthed salamander. Most are restricted to eastern Texas, where they breed during the cooler months of the year. In contrast, the barred tiger salamander is found in West

Texas and South Texas, where they are most active when late summer monsoon rains appear. Ambystomatid salamanders are not well-studied in Texas, and concern has been raised regarding the status of the long-lived eastern tiger salamander.

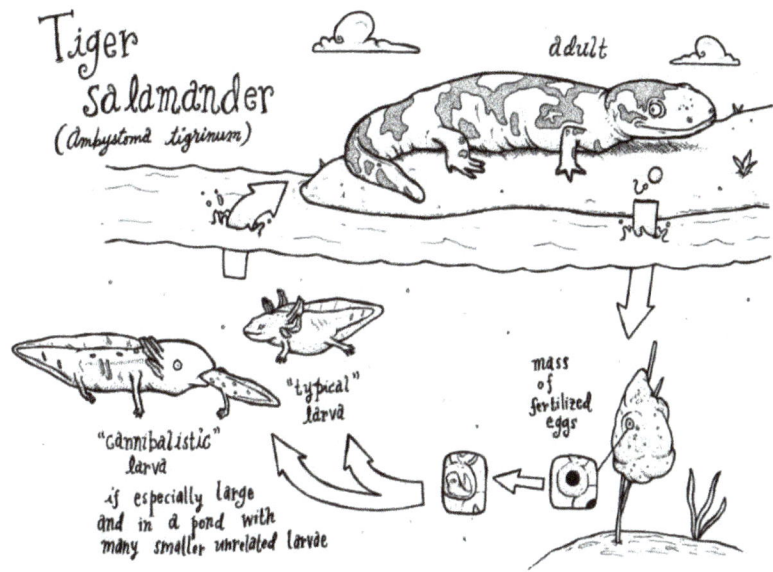

Figure 14.12. Typical salamander life cycle. Source: Dakuhippo, CC BY-SA 3.0.

Family Salamandridae – The Newts

The newts, made famous by Shakespearean witches, are common in Europe, but only two species are found in Texas—the red-spotted newt in East Texas and the threatened black-spotted newt in South Texas. Newts exhibit a variation on the typical salamander life cycle (Fig. 14.13). Eggs are laid in water. The larvae have gills and are aquatic, but when they metamorphose, they enter a terrestrial subadult phase called the eft. Efts have rough, granular skin and disperse away from the water. After one to three years, they reach adulthood, return to the water to live and breed, and change body form again—developing smooth skin and a compressed finlike tail.

Figure 14.13. Life stages of newts, from water to land to water again.

Family Plethodontidae – The Lungless Salamanders

Plethodontidae is the largest family of salamanders in Texas and the world. At least 18 species, comprised of two primary forms, are found in the state. The more typical form includes the Holbrook's southern dusky salamander, southeastern dwarf salamander, western slimy salamander, and southern red-backed salamander. These species live on land as adults but, true to their name, do not have lungs. Instead, they absorb all the oxygen they need through their skin and mouth cavity. Some species return to water to breed, but others lay egg clusters on the ground. Larvae may then move into water. The western slimy salamander produces larvae that live on land immediately.

The remaining Texas plethodontids are all in the genus *Eurycea*. All but one are found in the aquifers and springs of the Edwards Plateau. These spring species have a **neotenic** adult form—retaining the appearance of a larva. They remain aquatic for life, never living on land. The neotenic *Eurycea* also do not have lungs but retain fine-textured external gills. Cave-dwelling species, such as the Texas blind salamander, may have other bizarre adaptations, such as a lack of eyes and pigment, very long legs, a flattened head, and the ability to sense small changes in water movement. Taxonomists are still classifying new species of *Eurycea*, finding that species isolated in different spring systems often differ genetically from other populations. Nearly all spring-dwelling *Eurycea* are considered threatened or endangered due to threats to water quality and quantity in Edwards Plateau springs.

Figure 14.14. The Texas blind salamander has some remarkable adaptations for living its life in the underground waters of the Edwards Aquifer. Photo: brian.gratwicke, CC BY 2.0.

Families Proteidae, Sirenidae, & Amphiumidae – The Aquatic Salamanders

Each of the remaining three families of Caudata in Texas have only one species in the state. All three species are fully aquatic, dwelling in water their entire lives. Though they may move across the land for short periods, they always return to the water. The Gulf Coast waterdog, found only in East Texas, is spotted with four legs and very prominent gills. The lesser siren, which includes two subspecies in Texas, is characterized by external gills and no hind legs. The subspecies in South Texas, the Rio Grande lesser siren, is considered threatened due to threats to wetlands in that region. The three-toed amphiuma is eel-like, with four tiny limbs and internal gills behind a gill slit. It occurs only in East Texas in slow-moving waters with abundant vegetation and debris.

Figure 14.15. Gulf Coast waterdogs, like sirens and amphiumas, are salamanders that do not have a land-dwelling stage. Photo: Brad M. Glorioso; USGS, CC BY-SA 4.0.

THREATS

Amphibians, because of their two-stage life cycle and water-permeable eggs and skin, are sensitive to climatic factors, habitat changes, and a wide range of environmental degradation. Due to their unique characteristics, amphibians can serve as excellent bioindicators of the environmental health of terrestrial and freshwater aquatic ecosystems worldwide. Unfortunately, they are revealing the same message as many other species of animals and plants: ecosystems worldwide are changing faster than many organisms can adapt, and the current extinction rate is significantly higher than the normal background levels throughout the long history of life on earth. Add in the alarming rate of malformations in some amphibian populations, and it seems amphibians have much to reveal about the quality of the natural environment.

A 2004 report by the Amphibian Specialist Group of the IUCN (the International Union for the Conservation of Nature) determined that 43% of amphibian populations around the world are declining. Nearly one-third of the world's amphibian species are considered threatened (Fig. 14.16). Since 1990, more than 165 species are believed to have gone extinct. Many factors may have contributed to the global decline of amphibians. Causes for decline include:

- Habitat loss – As for all wildlife, habitat loss is considered the primary concern. In addition to the direct destruction of wetlands and adjacent uplands, land use can play a role. A study found that salamanders are five times less abundant in clear-cut forests than in intact forests. Overgrazing can lead to erosion and loss of vegetation cover near the water's edge. Fragmentation can affect the ability of amphibians to move between habitats and may isolate small populations. A study of the endangered Houston toad showed that toads may move up to 0.8 miles from their breeding ponds. However, they rarely move across unforested habitats.

- Climate change – Amphibians' life cycle, semi-permeable skin, and unprotected eggs make them vulnerable to changes in temperature and moisture. Although

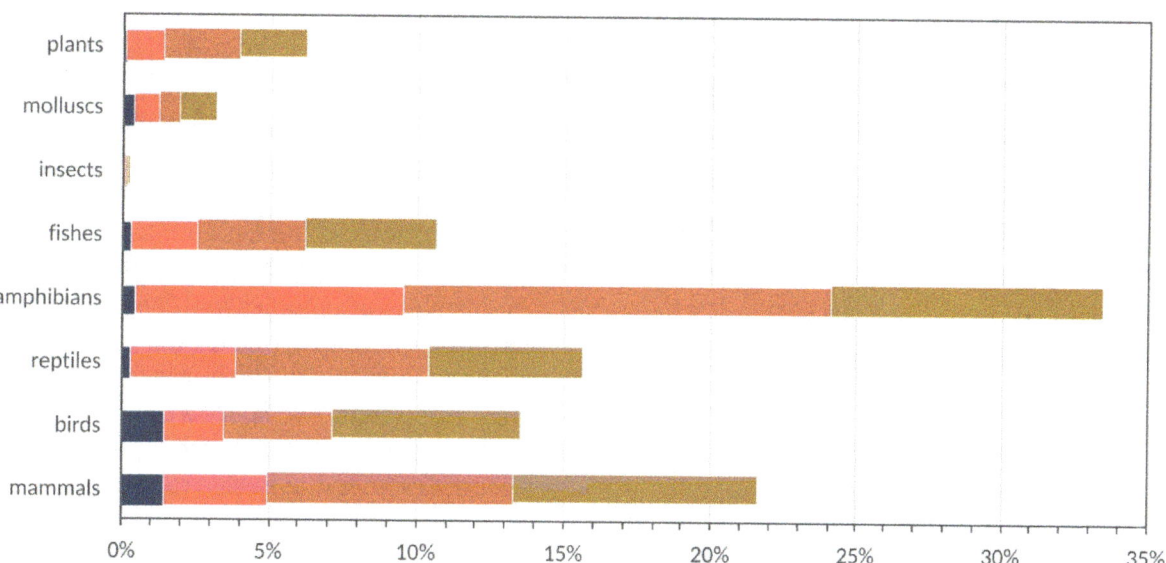

Figure 14.16. Of all animal groups, the IUCN estimates that amphibians have the greatest percentage that can be considered extinct, critically endangered, endangered, or vulnerable. Source: Erkcan, CC BY-SA 4.0.

periods of drought always impact amphibian populations, long-term climate change may have even more damaging effects. Amphibians cannot migrate long distances, so they are unlikely to be able to shift their distributions as mammals or birds might. The Houston toad is already restricted in its distribution due to millennia-long climate shifts in its former range. If modern climate change results in more extended periods of drought in the Post Oak Savanna, then it may not be able to survive.

- Introduced species – Introduced species have imperiled amphibians in other states and countries. Mosquitofish (*Gambusia*), widely introduced to reduce mosquito populations, decrease the survival of anuran eggs and tadpoles. Introduced trout have been associated with the disappearance of the southern mountain yellow-legged frog from 80% of its former range. Ironically, even introduced frogs can imperil other frogs. Bullfrogs were introduced across the western U.S. because of the popularity of eating frog legs. However, they prey on other frogs and have endangered several other ranid species. The South American cane toad, native to Central and South America, has been widely introduced, competing with and preying on local species.

- Chemical pollutants – With their semi-permeable skin, it is easy to imagine that chemical pollutants could adversely affect frogs. The effects on frogs vary, with different species sensitive to different chemicals. Long-term exposure to carbaryl, an insecticide, increases larval mortality and deformity rates. Atrazine, an herbicide, produces a high rate of hermaphroditic frogs (frogs with both male and female characteristics), with unknown impacts on their ability to reproduce. Studies have shown that heavy metals, fertilizer runoff, and acid rain also have the potential to impact amphibians negatively. As detailed in Highlight 14.2, pollution has also affected frogs indirectly through changes to the aquatic environment.

- UV radiation – Like their parents, the eggs of amphibians are covered with a semi-permeable membrane. Because they float on or near the surface of the water, they are especially vulnerable to damage in locations where declines in the ozone layer allow increased UV radiation. It is also thought that increases in UV radiation may make adult amphibians more susceptible to disease.

- Overcollection – Amphibians are collected for food, research, and pets. The 2004 Global Amphibian Assessment identified over-exploitation and habitat loss as two of the top threats to amphibians.

- Disease – All the above factors have played a role in amphibian decline. However, disease has been the most critical threat to amphibian populations in the last three decades. Chytridiomycosis, caused by chytrid fungus *(Batrachochytrium dendrobatidis)*, has impacted amphibian populations at many locations around the world. The fungus wreaked havoc as it was accidentally introduced in Australia, Central America, South America, and some portions of the United States and Europe. Chytrid fungus infects the mouthparts of tadpoles and the skin of adult frogs, leading to a hardening of those tissues and the ultimate death of the frog. Australia's two species of unique gastric brooding frogs are thought to have become extinct due to chytrid fungus. The spread of the fungus has been so extensive and deadly in Central America that a rescue program called the Amphibian Ark has been implemented to take species into captivity before they are lost to extinction. In addition to chytridiomycosis, a viral pathogen called *Ranavirus* has emerged recently. *Ranavirus* has been documented in at least 16 species of frogs, one species of toad, and six species of salamanders in the United States. Mortalities have been reported in 20 amphibian and turtle species in 25 states.

HIGHLIGHT 14.2 - MALFORMATIONS – "A PLAGUE OF FROGS"

In 1995, a group of middle school students was on a field trip to a farm in Minnesota. As the students began their hike, they came to a pond with northern leopard frogs leaping all around. Quickly, the students turned their attention away from the hike and toward capturing frogs. Then they started to notice something strange—many of the frogs, eventually a total of 22 frogs, were missing legs or had malformed legs. Shocked and even a little frightened, the students and their teacher, Cindy Renitz, decided they needed to tell someone. They contacted the Minnesota Pollution Control Agency and sounded an alarm that would spread nationwide.

Occasionally, amphibians, like all species, exhibit abnormal development. Mistakes in development are easy to imagine in a species that goes through a complicated process like metamorphosis. However, the findings in Minnesota soon led to many reports of frogs with extra limbs, wholly and partially missing limbs, skin webbings, bony protrusions, and even eye malformations. Ultimately, more than 50 species in 44 states were reported with malformations.

Scientists began investigating many potential causes, including chemical contamination, UV-B radiation, predation injuries, and parasite infections. Ultimately, the most frequent cause was identified as a parasite—specifically, a parasitic flatworm called *Ribeiroia ondatrae*. *Ribeiroia* is a naturally-occurring aquatic parasite. Its life cycle requires three hosts: ramshorn snails, larval amphibians or fishes, and birds or mammals. Snails serve as the first host and release free-swimming larvae called cercariae. The cercariae imbed themselves in the developing limbs of tadpoles, forming a cyst. That cyst can cause abnormal limb development. Ultimately, malformed amphibians are less likely to survive and reproduce. Birds may eat them, and the cycle continues.

Even though *Ribeiroia* occurs naturally, human impacts can cause an increase in its abundance. Snails are grazers. When fertilizers run off into ponds, there is an increase in algae and snail populations. In response, parasite numbers may increase up to eight times, resulting in unnaturally high rates of malformations. As data accumulated, it seemed to indicate that land use practices contributed to what author William Souder called a *Plague of Frogs* (University of Minnesota Press, 2000). Malformed frogs connected to snails connected to fertilizers in agricultural systems illustrates what Native American Chief Seattle once said, "All things are bound together. All things connect."

Figure 14.17. Malformed bullfrog found in the San Marcos River, Caldwell County, Texas. Photo: Melba Sexton.

How do Texas amphibians fit into this picture of declining amphibians? The short answer is that no one completely knows. One species native to the El Paso area, the northern leopard frog, disappeared from Texas due to habitat alteration. Many Texas species are naturally adapted to drought. Still, the combination of habitat loss and climate change may hasten their demise. Data on the endangered Houston Toad strongly suggests these as the primary reasons for this species' disappearance from its occurrence in the Houston area in the 1950s. Chytrid fungus, *Ranavirus*, and malformations have been documented in the state, although no major die-offs have been reported. More monitoring is needed. More information is also needed about the sustainability of amphibian harvests in Texas. In 1999, more than 45,000 herps (reptiles and amphibians) of more than 100 species were collected in Texas. Tiger salamanders, Couch's spadefoot toads, and Chihuahuan green toads were among the top 12 species collected. TPWD lists 23 amphibian species as Species of Greatest Conservation Need (12 salamanders and 11 anurans). Still, the data to verify population trends for most species do not exist.

CONSERVATION AND MANAGEMENT

The best hope for amphibian populations in Texas is maintaining healthy ecosystems. Endeavors to conserve wetlands, such as Joint Ventures, can provide habitat for amphibians. Protecting habitat quality is also essential. Some tools exist. The Texas Commission on Environmental Quality sets up rules regarding the discharge of pollutants into streams. It also provides funding for locally-implemented Watershed Protection Plans and guidance for voluntary prevention of non-point source pollution, such as nutrient runoff from farms and feedyards. Zoning rules to reduce runoff can also help protect water quality in aquifers for spring-dwelling salamanders, such as in the Barton Springs watershed in Austin. Guidelines developed by local groundwater districts can help maintain water quantity in aquifers and springs as well.

TPWD provides some protection against the overharvest of amphibians in Texas. It requires a hunting license or collection permit to collect herps in the wild, sets limits on the number of organisms and collection on public roadways, and requires commercial permits to sell Texas wildlife.

Preventing the spread of disease is also crucial to maintaining amphibian populations. Wetland enthusiasts should clean boots, nets, and other equipment used in amphibian habitats. Mud and debris should be removed, and the items treated with a 3% bleach solution or other suitable disinfectant before being used in another location. Because of disease risks, pet amphibians should never be released into the wild. In Europe *Batrachochytrium salamandrivorans*, a chytrid fungus that infects salamanders, has now been detected in captive populations. Biologists hope to prevent its spread to the U.S., which hosts a high diversity of salamander species.

There are also opportunities for individuals to take action to help amphibians. Several citizen science opportunities exist, such as FrogWatch USA, a program that invites individuals to conduct call count surveys. The Texas Herpetological Society provides opportunities for amateurs and professionals to survey amphibians and reptiles during field meets. TPWD recruits herp sightings to be shared with the Herps of Texas Project on iNaturalist.

Landowners can protect wetlands, springs, and sinkholes through good grazing management and appropriate use of fertilizers. Individuals can avoid introducing non-native plants or wildlife. One popular strategy is to build a frog pond. Whether a larger stock pond or a small backyard ornamental pond, if the pond provides shallow areas with native wetland plant cover and protection from fish predation, then amphibians will use it. Finally, it's even possible to create a "toad house" by knocking a hole in a flowerpot and turning it upside down near a water faucet. Providing even a tiny habitat patch can produce a lovely symphony of frogs on a warm spring evening.

Figure 14.18. Small ornamental ponds with shallow areas and vegetative cover can provide habitat for multiple frog species.

CHAPTER 15 – TEXAS REPTILES

Reptiles—some people love them; some people hate them. With Texas hosting more than its fair share of reptiles, Texans are bound to have had some reptile encounters. And yet, whether one appreciates or fears reptiles, the creativity of class Reptilia offers much to be admired. Texas reptiles come in all sizes—from the appropriately-named threadsnakes to 900-pound alligators. Some reptiles live in deserts and are water-conserving champions. Some live in the ocean and can hold their breath for hours. The body forms boggle the mind. Turtles carry a shell. Snakes speed along on their bellies without legs. Some lizards run on hind legs; others shoot blood from their eyes. Some have no ears, some have no legs, and some have no males!

Fascination with reptiles can lead to both conservation challenges and opportunities. People love to collect reptiles. Some people love to eat them. Some like the adrenaline thrill of hunting them. Others are just willing to spend hours looking for them. All these characteristics create a complex and rich picture of reptile conservation in Texas.

WHAT IS A REPTILE?

The class Reptilia is divided into four extant orders (groups that exist today).
1. Crocodilia is composed of crocodiles, alligators, and gharials. These are the largest living reptiles. Their powerful legs allow them to move quickly on land, but their primary habitat is water, where they use their muscular tail to swim effectively.
2. Squamata (which means "scaly") is divided into two suborders—Sauria (the lizards) and Serpentes (the snakes). Lizards have legs (usually), eyelids (usually), and external ears (usually). In contrast, no snakes have legs, eyelids, or external ears. They sense vibrations through ear bones connected to the jawbone.
3. Testudines includes the turtles. They are characterized by a shell composed of bones with overlying scutes. Turtles occur in fresh water, salt water, and on land.
4. Sphenodontia is an order of reptiles that includes only one species—the tuatara. This medium-sized, lizard-like reptile is found only in New Zealand. Its most unique characteristic is an eye-like structure on top of the head.

Reptiles and amphibians are both considered "herps." There are some similarities between the two classes, but there are some significant differences that allow reptiles to better explore life on land. Reptile characteristics include:

- Ectothermy – Like amphibians, reptiles are ectothermic. The external environment influences their body temperature. Most species actively **thermoregulate**, warming up by basking in the sun and cooling by panting or seeking shade and burrows. Some species can also change skin color to increase heat collection. Large-bodied reptiles like turtles and crocodilians have circulatory systems that let them redirect blood flow to the extremities if they need to warm up or cool down. They can also shunt it away from the lungs to avoid losing heat in cold temperatures.

Most reptiles prefer external temperatures between 82 and 104 F. For this reason, they are more restricted in distribution than amphibians. Turtles and alligators are poikilothermic, with their body temperatures varying greatly depending on water temperature. In contrast, most lizards take actions to stabilize their body temperatures to be more homeothermic. During cold periods, reptiles, like amphibians, brumate.

(opposite) **Figure 15.1. Texas horned lizard, the Texas State Reptile.**

Figure 15.2. Turtles, like many other reptiles, bask to bring their body temperatures up to ideal levels. Here, Texas river cooters bask along the San Marcos River.

- Lungs – In contrast to amphibians, reptiles breathe with lungs throughout their life cycle. They lack a diaphragm. Most species breathe by expanding their rib cage, but turtles must flex their shoulder and hip bones to draw in air. Because they are ectothermic, their oxygen requirements are lower than mammals, and many species can go for long periods between breaths.

- Scales – Reptile skin is covered with scales, a protective covering that develops from the epidermis. In lizards and snakes, the scales are distinctive and overlapping. In alligators, scales do not overlap. In addition, the scales on the back of crocodilians have bony structures under them called osteoderms; these form the pointed ridges on the alligator's back. Turtle have shells formed from fused bones covered with patterned scutes composed of keratin.

- Amniotic eggs – Unlike amphibians, reptiles can lay eggs on dry land. This is possible because they possess a more complex egg with membranes that help protect the developing embryo (Fig. 15.3). Reptiles do not actively incubate their eggs, so the external temperature affects the length of incubation in reptiles. A few reptiles, such as pit vipers, water snakes, garter snakes, and some lizards, give birth to live young. In this case, the eggs lack shells and are held inside the mother until hatching.

- Heart – Amphibian hearts have only three chambers, allowing the mixing of oxygen-rich blood with oxygen-poor blood in the single ventricle. Most reptiles also have only one ventricle, but a membrane called a septum partially separates the oxygenated from the unoxygenated blood. Crocodilians have a complete 4-chambered heart with two separate ventricles.

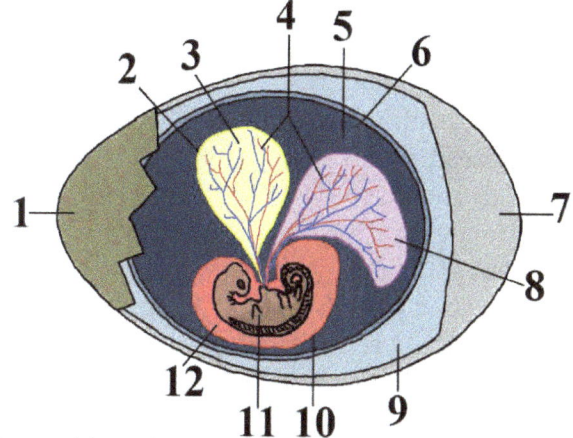

The eggshell (1) protects the crocodile embryo (11) and keeps it from drying out, but it is flexible to allow gas exchange. The chorion (6) aids in gas exchange between the inside and outside of the egg. It allows carbon dioxide to exit the egg and oxygen gas to enter the egg. The albumin (9) further protects the embryo and serves as a reservoir for water and protein. The allantois (8) is a sac that collects the metabolic waste produced by the embryo. The amniotic sac (10) contains amniotic fluid (12), which protects and cushions the embryo. The amnion (5) aids in osmoregulation and serves as a saltwater reservoir. The yolk sac (2) surrounding the yolk (3) contains protein and fat-rich nutrients that are absorbed by the embryo via vessels (4) that allow the embryo to grow and metabolize. The air space (7) provides the embryo with oxygen while hatching. This ensures that the embryo will not suffocate while it is hatching. Source: "Crocodile Egg Diagram" by Amelia P. User: Catsloveme207 is licensed under CC BY-SA 3.0.

Figure 15.3. Diagram of a crocodile egg.

IDENTIFICATION AND MONITORING

Field guides are essential in reptile identification. Some species are quite distinctive, but others may be distinguishable only by knowing certain key features as pointed out in a field guide. Identifying some species may require examination in the hand to count scales or scutes. The pattern of head scales, the characteristics of the ventral (belly) scales, and the count of the dorsal scales on the back may be needed to distinguish between some snake species. (Care should always be taken when examining snakes up close.) For turtles, the pattern of the scutes on the **carapace** and **plastron** sometimes helps with identification.

As taxonomists learn more about different species, the names of those species and their subspecies may change. In addition, different subspecies are often given different common names. The Society for the Study of Amphibians and Reptiles (SSAR) regularly updates a publication called "Scientific and Standard English Names of Amphibians and Reptiles of North America," a helpful reference for name changes. Still, no one needs to let worries about taxonomy diminish their enjoyment of herps; thankfully, a good

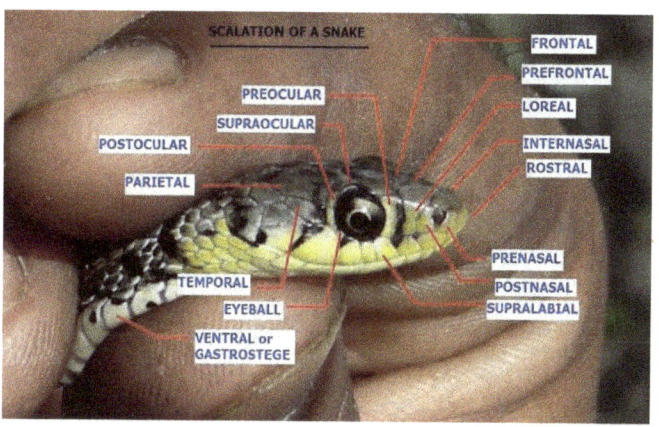

Figure 15.4. An example of different scales on a garter snake's head. Source: AshLin, CC BY_SA 2.5.

photo uploaded to a website like iNaturalist can provide identification help and links to current names.

Many monitoring techniques for reptiles take advantage of the fact that these ectotherms bask in the sun. Basking transects can be conducted on land or water to spot lizards, turtles, or alligators that emerge in the open to increase their body temperatures. Since visibility may be limited on these transects, distance sampling is usually not possible. Instead, results are typically expressed as the number of animals per length of distance traveled on a road, trail, or waterway. Basking surveys are best conducted at temperatures in the 60s or 70s when sunshine is present. In West Texas, roadside surveys are often performed at night when reptiles come to the warm road surface as desert temperatures drop. Night-time surveys using a spotlight are also used to count alligators on waterways. A bright spotlight reflects a red color from the tapetum lucidum layer in alligator eyes. (The tapetum lucidum is well-developed in some animals and helps them reflect more light to the retina at night.) Even size can be estimated in these night-time surveys. The distance from the eyes to the nostrils in inches approximates the length of the overall alligator in feet.

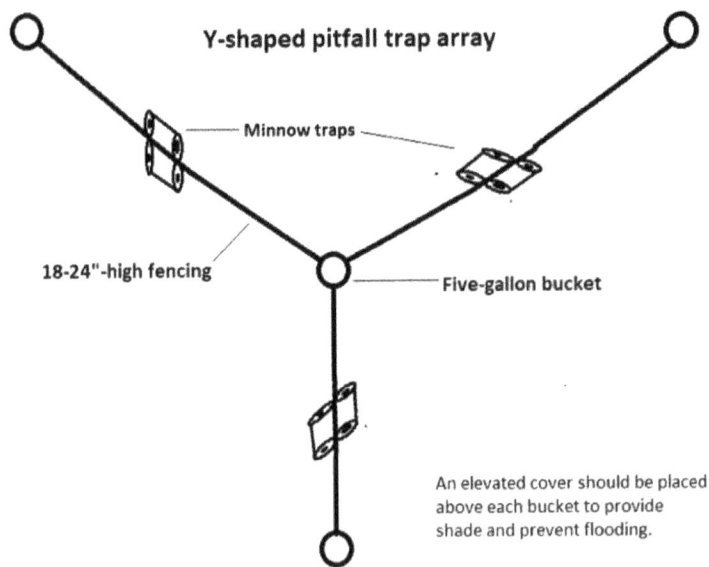

Figure 15.5. Terrestrial pitfall trap array with minnow traps used to collect snakes.

Time-constrained searches are also useful for reptiles. Rocks and logs can be turned to look for snakes and lizards underneath. Care should be taken, both in lifting the object and in replacing it. A snake hook can help turn objects to avoid the risk of a bite to the hand. (These hooks are also used to capture snakes by lifting them off the ground.) As with other time-constrained searches, the number of searchers is multiplied by the amount of time spent searching and then divided by the number of animals found to provide an abundance index for each species. Similar to mammals, signs can also indicate the presence of a species. Lizards and snakes concentrate nitrogenous waste, often producing a white uric acid tip on the scat. If that scat is from a horned lizard, it will be full of ant body parts (see Highlight 15.1). Shed snake skins can be used to identify snakes.

Coverboards and pitfall traps are also effective for reptile inventories. In a pitfall array, some buckets can be replaced by minnow traps for capturing snakes (Fig. 15.5). Baited minnow traps placed in water are used to catch aquatic snakes (just be sure a portion of the trap is left above water level). Lizards are often captured by hand or with a monofilament noose placed on a pole. For mark-recapture, PIT tags are frequently injected. Toe-clipping is often used on lizards. Turtle shells may be notched. Ear tags can be attached to the fin-like scutes on an alligator's tail.

TEXAS REPTILE DIVERSITY

More than 12,000 reptile species exist worldwide, and new species are discovered every year. About 440 reptile species live in the United States. Although different databases provide different numbers, Texas has about 154 native species of reptiles, about 30 more introduced species, and numerous subspecies. That's a lot of reptilian diversity!

Table 15.1. Reptile diversity in Texas, according to reptile-database.org.

Order/Suborder	Common Name	# Families	Native Species	Introduced Species
Crocodilia	Alligators & Crocodiles	1	1	1
Squamata/Sauria	Lizards	8	45	16
Squamata/Serpentes	Snakes	5	76	11
Testudines	Turtles	7	32	2
		Totals:	**154**	**30**

ORDER CROCODILIA – THE AMERICAN ALLIGATOR

Texas has one native crocodilian—the American alligator. (There have also been reports of South American spectacled caiman found in the state, probably from pet releases.) Alligators are Texas's largest reptile, with the largest specimens reported in the state over 14 feet long. They are found in freshwater bodies from the northeast corner of the state down through the Lower Rio Grande Valley, being strongly associated with slower-moving waters in the Western Gulf Coastal Plain.

Crocodilians are thought to be the most intelligent of the reptiles and provide more care for young than other species. Adult female alligators construct nests made of dirt and marsh vegetation where they lay 30-40 eggs covered by vegetation. The decomposition of the nest provides warmth. The sex of the alligator young is determined by the incubation temperature, with high temperatures producing more males and lower temperatures producing more females. The female alligator stays nearby, sometimes excreting liquid waste on the nest to moisten it. When the young hatch, they make a low-pitched chirping sound. In response, the female alligator begins to open up the nest. The young remain near the mother for up to three years, with her providing some protection from predation.

As described in Chapter 11, alligators are long-lived. Populations were once decimated by overhunting. As harvest has been regulated, alligators have bounced back and are now classified as a game species. In addition to spotlight transects on waterways, alligator numbers are monitored by nest transects conducted by helicopter. With an understanding of alligator life history tables, those nest counts are used to produce population estimates and set harvest limits.

Figure 15.6. This alligator is lowering its body temperature by "gaping," allowing a glimpse of the palatal valve that can close to prevent water entering the lungs and stomach. Note the scutes along the back. Photo: Mark Alan Storey.

ORDER SQUAMATA/SUBORDER SAURIA – THE LIZARDS

Texas has eight families and about 45 species of native lizards. Known for their quickness and variety of colors, all lizards are carnivorous. Most Texas lizards are **oviparous** (reproducing by laying eggs in the ground), although a few are **ovoviviparous** (holding eggs that hatch within the body, leading to birth of live young). In contrast to alligators, in some lizards high incubation temperatures produce more female lizards; low temperatures produce males. In other species, sex is determined genetically.

Figure 15.7. Slender glass lizards are lizards with no legs! Photo: 2ndPeter, CC BY 2.0.

Family Anguidae includes the slender glass lizard and the alligator lizard. Although not the most frequently seen species, they have several unique attributes. Both species have large, stiff scales. Because their scales are so rigid, they have a scale-less groove along the side of the body that allows for more flexibility. Slender glass lizards are lizards with no legs. However, a visible external ear opening marks them as lizards. They are called glass lizards because they readily lose their tail. When the broken tail regrows, it is a hardened sharp point in a different color than the rest of the original tail, leading to the nickname "horn snake." Texas alligator lizards are found only in the Edwards Plateau region of the state, where they hide in rough and rocky habitats. They have legs, and their tail is somewhat prehensile (able to wrap around objects). They are slower-moving than many lizards.

Family Dactyloidae includes the anoles. Texas has two species of anole, both of whom are good climbers. The green anole is a native species, but the brown anole was recently

Figure 15.8. Male anoles are well-known for their territorial displays. Photo: Mark Alan Storey.

introduced from the Caribbean. The green anole, often found around houses and house plants, is sometimes mistakenly called a chameleon because of its ability to change its color. Interestingly, scientists think the shifts from green to brown and back are not designed to match the environment but demonstrate the lizard's stress level or relative dominance. In fact, anoles are famous for their territorial displays—the males bob up and down, do push-ups, and expand a colorful throat fan, or dewlap, as they try to drive other males away and draw females in.

The family Crotaphytidae includes two collared lizards and one leopard lizard in Texas. These lizards are boldly marked, and their behavior is quite bold, often biting when handled. All three are fast. The eastern collared lizard can reach speeds at which it rises up and runs on its two hind legs. Males in this group also demonstrate head bobbing, nodding, and push-ups to declare territory.

Family Eublepharidae includes the two Texas native gecko species. The Texas banded gecko occurs in South and West Texas; the reticulate banded gecko is restricted to the Trans-Pecos. These well-marked lizards are known for their ability to drop their tail quickly, perhaps an adaption to escaping predation by snakes. Texas also has three introduced gecko species in the family Gekkonidae. Members of this family are known for their ability to climb vertical surfaces using their expanded toes covered by pads with tiny bristles and suction cups. The introduced Mediterranean gecko is very common on and in houses across the state. Geckos are unusual among the lizards because they make sounds.

The family Scincidae includes eight skink species found in Texas. Skinks have smooth scales that give them an almost wet appearance, appropriate for their moist habitats. They hunt actively among leafy debris and ground cover. They lose their tails easily. Some female skinks stay with their eggs, turning them and perhaps providing warmth.

The family Teiidae includes whiptails and racerunners. There are eight species in Texas. These species are fast, long, and slender, and they can be challenging to tell apart. The common spotted whiptail has the broadest distribution, including most of the state. At least five of the eight species of teiids in Texas are unisexual; they reproduce through parthenogenesis from an all-female population. (In parthenogensis the embryo develops directly from an egg without needing fertilization.) The parthenogenic species are primarily found in West and South Texas, where parthenogenesis may help ensure reproduction in a harsh, arid environment.

Figure 15.9. Collared lizards are large and boldly colored. Photo: 2ndPeter, CC BY 2.0.

Figure 15.10. Male four-lined skinks develop a blue tail during breeding season. Photo: Mark Alan Storey.

The family Phrynosomatidae is the largest and most diverse family of lizards in Texas. Most of the 20+ species are sit-and-wait foragers, holding very still while waiting for prey to pass. As a result, they are often well-camouflaged in their environment. Three genera are especially notable.

Figure 15.11. A spotted whiptail in the background eyes a Texas spiny lizard as a possible meal.

The earless lizards in the genus *Holbrookia* lack external ears, although they can still detect sound with internal ear structures. Many of them burrow into the sand, so the lack of an ear opening may be an advantage in those soils. There are four species in Texas. The males are often marked with a pair of bold black bars on their sides. The spot-tailed earless lizard is a species of concern in Texas with a limited number of records.

The genus *Sceloporus* includes the spiny lizards, a group of well-camouflaged lizards with strongly keeled and pointed scales (Fig. 15.11). Eight species occur in Texas. All are good climbers. Males often have bright blue markings on either side of the belly and sometimes under the throat. The dunes sagebrush lizard in West Texas is a species of conservation concern recently listed as endangered.

Finally, the horned lizards in the genus *Phrynosoma* are appealing and enigmatic. Texas has three horned lizard species—the greater short-horned lizard in the mountains of the Trans-Pecos, the round-tailed horned lizard that occurs in open habitat in the western half of the state, and the Texas horned lizard that once occurred nearly statewide. All three of these wide-bodied, slow-moving lizards are primarily **myrmecophagous** (feeding on ants) and bear pointed keratin structures on their heads (the "horns"), but each has its own unique adaptations. The short-horned lizard is a live-bearer whose internal body cavity is black. Herpetologists hypothesize that, in mountain habitats, ovovivipary and the black tissues enable the females to maximize warmth for the developing young. The round-tailed horned lizard has a unique form of camouflage. It bears dark, shadowy marks beside each leg. It hunches its back when frightened, assuming a round, rock-like shape further accentuated by the dark markings. Finally, the much-beloved Texas horned lizard is associated with unique legend and lore and has earned the designation of the Texas State Reptile (Highlight 15.1).

HIGHLIGHT 15.1 – THE HORNED "TOAD" – THE STATE REPTILE(?) OF TEXAS

In 1993, a group of schoolchildren discovered that Texas did not have an official state reptile. The state had a state flower (bluebonnet), a state tree (pecan), a state bird (northern mockingbird), a Texas state gem (blue topaz), a state grass (sideoats grama), a state fish (Guadalupe bass), and even a state dish (chili), but no Texas State Reptile. The students wrote a letter to the state legislature suggesting that the Texas horned lizard be designated the official state reptile. With overwhelming support, it was awarded the title.

Undoubtedly, there was much affection for horned lizards at the state legislature. Many Texans grew up playing with these fierce-looking yet docile lizards that were everywhere, and they were common characters in Texas legend and lore. It was reported that Texas Boy Scouts regularly took hundreds of horned lizards to the National Scout jamborees to exchange as gifts. The lizards were so abundant on the original Texas Christian University football field in Waco that the school adopted the "Horned Frog" as their mascot. Eastland, Texas, lays claim to the tale of an infamous horned lizard, "Old Rip," that reportedly survived for 31 years in a time capsule in the cornerstone of their courthouse. And even long before Texas was Texas, horned lizards were symbols of health and happiness for Native Americans.

But how did a scaly lizard with two big horn-like projections on its head come to be affectionately known as a horned toad? The mixed-up identity seemed to exist even among early taxonomists. When the species was first described by European naturalists exploring North America in the 1820s, it was given the genus name *Phrynosoma*, which

Figure 15.12. Horned lizards mate soon after emerging from hibernation. The female digs a hole and lays one to three dozen eggs. Hatchling emerge 40 to 60 days later. Photos: Chip Ruthven, TPWD.

translates as "toad body." Its mild demeanor and plump body form just seemed un-reptilian.

Texas horned lizards utilize a variety of habitats. They are usually found in flat, open terrain, with sparse plant cover allowing movement on the ground. They use mammal burrows, rock piles, or clumps of vegetation for cover and may excavate shallow burrows in loose soils or scoop out a shallow depression in which to hide. They can survive in both moist and arid environments. They get water from their diet and from rain harvesting—arching their back during rain showers and allowing their scale pattern to direct water flow to their mouth. They may also absorb some rainwater through the skin.

Horned lizards primarily eat harvester ants (sometime called big red ants)—up to 70 to 100 per day—but their diet also includes termites, other native ants, other insects, spiders, and sowbugs. Their body is built around their ant diet. They have a sticky tongue for lapping up ants and teeth and jaws designed for crunching them. They have a compound in their blood that detoxifies the venom of harvester ants. Their stomachs are enlarged to manage the large amount of chitin found in ant exoskeletons, leading to their plump body shape. (The indigestible ant exoskeleton is visible in horned lizard scat.)

Texas horned lizards also use their ant diet as a part of their defense system against predators. Because they are slow-moving lizards, they first defend themselves through camouflage and flattening the body. If discovered by a predator, they may take in air to inflate their body and erect their spiny scales, or they may "shield," tipping their body sideways to look broader to a potential predator such as a snake. But they have an even more bizarre form of defense. When grabbed by a predator they can shoot a stream of blood from the sinuses in their eyelids! The blood they eject contains a compound especially distasteful to canine predators such as coyotes and foxes. The distasteful compound may be derived from the venom in the ants they eat.

Although many people remember growing up with horned toads, those are distant memories in many parts of the state. Texas horned lizards can still be readily found in West and South Texas. Unfortunately, only small, isolated populations remain in the state's eastern third. Many factors have been proposed as culprits in their disappearance, including collection for the pet trade, the spread of the red imported fire ant, changes in agricultural land use, habitat loss and fragmentation due to urbanization, and environmental contaminants, including pesticides used to kill ants.

Much public interest in saving horned lizards exists. The Texas horned lizard was featured on the first TPWD specialty license plate. The Horned Lizard Conservation Society started in Austin in the 1990s to call attention to the plight of horned lizards across the Southwest. Kenedy, a city in South Texas, actively promotes horned lizard conservation and viewing and received designation by the Texas legislature as "The Horned Lizard Capital of Texas." Several zoos now raise horned lizards in captivity. They work with TPWD and private landowners to restore habitat and reintroduce horned lizards to the wild. Reducing pesticide use and reducing red imported fire ant populations are essential components of habitat restoration. With support and science, perhaps horned toads can find new vigor and reclaim their place as the Texas State Reptile in *all* of Texas.

Figure 15.13. Like the Texas horned lizard, short-horned lizards, found in the mountains of the Trans-Pecos, can squirt blood from the eyelid to escape predators. Photo: USFWS, CC BY 2.0

ORDER SQUAMATA/SUBORDER SERPENTES – THE SNAKES

Probably no reptile group elicits more complex emotions than snakes. Despite some people's aversion, they are fascinating animals that play critical roles in the ecosystems they inhabit. Texas is the champion for Serpentes, hosting more snake species (76 native and 11 introduced) than any other state.

Snakes have many unique adaptions associated with their long body form, lack of limbs, and predatory lifestyle. Body organs in snakes are elongated. They have just one very long single lung. Like other amphibians and reptiles, snakes have a Jacobson's organ on the roof of their mouth. Snakes flick out their forked tongue to pick up chemicals from the ground and the air and then press it against the Jacobson's organ to provide detailed cues about their surroundings. Snakes eat infrequently, but they can consume large prey. Their stomach is large and elongated. Their skull bones are reduced and connected with multiple joints to make the skull more flexible. Snakes also can separate the two sides of the lower jaw in the middle to aid in swallowing prey. Some snakes kill their prey by constriction, some use venom, and some simply swallow their prey alive.

Snakes have between 120 and 400 vertebrae. The ribs from those vertebrae attach to the belly scales, allowing snakes to move through flexion of the body and gripping with the scales. As snakes grow, they must shed the skin layer containing their scales. In snakes, this is accomplished when a new skin layer grows underneath. The snake develops a dull, cloudy color while the new layer forms, and then the old skin is loosened. When the loosening is complete, the snake crawls across a rough surface and essentially crawls out of its skin. Sometimes, it is possible to identify a snake by its shed skin.

Texas Snake Families

In the U.S., the family <u>Viperidae</u> are the pit vipers. The family includes about a dozen species of venomous snakes in Texas, including two copperheads (eastern and broad-banded), the northern cottonmouth (sometimes called water moccasin), and nine species of rattlesnakes. (Note that the proper term for a species that injects venom is "venomous," not "poisonous.") The pit vipers get their name from a pair of heat-sensitive organs, one on each side of the head. These pits can detect infrared radiation from warm-blooded animals, even in the dark. Vipers have moveable hollow front fangs. A set of muscles in the head pump venom from venom glands to immobilize their prey. Vipers may have a blend of several types of venom. The dominant venom is a hemotoxin, which causes changes in blood cells, prevents blood from clotting, and damages blood vessels, causing them to leak. The family is ovoviviparous, with the females giving birth to 3 to 16 live young. Their scales have strong keels (ridges down the middle of the scale).

Many people have come to associate the appearance of vipers—the broad triangular head, the elliptical (vertical) pupils, and the rattles on a rattlesnake tail (a rattler adds one segment to its rattle each time it sheds its skin)—with fear. However, pit vipers predominantly use their venom to take prey, primarily rodents. In this way, they provide a valuable ecosystem service. Bites to humans happen when the snake is surprised or threatened. Caution is always recommended in habitats and weather favorable for snakes, but in the United States fatalities from venomous snakes average less than 15 per year. Still, as discussed at the end of this chapter, there are many passionate opinions concerning the management of vipers.

Broad-banded copperhead
Photo: John Jefferson

Northern cottonmouth
Photo: U.S. Dept of Energy, CC0 1.0.

Western diamond-backed rattlesnake
Photo: Mark Mitchell, TPWD.

Figure 15.14. Three of the pit viper species found in Texas.

The family Elapidae contains one species in Texas—the Texas coralsnake. Coralsnakes occur in East, Central, and South Texas and are usually found in well-vegetated habitats feeding on lizards and small snakes. They are a colorful species with repeating bands of yellow, red, yellow, and black (Fig. 15.15). There is variation among individuals, and colors can sometimes be muted or expanded. Still, the old folklore saying—"Red next to yellow, kill a fellow. Red next to black, venom lack."—provides some safety guidance:

The coralsnake is venomous and is related to highly venomous species, such as cobras, mambas, kraits, and tiger snakes. The neurotoxic venom is delivered through two small hollow fangs fixed at the front of the upper jaw. With a small mouth size, however, bites are rare. Only two human fatalities have been recorded in the United States in the past 100 years.

At the other extreme, perhaps no one is afraid of the tiny blind snakes and threadsnakes in the families Typhlopidae and Leptotyphlopidae. These four native species and one introduced species are small, earthworm-like snakes that spend most of their time underground, occasionally emerging after rain. Even these tiny pink or brown eyeless snakes provide a service, feeding on termites and larval ants. They secrete substances to protect themselves as they feed in ant colonies.

Figure 15.15. Typical color pattern in coralsnakes; however, variations exist. Photo: Mark Mitchell, TPWD.

Colubridae is by far the largest family of snakes in Texas, with 68 species present. The family is quite variable, from the 12" DeKay's brownsnake to the vivid, tree-climbing green snakes to the massive bullsnake and Texas indigo (the record length for these species is over 8 feet). A few members of this family have rear fangs (the hog-nosed snakes, Texas lyresnake, Taumalipan black-striped snake, and Chihuahuan nightsnake), but their venom is not dangerous to humans. Several genera of Colubrids are quite familiar.

Figure 15.16. Great Plains ratsnakes are found across much of Texas. Photo: Meghan Cassidy.

The ratsnakes in the genus *Pantherophis* are commonly encountered. Texas has at least four species, and some, such as the Great Plains ratsnake and western ratsnake (formerly called Texas ratsnake), are widespread in their distribution. These are large-bodied snakes with a flat belly that look like a loaf of bread in cross-section. They are constrictors and eat many rodents.

The genus *Coluber* includes the coachwhips, whipsnakes, and racers. They are slender, fast-moving snakes, usually with plain markings. These quick snakes often take lizards and other vertebrates as prey. They may pin down their prey using a loop of their slender body. Coachwhips are named for their large diamond-shaped scales that look like the braided leather straps of a whip. At least four species with several subspecies occur in Texas.

Hog-nosed snakes make up the genus *Heterodon*. The unique attributes of this genus are numerous. These boldly-marked snakes have an upturned snout reminiscent of a hog. Hog-nosed snakes feed on a variety of prey and are often encountered feeding on toads in moist environments. They have a seemingly infinite array of strategies to deter their own predators. If threatened, they may

Figure 15.17. Hog-nosed snakes are boldly-marked and are sometimes called "spreading adders" because of their tendency to flatten their necks in a threat display. Eastern hog-nosed snake photo: Mark Mitchell, TPWD.

spread the skin on their neck cobra-like, fill their body with air, and hiss loudly (they get their nicknames—hissing adder, puff adder, or spreading adder—from this behavior). If that behavior fails, then they may roll onto their back, convulse, throw their mouth open, emit foul-smelling musk, regurgitate their food, and feign death. Even if one rolls them back over, they often flop over to their back again. Texas has several species of hog-nosed snakes, so nearly everyone in the state has a chance to see this dramatic performance.

Figure 15.18. In the non-venomous western milksnake, red color lies next to black.

Figure 15.19. The diamond-backed watersnake shares habitat with water moccasins, but is non-venomous. Note the round pupils and more slender head. Photo: pondhawk, CC BY 2.0.

Figure 15.20. Western ribbonsnakes, along with gartersnakes, are usually found near water. Photo: calinsdad, CC0 1.0.

The genus *Lampropeltis* includes the kingsnakes and milksnakes. There are at least six species in Texas. Several species of kingsnakes are known to feed on other snakes, including venomous snakes, as they are immune to the venom. Milksnakes are significant predators of rodents. Many milksnakes are tri-colored, with red, black, and yellow/cream; however, the pattern is different than the coral snake, with red bands always touching black bands (Fig. 15.18). The gray-banded kingsnake is highly variable in color and is quite popular as a pet in captivity.

Nerodia are the watersnakes. At least seven species occur in aquatic habitats in Texas, where they feed on fish, amphibians, and other aquatic organisms. They can often be seen basking on trees and debris near water and can be trapped in minnow traps. They are live-bearers with a bad reputation. They bite readily and give off a musk when captured. They are often persecuted because they are misidentified as cottonmouths; however, the pupils in watersnakes are round, their heads are not enlarged, and, when swimming, they tend to ride lower in the water than cottonmouths.

Thamnophis are the gartersnakes and ribbonsnakes. At least five species occur in Texas. Long and thin with stripes running the length of their bodies, the gartersnakes are named for the fancy garters that once supported men's stockings. These snakes are live-bearers, occur near water, and feed on aquatic organisms. The various species can be similar in appearance, and it is sometimes necessary to count the row of scales where the stripes appear.

ORDER TESTUDINES – THE TURTLES

Texas is home to 35 species of turtles, including five sea turtles, three land turtles, one estuarine turtle, and at least 26 freshwater turtles. Aquatic turtles have physiological adaptations to reduce oxygen use and absorb oxygen from their mouth cavity, allowing these vertebrates to extend their time under water. In fact, snapping turtles have been shown to survive up to 17 hours without access to air. All turtles lay their eggs on land. The temperature of incubation determines the sex of the young. High temperatures usually produce females; low temperatures usually result in males. These species have a long lifespan but also have high hatchling mortality. As a result, many species have been endangered by overharvest, and conservation strategies must take a long-term approach.

Family Chelydridae – The two species of snapping turtles in family Chelydridae are probably Texas's most fearsome turtles. They are known for their large heads, long tails, aquatic habits, and ferocious temper. The alligator snapping turtle, found in East Texas, is one of the largest freshwater turtles in the world. It can reach weights up to 150 pounds but has a surprisingly passive feeding strategy. It has a small, worm-like appendage inside its mouth that it wiggles to lure fish into its mouth as it sits in the bottom of the stream or lake. The North American snapping turtle, found across most of the state, is smaller but equally prehistoric-looking. Its tail has serrations reminiscent of dinosaur depictions. A folklore myth states that a biting snapping turtle will not release its victim until it thunders. Though there is no

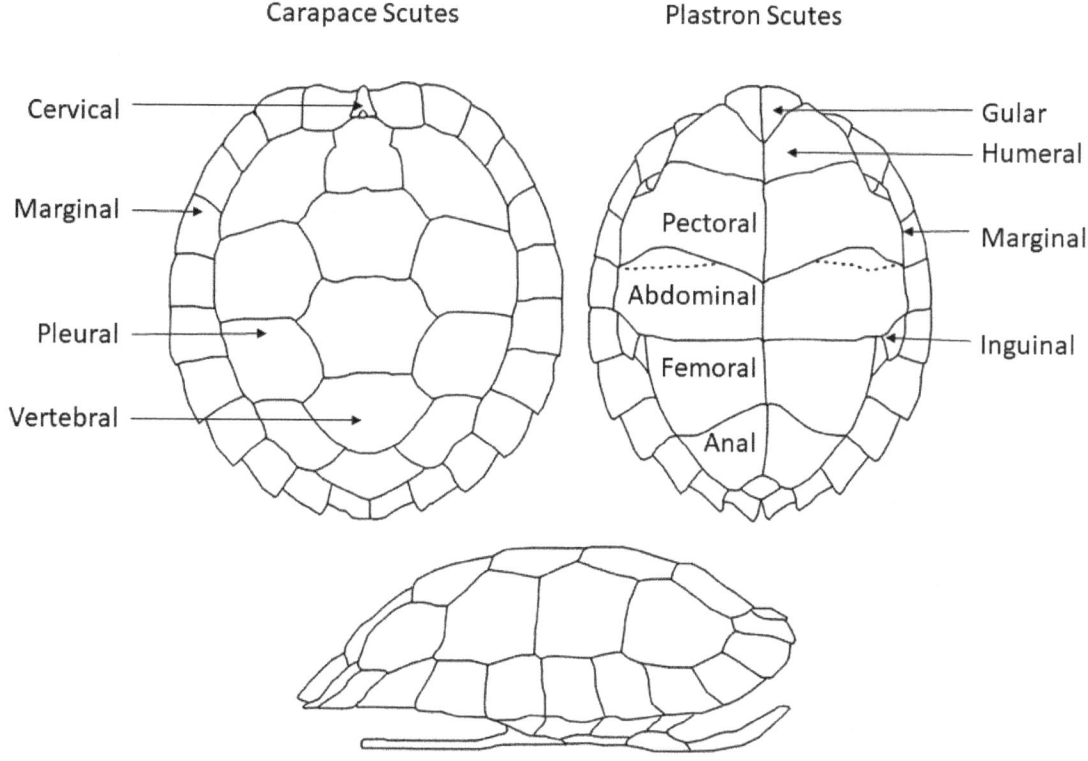

By Obsidian Soul - Own work, CC0, https://commons.wikimedia.org/w/index.php?curid=14731046

Figure 15.21. Scutes on a turtle's upper shell (carapace) and lower shell (plastron) can help in identification of the species. Scutes are named differently, based on their location on the shell.

basis for that belief, a bite from the powerful jaws of a frightened snapper certainly is painful!

Family Trionychidae – Texas has two species of softshell turtles with leathery, flexible shells and snorkel-like snouts. These pancake-shaped aquatic turtles do not look threatening but are prone to biting. Softshell turtles are frequently seen basking on the shore, quickly scooting into the water when approached. They sometimes bury themselves in sediments in the river bottom. In the past, softshell turtles were heavily harvested for food.

Family Kinosternidae – Texas is home to at least five species belonging to Kinosternidae, the mud and musk turtle family. Known for their musky odor (one musk turtle in particular, *Sternotherus odoratus*, has a nickname "stinkpot"), these are all fairly small turtles with domed shells and barbels (fleshy "whiskers") on the chin and neck. They are strongly aquatic and rarely bask outside of the water. The shape of the carapace and plastron can help to differentiate between species.

Family Emydidae – This family, referred to as the water and box turtles, is large, with many diverse habitats and life history strategies among its members. There are 18 species in Texas. A quick look at some of the genera gives an appreciation of the family's diversity.

The diamond-backed terrapin (genus *Malaclemys*) is a widely distributed estuarine turtle found in salty and brackish waters from New England to Mexico. It was once harvested heavily as the traditional source of

Figure 15.22. The alligator snapping turtle is associated with large rivers in East Texas. Photo: Peter Paplanus, CC BY 2.0.

Figure 15.23. Softshell turtles have reduced bony plates in their carapace covered with a leathery skin instead of hardened scutes. Photo: 2ndPeter, CC BY 2.0.

turtle soup, leading to a threatened status. Diamond-backed terrapins are strikingly marked, with spotted gray bodies and strong concentric rings on their shell.

Texas has five species of map turtles (genus *Graptemys*). These beautiful turtles are known for their topographic map-like markings on their shell and bodies. Several of them have very restricted ranges, including the Texas map turtle, found only in the Colorado River drainage, and the Cagle's map turtle, located only in the Guadalupe River drainage. Because these species need diverse aquatic habitats, such as both riffles and pools, they have been impacted by dams and channelization. They have also been collected for the pet trade.

Several types of turtles in the family Emydidae might be called basking turtles. These include the sliders, cooters, painted turtles, and chicken turtles. All these species spend a fair amount of each day basking in the sun to raise their body temperatures. They have relatively flattened shells and can be distinguished by patterns on the heads or shells. The red-eared slider in this group is probably the most widespread and familiar turtle in Texas. It is often sold in the pet trade.

Figure 15.24. The red-eared slider is one of the most commonly-encountered turtles in Texas. As sliders age, the red head marking often disappears. Photo: Mark Alan Storey.

In addition to these aquatic turtles, the terrestrial box turtles (genus *Terrapene*) are in the family Emydidae. Texas has two species—the three-toed box turtle found in East Texas and the ornate box turtle found across the entire state. Box turtles get their common name from the hinge on the plastron that allows the two lobes of the plastron to close completely like a box. They have a domed carapace and are strictly terrestrial, not using water as a habitat. Unlike most turtles, their diet is omnivorous. They may live up to 25 to 35 years in the wild. They are frequently kept as pets, which poses a couple of concerns. Collecting these long-lived turtles may deplete wild populations. In addition, the release of pet turtles has significant potential to spread diseases from captive to wild populations.

Figure 15.25. The ornate box turtles is often recognizable by the starburst pattern on its carapace. Eastern box turtles often sport red coloring on the skin, and males have red eyes. Photo 1 (ornate box turtle): Patrick Feller, CC BY 2.0. Photo 2 (three-toed box turtle): Mark Alan Storey.

Family <u>Testudinidae</u> - The Berlandier's tortoise (also commonly called Texas tortoise) is a member of this family, which includes the massive Galapagos and Aldabra tortoises. Texas tortoises share characteristics of those species, including a domed carapace with ridges on the scutes and heavy, elephant-like legs. Restricted to South Texas, they are primarily herbivorous, feeding on cactus and forbs. However, they may occasionally take meat or invertebrates. Texas tortoises can reach ages over 50 years. Because of its longevity and interest in the pet trade, the Texas tortoise is listed as threatened by TPWD.

Figure 15.26. The Texas tortoise, like box turtles, is strictly terrestrial. It is recognized by its thick strongly-ridged carapace and heavy legs.

THREATS

TPWD considers 49 species of reptiles to be Species of Greatest Conservation Need. All five sea turtles that occur in Texas waters are protected under the Endangered Species Act. One snake—the Louisiana pinesnake—is listed by USFWS as threatened and is thought to be extirpated in Texas; the dunes sagebrush lizard is listed as endangered. TPWD lists six more snakes, four turtles, and two lizards (the Texas horned lizard and the greater short-horned lizard) on the state list as threatened.

Habitat threats create concerns for reptiles, especially those with limited distributions or specific habitat requirements. For example, the dunes sagebrush lizard is restricted to shinnery oak dune complexes in West Texas and eastern New Mexico. Louisiana pinesnakes are associated with pocket gophers in natural longleaf pine forests on sandy soils—a scarce habitat now. The Concho watersnake and Brazos River watersnake are only found in the upper Colorado and Brazos River watersheds. They thrive best in natural riffle-and-pool habitats that are threatened by reservoirs. Alligator snapping turtles, map turtles, and cooters need clean, unaltered streams.

Introduced species also have the potential to threaten native reptile species. Data indicate that Texas horned lizards are more likely to be absent where red imported fire ants are present. It is largely unknown how introduced herp species compete with native species. Mediterranean geckos are now one of the most abundant lizards in Texas. The more robust brown anole uses the same habitats as the native green anole. More challenges could be on the horizon. In 1993, a brown tree snake was discovered and removed from a cargo ship that had arrived in Corpus Christi, Texas, just in time to prevent its escape into the state. The introduction of this species has caused the extinction of many species in Guam.

There is limited knowledge regarding the impact of diseases and toxins on reptiles. Over the last 20 years, biologists have become concerned by an increasing prevalence of snake fungal disease in the eastern U.S., including in East Texas. The disease causes a thickening of the epidermis and can lead to death. Box turtles in captivity are very susceptible to bacterial and viral respiratory diseases, diseases that can spread to the wild if captive turtles are released. One study found that oviparous snake species were absent from a study site with high levels of pesticide in South Texas.

Perhaps more than any other wildlife group in Texas, concern has been expressed about the effects of overharvesting reptiles for commercial trade. In 1999, more than 45,000 herps comprising 100 species were collected in Texas. A few selected species made up 90% of the harvest. These included the spiny softshell turtle, western diamondback rattlesnake, eastern collared lizard, common side-blotched lizard, red-eared slider, yellow mud turtle, and ornate box turtle. A total of 16,110 native turtles were collected that year. Texas accounts for 75% of the nation's trade in rattlesnake products. These concerns have led TPWD to revise regulations for the harvest of nongame wildlife in the state.

CONSERVATION AND MANAGEMENT

Reptiles can benefit from management actions taken to improve wildlife habitat in general. Strategies that prevent overgrazing and wildfires and create variety in habitat structure enhance habitat for reptiles. Moderate grazing, well-controlled prescribed burns, and selective clearing can help to develop the open structure that many reptiles prefer. Leaving dens, natural rock structures, and brush piles in place is essential. On the other hand, the rearrangement of rocks on land or water to create artistic rock stacks or cairns can cause a loss of herp habitat. These structures can also directly kill reptiles and amphibians if the piles fall.

As indicated earlier, TPWD has taken action to monitor and, in some cases, restrict the collection of reptiles from the wild. Although anyone with a hunting license can collect reptiles (as long as they are not listed as threatened or endangered), there is a possession limit of 25 animals, including no more than six freshwater turtles. There are also some restrictions regarding collection on public roads and the use of lights at night.

Anyone who wants to sell nongame must have a Commercial Nongame Permit and provide an annual report. Due to concerns about increasing overseas demand for turtle meat and the very long lifespan of adult turtles, no commercial sale of wild-caught turtles is allowed in the state. Two lists regulate the sale of other species. The White List includes those species that may be collected in the wild and sold. This list contains about 75 species of herps. The Black List consists of those species that cannot be collected in the wild for commercial purposes but may be bought and sold from captivity or other legal states. This list includes over 100 species of herps.

One issue that continues to arouse contention in the state is the appropriate harvest level for rattlesnakes. Springtime traditions in several small towns in Texas include "rattlesnake roundups"—festivals organized by local civic groups designed to draw tourists to observe the harvest of thousands of rattlesnakes. Originally organized with the intention of reducing rattlesnake populations, rattlesnake roundups have morphed into a fund-raising carnival-type event that engenders pride in many local communities. Rattlesnake roundups in about a half-dozen Texas towns like Sweetwater and Freer draw tens of thousands of tourists. They are estimated to raise thousands of dollars for local charities and add millions to local economies. Local organizers emphasize that roundups provide education about rattlesnakes and keep rattlesnake populations under control. Yet, many questions have been raised about these events. Because some activities seem to build fear and glorify butchering snakes, both animal rights activists and wildlife professionals express concern about the messages being conveyed. Little data exists to understand whether the harvests are sustainable, and environmental concerns exist because gasoline is often used to drive rattlesnakes from their springtime dens. TPWD organized a working group in 2016 to bring diverse interests together to try to develop agreements about rattlesnake roundups. Unfortunately, the participants found it hard to discover common ground, and no agreements were reached regarding harvest methods or limits.

Nevertheless, herps continue to have their heroes in Texas. About a dozen herpetological societies are scattered around the state, plus several groups devoted to specific interests such as turtles or horned lizards. Amateur and volunteer herpetologists gather valuable data while "herping" (looking for reptiles and amphibians in the wild). Some communities have begun to embrace alternatives to rattlesnake roundups. The city of Sanderson in West Texas offers a festival called Snake Days. The Snake Days website describes it as "an event that celebrates the wonderful reptile and amphibian diversity of West Texas and raises money for wildlife diversity conservation. It brings together enthusiasts from around the United States who are passionate about herpetology, field herping, herp photography, and building relationships with other like-minded individuals." Reptiles need more friends like that!

Figure 15.27. Rattlesnake roundups, such as the popular event held in Sweetwater, Texas, each year, draw large crowds. They also raise questions about the impact on snake populations and the messages they convey. Photo: WC (public domain).

CHAPTER 16 – TEXAS INVERTEBRATES

Invertebrates, animals that lack a vertebral column (or backbone), dominate the world. They make up more than 95% of all living animal species. In a given ecosystem, they may compose more than 90% of the living biomass. Invertebrates play crucial ecological roles through their diverse interactions with other organisms. They serve as decomposers, pollinators, parasites, and predators. More than 85% of the world's plants, including two-thirds of our food crops, require animal pollinators, mostly invertebrates. Invertebrates aerate and create soil. Seed-eaters, such as harvester ants, aid in plant dispersal. Invertebrates are key elements in food chains that support all other vertebrate species. Nearly all bird species (96%) feed on invertebrates at some point in their life. (One swallow chick can consume two million insects before it fledges!) Many invertebrates eat plants, but they also prey on pests that eat plants; one-quarter of all insect species are parasites or predators of other insects. Without their presence, ecosystems would crash. In fact, E.O. Wilson, the renowned late Harvard entomologist, once said that invertebrates are "the little things that run the world."

Invertebrates also provide direct benefits to people. Many cultures depend on invertebrates as a major source of protein. Horseshoe crabs are used for testing bacterial contamination in medical devices and vaccines. Spiders produce silk that is used in industrial applications, such as bulletproof vests and surgical sutures. Invertebrate communities are used as crime scene evidence in forensic investigations. Some insects spread human disease, but other insects help keep vector insect populations in balance.

E.O. Wilson was right—we need invertebrates!

INVERTEBRATE DIVERSITY

About 1,125,000 species of animals have been cataloged on Earth. Of these, 97% are invertebrates. When plants, fungi, & protists (simple, few-celled organisms) are included, invertebrates still make up 75% of all known species. Dr. Rob Wiedenmann, past president of the Entomological Society of America noted that, when it comes to insects, "Species diversity is not only greater than we imagine, but greater than we *can* imagine."

Table 16.1. Percentage of animal species by category worldwide.

Phylum name(s)	Description	Percent
Chordata	Animals with spinal cords	3%
Porifera, Cnidaria & Echinodermata	Marine inverts – corals, sponges, starfish, etc.	2%
Platyhelminthes, Nematoda & Annelida	Worms	4%
Mollusca	Mollusks	7%
Arthropoda	Jointed-legged invertebrates	84%

(overleaf) Figure 16.1. Texas State Insect, a male Monarch Butterfly. Photo: Mark Alan Storey.

In this chapter, we focus on terrestrial invertebrates found in Texas, notably worms, mollusks, and arthropods. With such diversity, identification can be challenging. Field guides, especially those with a dichotomous key, are helpful. In addition, a strong magnifying lens or a binocular dissecting scope may be required. Visits to university and museum collections can aid in identification. A good photo uploaded to a website such as www.iNaturalist.org can invite other experts to help with the identification of some species.

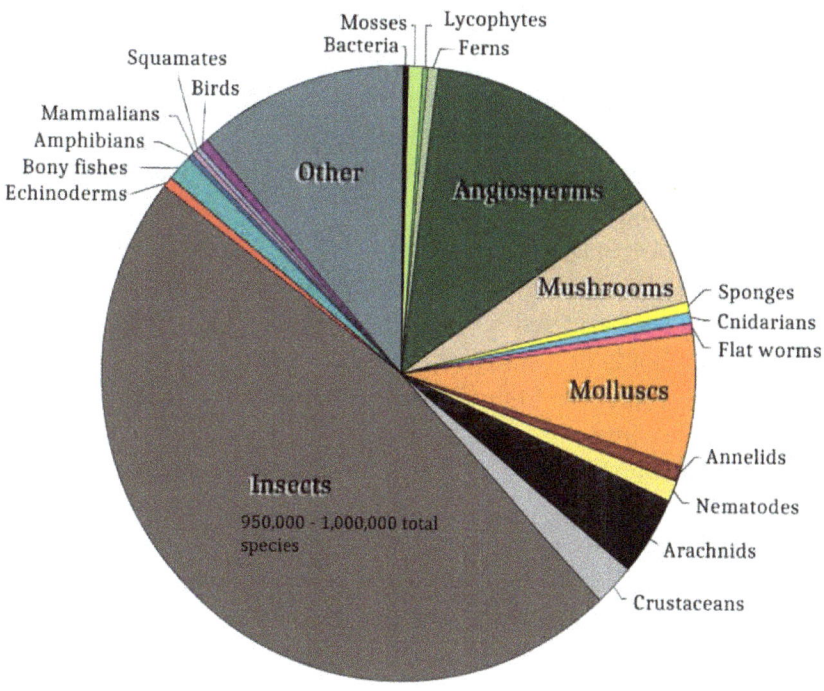

Figure 16.2. Invertebrates make up about three-fourths of all known species on earth. Source: Gretarsson, CC0 1.0.

WORMS

There are three phyla of worm-like organisms.

- Platyhelminthes are the flatworms. Many species are parasitic or marine; aquatic planarians in the order Triclada are frequently used in science classrooms. A few terrestrial species exist in moist habitats where they are predators of other invertebrates.

- Nematodes are roundworms. Tapered at both ends, most are tiny, although parasitic *Ascaris* can grow quite large. Some species such as hookworms are parasites of animals and plants, but many are free-living in soil or aquatic environments. Native nematodes are part of a healthy soil environment, feeding on a variety of plants, fungi, and grubs.

- Annelida includes the segmented worms, like earthworms and leeches. Earthworms are ecologically important as recyclers. They are decomposers, consuming soil and detritus and then excreting it in a

Figure 16.3. Most earthworms encountered in Texas are actually non-native. Photo: schizoform, CC BY 2.0.

more usable form. Other types of annelids may be predatory or omnivorous in their feeding habits. Leaches are infamous because some species can be parasitic blood feeders. They are studied in the medical community for their anticoagulant properties.

Little is known about the number of species of worms in Texas. One concern is that many non-native species have been introduced, presumably transported in soil. The two most familiar Texas earthworms, red wigglers and nightcrawlers, are both introduced species thought to originate in Europe. The effects of introduced earthworms on native species and soil ecology are not well understood. Some non-native species, such as predatory hammerhead flatworms, pose risks to other native invertebrates like snails, slugs, and earthworms.

MOLLUSKS

The phylum Mollusca is a diverse group including clams, snails, slugs, octopus, and squid. The snails and slugs are in the order Gastropoda, which translates as "stomach foot." Gastropods have a soft, slimy body. They secrete a mucus that protects them from drying out and enables them to glide over a variety of surfaces on their stomach. Snails have a coiled shell that grows as the animal grows. The shell protects them from some predators, and some shells have a door or "operculum" that snails can close during arid conditions to protect them from desiccation (drying out). Terrestrial gastropods lay globular masses of gelatinous eggs on plants or in the soil. Most terrestrial species feed on plants or detritus. Birds, ground beetles, garter snakes, turtles, lizards, frogs, and toads feed on snails and slugs.

According to the pre-eminent Texas malacologist Raymond Neck (a malacologist studies mollusks), Texas is home to at least 400 marine, 42 freshwater, 165 terrestrial, and 11 introduced gastropods. Terrestrial species tend to be associated with moist microclimates and may require specific soil types or vegetation communities. Some species of concern include endemic snails in the family Polygyridae (the pill snails) found in woodland habitats of the Edwards Plateau, land snails associated with specific mountain ranges in West Texas, a few South Texas species, and freshwater spring snails of the Trans-Pecos. Because snails cannot disperse widely, small isolated populations are especially vulnerable to disappearing. TPWD's citizen science program, Texas Nature Trackers, sponsors a Terrestrial Mollusks of Texas Project on iNaturalist.

On the other hand, several introduced mollusks have established large populations in the state with significant ecological threats. These species were brought in accidentally on plants or purposefully for escargot or the pet trade. The Texas Invasives Species Institute lists 18 species of non-native gastropods that occur in Texas. Species like the brown garden snail, Asian tramp snail, and chocolate-band snail pose threats to a variety of fruits and vegetables, as well as native plants. The decollate snail, usually recognized by the blunt tip of its conical shell, was introduced to Texas from the Mediterranean to try to control the brown garden snail. Several introduced slug species, including the black velvet leatherleaf slug, leopard slug, and yellow cellar slug, are also widespread in Texas. Slugs can be intermediate hosts for parasites that threaten humans—another drawback to their

Figure 16.4 Wolfsnails are predators on other snails. At least three species are native to Texas.

introduction and expansion. Introduced snails are also a threat to aquatic systems in the state. A notable example is the island (or channeled) apple snail, a very large freshwater snail that lays large clumps of bright pink eggs.

Identification help for snails is available online at the Field Guide to Texas Snail Shells, Mollusk Man Field Guide.

Figure 16.5. The chocolate-band snail is a common non-native species around homes and gardens. Photo: Holger Krisp, CC BY 4.0.

PHYLUM ARTHROPODA – THE JOINTED-LEGGED ANIMALS

Table 16.2. Percentages of arthropods by class.

Class	Description	Percent
Insecta	Insects	73%
Arachnida	Spiders, scorpions, ticks, mites, etc.	20%
Crustacea	Crabs, shrimp, crayfish, amphipods, isopods, etc.	7%
Diplopoda	Millipedes	0.1%
Chilopoda	Centipedes	0.1%

By far the largest group of invertebrates in Texas and the world is the phylum Arthropoda. Arthropods have bilaterally symmetrical bodies and a protective rigid outer exoskeleton made of chitin. As the animal grows, they must shed the exoskeleton. As the name implies, their legs are jointed ("arthro" means joint, and "pod" refers to feet or legs). In addition, the legs and body segments of Arthropods are usually arranged to form a few main sections. Arthropods are divided into five classes; the vast majority of species belong to the class Insecta (Table 16.2). The website www.BugGuide.net offers help in the identification of arthropod species.

Class Crustacea

The class Crustacea is very diverse, including crabs, shrimp, lobsters, crayfish, barnacles, amphipods, isopods, and tiny species such as copepods, krill, and even some parasitic forms. Crustaceans are characterized by two pairs of antennae and at least five pairs of legs. The mouthparts and legs may be specialized to help with feeding, such as the pinchers of lobsters or crayfish.

Crustaceans are primarily aquatic. In Texas, only the order Isopoda has terrestrial species. Isopods include the pillbugs and sowbugs, small segmented terrestrial crustaceans with seven pairs of legs. They usually occur in moist microhabitats where they feed on decaying wood, leaf litter, fungi, fallen fruit, dead animals, and even feces. Observers on iNaturalist have reported 24 species of isopods in Texas. Little is known about the status of isopods in the state. The most familiar species, the common pill woodlouse or roly-poly, is introduced from Europe.

pair of legs is modified to serve as venomous claws to grasp and immobilize their prey. Centipedes primarily feed on invertebrates, but large species occasionally take small vertebrates. Observers on iNaturalist have reported 27 species of centipede in Texas. Two familiar species are the Texas redheaded centipede (sometimes called giant desert centipede or *Scolopendra*, its genus name) and the introduced house centipede. *Scolopendra* are large (up to eight inches long), have 21 or 23 pairs of legs, and are infamous for their painful bite. House centipedes, native to the Mediterranean region, are much smaller with 15 pairs of very long legs. They prey on insect pests in the home.

Figure 16.6. The pill woodlouse, or roly-poly, is the most commonly-encountered terrestrial crustacean in Texas. Photo: Peter O'Connor, CC BY-SA 2.0.

Figure 16.7. The Texas redheaded centipede has a painful bite. They feed on invertebrate pests and even small vertebrates . Photo: LennyWorthington, CC BY-SA 2.0.

Myriapoda

The subphylum Myriapoda includes the classes Chilopoda (the centipedes) and Diplopoda (the millipedes). Myriapods (which translates as "many feet") have legs on multiple body segments and one pair of antennae.

Centipede means one hundred legs, although species may have as few as 15 or as many as 177 pairs of legs. Centipedes have one pair of legs per body segment. They are carnivorous, aggressive, and fast. The first

Millipede means one thousand legs, but the highest number of legs ever reported is *only* an impressive 752. Body segments are fused in millipedes, so they appear to have two pairs of legs per body segment (Diplopoda means "double foot"). Millipedes move more slowly, feeding on moist leaf clutter, fungi, and decayed plant material on the ground. They are docile and easy to handle, often coiling to protect themselves, but they may exude a noxious odor. They may live up to seven years. Observers on iNaturalist have reported 56 species of millipede in Texas. The pill millipedes resemble the woodlouse in the phylum Crustacea; however, woodlice have only one pair of legs per body segment.

Figure 16.8. Desert millipedes have a long lifespan and can grow up to nine inches in length. Photo: Panza Rayada, CC BY 2.0.

Figure 16.9. Golden orb-weaver spiders, also called banana spiders, produce intricate webs. They also use silk to wrap up their prey. Photo: HavannaMabel, CC BY-SA 4.0.

Class Arachnida

Arachnida includes spiders, harvestmen, ticks and mites, scorpions, and several scorpion-like species. Arachnids have eight legs, but other body features differ. Texas has about 1,000 species of arachnids.

Spiders are the most familiar arachnids. These carnivorous (and somewhat intimidating) creatures have two body parts—the cephalothorax and the abdomen. Most have six or eight eyes, although some have as many as 12 eyes and some cave species have none! All spiders have fangs at the end of mouthparts called chelicera. The venom in the fangs is used to subdue prey. Spiders then regurgitate digestive enzymes on the subdued prey, thus digesting it before eating it. Spiders have four spinnerets on the abdomen that produce silk. Silk is used to spin webs or snares to capture prey, build shelters for the spider, and wrap eggs. Spiders can also produce a line of silk called a drag line to catch itself if it falls from a surface. Young spiderlings produce extremely long strands that allow them to float on the wind to reach new habitats, a process called "ballooning."

There are 62 families of spiders in North America; 48 families and about 900 species have been reported in Texas. When people imagine spiders, many think of the large, bright garden spiders that weave intricate webs spanning spaces in gardens and woodlands—the orb-weavers. However, spiders vary in their web strategies. Like many other species, two notable spiders in Texas—the black widows and brown recluse—tend to build messy webs in corners. Active hunters like tarantulas and wolf spiders only use silk to line their burrows and wrap their eggs. Spiders may lay just a few up to several hundred eggs. Wolf spiders and a few other species show maternal care by carrying the egg sacs and then the young spiderlings on their backs. Five species of cave-dwelling spiders are listed as threatened or endangered in Texas.

Figure 16.10. Texas is home to at least eight species of tarantula. The Texas brown tarantula is the most commonly-reported. Photo: snowpeak, CC BY 2.0

HIGHLIGHT 16.1 - WHOM THEN SHALL WE FEAR? – GETTING TO KNOW TEXAS'S VENOMOUS SPIDERS

Texas has only two types of spiders with venom dangerous to humans—the brown recluse and black widows. The three species of black widow in the state are characterized by a bulbous black abdomen with red markings, often in the shape of an hourglass, on the underside. They build messy webs in dark, secluded areas, such as caves, hollow logs, and storage buildings. The effects of a black widow encounter can vary. Antivenins may be used to treat bites; however, a study of bite victims showed that 65% of the reactions were mild, and only 1% were life-threatening. Black widows have important ecological roles. They feed on ants, including red imported fire ants, weevils, scorpions, and other pests. In addition, mud dauber wasps prey on black widows. They sting widows to paralyze them and then place the hapless spiders in a chamber where they provide food for the emerging wasp larvae.

The brown recluse accounts for a greater number of medically-significant bites in the state. Still, extensive tissue damage only occurs in a small percentage of bite cases. This small, long-legged spider is frequently encountered indoors. Bites often occur when a spider becomes injured or trapped in clothing or bedding. Brown recluses are recognized by the presence of a dark violin shape on the top of the cephalothorax. They build loose irregular sheet webs under woody debris and often under furniture indoors.

Figure 16.11. Two types of Texas spiders with venom dangerous to humans. Black widow photo: tkksummers. CC BY-SA 2.0. Brown recluse photo: Annika Lindqvist, CC BY 4.0.

Harvestmen, such as the daddy-longlegs, are in a different Arachnid order. They have a round unsegmented body and only two eyes. Most have long spindly legs; many have scent glands that produce a smelly fluid. They feed on plant juices and other arthropods, but they lack venom and silk glands. Texas is home to at least 18 species. Three species of harvestmen found in Central Texas caves are listed as threatened or endangered.

Ticks and mites have two body segments. The species are parasites on other organisms. As ticks progress through their development they may require several different feeding episodes on different hosts. Ticks can be significant disease vectors in some parts of the country. Texas is home to blacklegged (deer) ticks associated with Lyme disease and dog ticks associated with Rocky Mountain spotted fever. Ehrlichiosis is another tickborne disease that is occasionally reported from Texas. Chiggers are another pesky member of this group. The parasitic larvae of these mites latch on to passing vertebrates, creating an itchy reaction to the digestive enzymes that they secrete at the bite mark.

Figure 16.12. American dog ticks, shown here before feeding, carry rocky Mountain spotted fever, a disease that has been reported in East Texas. Photo: Judy Gallagher, CC BY 2.0.

Scorpions are also arachnids. In addition to eight legs, they have two enlarged mouthparts called pedipalps with pinchers for catching prey. They also have an elongated tail with a stinger that injects venom into their prey. Scorpions give birth to live young, and the female scorpions provide care for the young, sometimes carrying them on their backs. Scorpions mature slowly, and some species may live for 20 to 25 years. Interestingly, their cuticle (exoskeleton) glows at night under UV light. Texas has 18 species of scorpion, with the greatest diversity in West Texas. The most widely distributed species is the striped bark scorpion.

Figure 16.13. Scorpions, such as the striped bark scorpion, give birth to live young that the mother carries on the back. Photo: Alejandro Santillano, UT-Austin (public domain).

Finally, there are three orders of scorpion-like arachnids. Only one species of whip scorpion, the giant vinegaroon, is found in Texas. It is easily recognized by its long, whip-like tail and the vinegar-like odor it secretes. Texas has 26 species of windscorpions. Also called sun spiders, they are fast—running like the wind! Finally, the pseudoscorpions lack the long tail of the scorpion. These small predators exist in a wide variety of habitats, including bird nests, where they prey on other arthropods. There are at least 33 species in Texas. One karst (cave-dwelling) pseudoscorpion in Central Texas is listed as endangered.

CLASS INSECTA

Insects dominate our planet. Over 60% of all animal species on earth are insects. There are estimated to be between 950,000 and 1,000,000 described species of insects; many more are likely to be discovered. About 100,000 species occur in the U.S., with an estimated 30,000 species in Texas. The class Insecta is divided into about 30 orders, but 90% of the species are found in just five orders. *Common Insects of Texas and Surrounding States, A Field Guide* by John C. Abbott and Kendra Abbott is a useful reference for the state.

Adult insects have 6 legs and usually two pairs of wings. Their body segments are fused into three main parts—the head, thorax, and abdomen. The head of adult insects bears one pair of antennae and one pair of compound eyes. Some insects also have additional simple eyes located on the head called ocelli.

Insects may go through several stages of development as they mature. In incomplete metamorphosis, also called simple metamorphosis, the immature look much like the adults. The terrestrial larval forms pass through several growth stages called instars, molting or shedding the exoskeleton between each instar stage. They gradually develop their wings with each molt (Fig. 16.14).

In the process of complete metamorphosis, the larva that hatches from the egg looks different than the adult form. The larval forms molt as they grow. Eventually, temperature, day length, and growth stage combine to trigger the larva to enter a pupal stage that involves forming a protective coating. During the pupal stage, the tissues of the larva are broken down, and small groups of cells develop into the adult. The former larva emerges from the pupal covering as an adult, usually with wings. The lifespan of adults varies considerably, but it can be short, often only allowing time for the adult to breed.

Insects are adapted to a wide variety of habitats and climatic conditions. They inhabit every imaginable environment, even including polar regions. They occur in temperatures ranging from -40°F up to 150°F. Insects aestivate during cold or drought and can enter **diapause**, a period of delayed development due to unfavorable environmental conditions. They are specialized to feed on nearly everything—plants, pollen, nectar, seeds, wood, blood, carrion, hair, horn, bone, dung, fungus, wax, and even petroleum (mealworms are used to help degrade plastics!). Exploration of a few orders engenders an appreciation of their remarkable diversity and adaptations in Texas.

Table 16.3. Insect orders with the most species worldwide.

Order	Description	# of Known Species
Coleoptera	Beetles	386,500
Lepidoptera	Butterflies and Moths	157,338
Diptera	Flies	155,477
Hymenoptera	Bees, Ants, & Wasps	116,861
Hemiptera	True Bugs	103,590

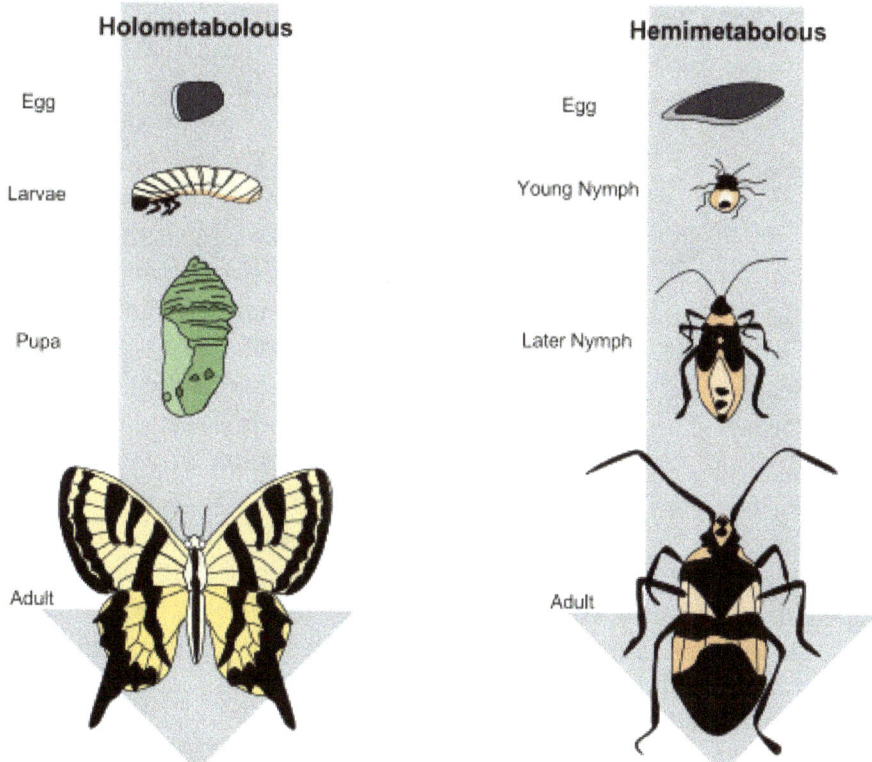

Figure 16.14. Complete metamorphosis (holometabolous) and incomplete metamorphosis (hemimetabolous) in insects. Source: Username1927, CC BY-SA 4.0.

Order Coleoptera – Beetles

When it comes to species diversity, beetles dominate. About 40% of insects are beetles. One-third of all animal species are beetles. Nearly one-quarter of every living species on earth is a beetle. There are more species of weevils (one beetle family) than there are in most animal orders. Geneticist John B.S. Haldane once said, "If one could conclude as to the nature of the Creator from a study of creation it would appear that God has an inordinate fondness for stars and beetles."

The term "Coleoptera" means "sheath wing." Beetles are named for their hardened forewings called elytra. When not in flight, the elytra give beetles their domed, often colorful, appearance. Beetles range in size from less than one millimeter to over three inches. They are specialized for many food sources and habitats. Some coleopterans, such as weevils, are pests, but others, such as ladybugs, are used as biological control agents. Beetles have complete metamorphosis. The larvae of beetles also have a diverse array of body shapes, feeding preferences, and habitats. Some larvae are called grubs and may be found sheltering in the ground, in the stems of plants, and under the bark of trees.

There are 131 families of beetles in North America and about 7,800 species of beetles in Texas. Nine Texas species are listed as threatened or endangered. Most imperiled species are associated with spring or cave habitats, such as the Comal Springs riffle beetle found only in San Marcos and Comal Springs. The American burying beetle is a large species once found in the grasslands of Northeast Texas that raises its young in carrion (dead animals). Concern also exists for many tiger beetles—small, fast, colorful predatory beetles found in sandy habitats—

and dung beetles—important waste recyclers in many ecosystems.

Figure 16.15. Dung beetles, such as this rainbow scarab, use feces to feed themselves and their young. They play an important role in breaking down waste in the environment. Photo: Curtis Burke, CC BY 4.0.

Order Lepidoptera – Butterflies and Moths

Lepidoptera, translated as "scale-winged," is named for the colorful scales that cover the wings of these beautiful insects. Butterflies, moths, and skippers (smaller butterflies with wings held over their backs like a fighter jet) are known for their complete metamorphosis life cycles. Lepidopterans lay their eggs on host plants, plant species that provide the specific food that their larvae need. The larvae, called caterpillars, have chewing mouth parts designed for their herbivorous and voracious feeding habits. Most larvae have silk glands. When the larvae complete their growth phase, they secrete a case as they enter the pupal stage (the pupa of butterflies is a **chrysalis**; the pupa of moths is a silken cocoon). The transformed pupae emerge as nectar-feeding adults with mouthparts that form a tubular **proboscis**.

Texas is home to at least 5,088 species. Lepidopterans are popular with citizen scientists who are eager to contribute trend data for the group. The National Butterfly Center in the Lower Rio Grande Valley of Texas hosts an annual butterfly festival, and several websites recruit sightings. Over 50,000 observers have submitted over one million sightings of nearly 4,000 species of Lepidopterans in Texas to iNaturalist. The most frequently reported butterfly is the enigmatic monarch (Highlight 16.3). The fiery skipper, whose larvae take advantage of non-native grasses for feeding, is the most frequently reported skipper species. The white-lined sphinx, also called the hummingbird moth, is the most frequently reported moth. This large striped moth with an extremely long proboscis feeds both day and night, hovering in front of flowers. Its size and feeding behavior bear a remarkable resemblance to hummingbirds (Fig 16.16).

Figure 16.16. The white-lined sphinx moth resembles hummingbirds with its fast wingbeats and long proboscis. Larvae most commonly feed on evening primroses. Photo: Mark Alan Storey.

Order Diptera – Flies

Texas has about 5,100 species of true flies. Diptera means "two wings." Flies, unlike other typical insects, have only one pair of wings. Many have specialized mouthparts designed for piercing, lapping, or sponging. Some flies use external digestion like spiders. The life cycle of flies also demonstrates complete metamorphosis. Some species, such as mosquitoes and black flies, have aquatic larvae; other flies are well-known for their maggot larvae that develop in decaying organic matter. Many fly species have a close association with human habitats. Mosquitoes, a type of fly, are recognized as the most dangerous animals in the world because of the number of human diseases they can transmit, including yellow fever, malaria, and dengue, among others.

Figure 16.17. Some flies, such as this margined calligrapher, resemble bees; however, they bear only one pair of wings. The adults eat nectar, pollen, and small insects. Photo: Judy Gallagher, CC BY 2.0.

Order Hymenoptera – Social Insects

Texas has about 5,000 species of Hymenoptera, the "membrane-winged" insects. The order includes ants, bees, and wasps. They are often called the social insects because some species live in complex cooperative colonies. Although many people associate bees with beehives, of the more than 800 species of bees in Texas, only honey bees and bumblebees are colonial. Others, such as carpenter, mason, and sweat bees, are solitary.

For species that are colonial, individuals in the colonies have unique roles associated with their gender and size. Reproduction is usually based on one or a few specialized females called queens. The young larvae, tended by other colony members, experience complete metamorphosis. Colonial hymenopterans collaborate in producing food for the colony. Leaf-cutter ants in the genus *Atta* live in huge underground colonies that can be 16 feet deep and 100 feet in diameter. They harvest leaves that they carry back to the colony. They use those leaves to grow fungi to feed their young. Honey bees feed on nectar and pollen collected by worker bees called foragers. Honey is produced from the nectar that is collected, and pollen is converted to royal jelly, a food for the larvae, queen, and other colony members.

Hymenoptera are famous, or perhaps infamous, for their stinging ability. Only the females sting. The purpose of stinging is usually to paralyze prey, such as the impressively-sized tarantula-hawk wasps that use their stingers to immobilize tarantulas so that they can lay eggs in the spider's abdomen. Some species also use their stingers in defense of the colony or themselves. Baldfaced hornets, western honey bees, paper wasps (red wasps), and yellowjackets are known to be especially defensive of threats to their colony sites.

Figure 16.18. Bumble bees, such as this Sonoran bumble bee, are distinguishable by the pattern of yellow and black. Texas has several bumblebee species. Photo: Monkeystyle3000, CC BY 2.0.

Despite their prickly reputation, Hymenopterans are extremely beneficial to humans. Many wasps are parasites of pest species. In addition, bees are the most important pollinators on earth for native plants and many food crops. Many native plant species demonstrate colors and shapes designed to perfectly fit their bee pollinators.

Texas Nature Trackers has sponsored a Bees and Wasps of Texas Project on iNaturalist to track these important species.

Order Hemiptera – True Bugs

This large group, with about 3,000 species in Texas, is very diverse. Members have piercing, sucking mouthparts that they use to feed on plants and animals. Reduviidae, the assassin bugs, feed on a variety of prey items, including birds and mammals, and can also inflict a painful defensive bite. The genus *Triatoma* in this family, sometimes called kissing bugs, can carry Chagas disease, a major health concern in Central and South America and occasionally in South Texas. Other species excrete noxious chemicals, giving them the common name "stink bugs."

Figure 16.19. Public health specialists have noted a recent expansion in kissing bug distribution in Texas, raising concern about Chagas disease. Photo: xpda, CC BY-SA 4.0

Members of the order have three primary wing patterns. The "half-wing," for which the order is named, forms a triangular pattern seen in stink bugs and assassin bugs. Plant eaters, such as cicadas, have a peaked roof-like wing pattern. Other members of this order, such as aphids, are wingless. The order demonstrates incomplete metamorphosis. The young larvae, called nymphs, look similar to adults.

Order Orthoptera – Grasshoppers

Orthoptera means "straight wing" and is named for the long forewings on many species. This order includes grasshoppers, crickets, katydids, lubbers, cave crickets, and mole crickets. Although not the most abundant order in Texas, they are extremely valuable as wildlife food. Orthopterans are primarily grazers with long hind legs used for jumping. The larval nymphs undergo simple metamorphosis. Orthopterans are known for the sounds they produce. The males "sing" to attract females by rubbing their wings together or their legs and wings. In turn, they can hear with a tympanum located either on the front leg or abdomen. Texas has about 525 species in 8 families.

Figure 16.20. Snowy tree crickets, like many orthopterans, call at night. Their rhythmic calls can be used like a thermometer! Add 40 to the number of chirps heard in 15 seconds to calculate the temperature in Fahrenheit. Photo: Michelle, CC BY 4.0.

Order Odonata – Dragonflies and Damselflies

The Greek word "odontos" refers to teeth, so the odonates are named for their large, strongly-toothed mouth parts that enable their predatory habits. They are a beneficial group, feeding on mosquitoes and other flying insects. Odonates are strong flyers with large membranous wings. When at rest, dragonflies spread their wings out to the side; the smaller damselflies hold their wings over their backs. The pattern and coloration of cells in the wings can aid in identification of adults. Metamorphosis is incomplete, although the predacious aquatic juveniles called nymphs look quite different from the winged adults. Many odonates are habitat specialists and can serve as indicator species for aquatic habitats.

Due to their size, colors, and visibility, dragonflies are popular among wildlife watchers. Many odonates travel long distances; some species are migratory and concentrate along the Texas Gulf Coast in the fall. Texas has recorded 246 species in 10 families, representing over half of all North American species. In addition to iNaturalist, many people submit sightings to the website Odonata Central.

Figure 16.21. Dragonflies, such as this widow skimmer, are important predators as larvae in water and as flighted adults. Males, females, and young adults often have different coloration. Photo: Mark Alan Storey.

MONITORING

Transect-based sampling can be conducted for some invertebrates, especially highly visible species such as butterflies or dragonflies. However, monitoring or inventory of most invertebrates usually requires capturing the organisms so they can be identified accurately to species. The amount of time sampled and the number of insects encountered during that time can be used to develop a measure of insect abundance. A variety of techniques are described in the *Texas Master Naturalist Curriculum*, including:

- Rocks and logs on land and in water can be overturned to search for invertebrates.

- Sweep nets made of canvas can be swept through herbaceous vegetation to sample a wide variety of adult insects. Lighter-weight nets can be used for capturing aerial insects such as butterflies and dragonflies.

- Aquatic insects and other invertebrates can be collected from streams and ponds using aquatic dip nets or similar collecting tools.

- Pitfall traps can be constructed from buried cups or bottles. A small amount of alcohol placed in the cup or bottle traps the insects that fall into the trap while walking on the ground.

- Malaise traps are a tent-like structure that funnels flying insects into a collection jar; flight intercept traps are a vertical structure that cause insects to fall into a pan placed beneath the trap.

- Berlese funnels use light to drive soil insects downward through a screen into a jar filled with alcohol placed beneath the funnel.

- A hanging bait trap can be constructed from a milk jug. If a smelly liquid with rotting produce is placed in the bottom of the jug, it attracts smell-sensitive insects, such as beetles.

- A very simple trap can be built from a bright yellow plastic bowl filled with a little soapy water. Some insects, especially hymenopterans, are attracted to the color and then fall into the water.

- In some instances, larvae and pupae can be collected in the field and placed in rearing chambers fitted with screened tops. When the adults emerge, they can be identified and catalogued.

- Light traps are an enjoyable way to sample nocturnal insects. They use a black light and/or mercury vapor light shining on a large white sheet hung outdoors at night. Lights can also be placed above a pan of alcohol.

- Finally, a sheet can be stretched or held beneath a tree or bush. If the branches are struck with a stick or other implement, insects fall onto the sheet where they can be collected. This technique can be used during the day or night.

Insects that are collected live can be photographed and released. Photos can then be shared on iNaturalist to increase knowledge about local species. Insects that are killed may be preserved in alcohol or pinned and then placed in a museum or other collection so that the data may be beneficial to others.

THREATS

Unfortunately, the status of most invertebrates is not well known. William Godwin, an entomologist at Stephen F. Austin State University, notes that there is little understanding of insect imperilment. Several studies around the world show cause for concern. A recent German research study found that the total mass of flying insects there had fallen by 80 percent in three decades. Some scientists have measured the downward trends in insect numbers by counting splats on windshields or license plates in standardized annual surveys. This approach examines what is called the "windshield phenomenon." Decades ago, it was difficult to clean car windshields because of the number of insects that would be smashed there while driving, especially at night. Now, most windshields are eerily clean.

TPWD lists 418 invertebrates as Species of Greatest Conservation Concern—about the same number as the vertebrates on the list. The Texas Entomology website compiled by Mike Quinn lists 27 lepidopterans, 4 beetles, 14 orthopterans, 3 stoneflies, 18 odonates, 9 mayflies, and 11 crayfish as invertebrate species of concern, noting that more assessment is needed. A recent review published in the journal *Biological Conservation* noted that some of the greatest declines have been recorded in Lepidoptera, Hymenoptera, Odonata, and dung beetles. The authors cite evidence suggesting that the major causes in order of importance are habitat loss, pollution from pesticides and fertilizers, introduced diseases and invasive species, and climate change (especially in tropical areas).

Figure 16.22. Non-native insects play a significant role in the ecology of Texas ecosystems. Western honey bees are often thought to be beneficial. (Photo: Smudge 9000, CC BY-SA 2.0.) Red imported fire ants are agreed to have caused ecological damage. (Photo: Judy Gallagher, CC BY 2.0.)

HIGHLIGHT 16.2 - THE GOOD AND THE BAD OF NON-NATIVE INSECTS

<u>Western honey bee</u> - Think of pollinators, and everyone thinks of honey bees. However, the most prolific honey bee species in North America is not a native species. Western honey bees (often called European honey bees) were brought from Europe in the 1600s for honey production and crop pollination. As a generalist species, honey bees are excellent at pollinating a wide variety of crops and native plants. Now, about 40% of crops in the United States depend on European honey bees for pollination, and bee colonies are transported across the country to provide pollination services at different times and places. European honey bees have been a welcome introduction.

Beekeepers have faced challenges with these popular honey producers. In 1956 African honey bees were brought to Brazil to produce honey in a more tropical climate. The African bees, which are more aggressive and defensive, soon formed hybrids with the European honey bees. The hybrids quickly moved north from South America, reaching Texas in 1990. Africanized bee colonies can still produce honey, but beekeepers are advised to manage them carefully due to their aggressiveness and may choose to replace Africanized queens with European queens.

Over the past 20 years, additional threats to honey bees have emerged. Fewer honey bee colonies are surviving winter due to a phenomenon called Colony Collapse Disorder. Causes are not yet fully understood, but the U.S. Environmental Protection Agency lists a variety of causes that are being investigated, including infestations by varroa mites, pesticide poisoning in fields or hives treated for mites, reduced foraging habitat and nectar plants, and stress from seasonal hive movements.

<u>Red Imported Fire Ant</u> – Although honey bees have their supporters, no one is a fan of the red imported fire ant (RIFA). This South American species was accidentally introduced in Mobile, Alabama, in the 1930s, probably arriving in soil carried by a ship. RIFA, who have a high reproductive rate and readily invade disturbed habitats, spread quickly across the South. They reached Texas in 1953 and now occur in about two-thirds of the state. RIFA cause damage to electrical equipment, and their bites and stings are a constant nuisance to humans, livestock, and pets. In addition, the introduced fire ants (Texas does have native fire ant species) have been an ecological threat to native wildlife. They are thought to have caused declines in fireflies, other native insects, bobwhite quail, Texas horned lizards, and perhaps other wildlife species.

Pesticides have been widely applied while trying to eradicate RIFA (myricides—ant poisons—were even dropped from airplanes in some counties), but they have not been successful. Instead, the widespread application may have led to the decline of native ant species. A less toxic approach used in smaller habitat areas is to inject steam or boiling water into RIFA beds to kill the queen and larvae. In addition, a RIFA predator from South America—a tiny parasitic phorid fly—has been experimentally introduced in some areas. The phorid flies sting the fire ants, laying an egg in the ant's abdomen. When the larva hatches, it grows inside the RIFA, finally killing the ant and causing its head to fall off. The phorid flies cannot wipe out all the fire ants in a colony, but the decapitation threat causes ants to be less active and aggressive in their environment.

CONSERVATION AND MANAGEMENT

With so many diverse species and such broad threats, designing conservation and management strategies for invertebrates is challenging. It would be impossible to write 30,000 species management plans, and that would just cover Texas's insects! Instead, most conservation biologists suggest taking an ecosystem management approach. Using fire and careful grazing to maintain diversity of native plants, ensuring spring and stream flows, and protecting riparian habitat can benefit invertebrates.

Another focus is to prioritize the conservation of high-value habitats—locations that support unique levels of biological diversity. Many of the threatened and endangered invertebrates in Texas are associated with caves and springs. Management of these environments requires a watershed strategy to prevent polluted run-off into caves, streams, and aquifers, as well as water use strategies that conserve water flow. Unique terrestrial ecosystems also harbor high numbers of unique invertebrates. The Monahans Sandhills in West Texas and the Wild Horse Desert in South Texas provide habitat for many endemic species of plants, vertebrates, and invertebrates. On a smaller scale, keystone species often provide important habitats for invertebrates. Pocket gopher burrows support unique flies, histerid beetles that feed on fly larvae, dung beetles, moths, cave crickets, and carabid beetles that eat cave crickets. Similar ecosystem assemblages exist in wood rat middens, prairie dog towns, beaver lodges, and squirrel and bird nests.

Other individual-level guidelines include leaving dead wood and stalks of forbs in place, allowing leaf litter to build up, building compost piles, maintaining natural spring and stream flow, leaving stream cover like rocks and logs, and careful use of pesticides. Although red imported fire ants are problems for cave invertebrates and many other ground-dwelling species, widespread application of insecticides further harms invertebrate communities and all the species that depend on them. Biologists managing caves instead use boiling water and steam injectors to control RIFA populations. Reducing outdoor lighting can also help insects. Many insects are drawn to artificial lights, disrupting their natural behaviors and making them susceptible to predation. Dark sky initiatives designed to help songbirds and other species can also help insects.

In agricultural settings, farmers can protect insect diversity by using Integrated Pest Management (IPM). IPM tries to prevent insect pest problems through activities like crop rotation or intercropping (growing more than one crop together). Maintaining ground cover and minimizing the amount of soil disturbance during planting (minimum tillage) can help ensure that insect predators remain in the ecosystem. Finally, if insect pests are detected, then trapping, weeding, or using targeted pesticides like **pheromones** (insect hormones) to drive away pests can help maintain beneficial insects.

But perhaps the most important key to maintaining invertebrate communities is to realize that the diversity of animals in a habitat is very closely linked to the diversity of plants in the habitat. Entomologist Doug Tallamy advocates that native plant diversity produces insect diversity which produces wildlife diversity.

Many insects are closely linked to the native plant species in their environment. It is a relationship that has been ongoing for many, many years. Plants fight to develop chemical defenses against insects. Insects fight to overcome them. Then plants develop more defenses. In the process, plants develop unique chemical compounds, many beneficial to human nutrition or medicine. Examples of these unique plant-animal relationships abound. Many species of leaf beetles are so specialized that they feed only on one type of skullcap forb species. Many flowers are designed in a certain shape and color so that they attract pollinator bees that are the perfect shape to transfer pollen efficiently. Many yucca species require one particular yucca moth species for pollination, and those moths require those yuccas for food. All the insects involved in these special relationships in turn feed other insects and other wildlife.

Not all plants are equal. Native plants support many more invertebrates than non-native plants, so increasing the dominance of native plant species can help conserve invertebrate biodiversity. Non-native plants like crape myrtle have no close relatives in the U.S., so they support few native invertebrates. In addition, the importation of non-native horticultural plants can harm native plants. Examples include introduced diseases such as Dutch elm disease and plant pests such as the bacterium that causes citrus greening disease. As noted in earlier chapters, escaped horticultural plants also wreak havoc in native ecosystems. Non-natives in native environments can alter fire frequency, deplete groundwater, decrease soil health, and degrade aquatic habitats by affecting run-off.

Tallamy and his graduate students at the University of Delaware were able to document that small tracts of habitat with native plants produced four times the insect biomass and more than three times as many insect species as tracts that were dominated by non-native species. Even more impressive, the caterpillar biomass—critically important as a bird food—was 35 times higher in on native plant species. Another study found 90% more caterpillars in native patches versus invaded patches.

Birds also respond to non-native versus native ecosystems and the insects they support. In South Texas, introduced grasses such as buffelgrass and Lehmann lovegrass dominate many rangelands. A study conducted by Texas A&M University-Kingsville found that arthropods were 60% more abundant in native grasslands compared to the non-native rangelands. Breeding grassland birds were 32% more abundant in the native habitats. A study of chickadees in suburban environments showed that they prefer yards with native plants. When they nest in yards landscaped with non-natives, they raise fewer chicks. Even when introduced plants provide berries for birds to eat, the nutritional value of the food is lower. Berries of native plants are high in fats, crucial for migration; berries of introduced plants tend to be high in sugars.

More native plans are just what our invertebrates need to run the world!

HIGHLIGHT 16.3 - WAKING UP TO THE PLIGHT OF THE MONARCH BUTTERFLY – A PERSONAL REFLECTION

It was one of the transformative experiences of my life. As I passed through the riparian woodlands of the Frio River that October day in 2008 the trees were alive. The air was alive. The sky above was blue, but everywhere else I looked was a flutter of orange and black. I was experiencing monarch migration.

The scientific name of the monarch butterfly is *Danaus plexippus*, meaning "sleepy transformation." The species, selected as the State Insect of Texas in 1995, has one of the most remarkable insect migrations in the world. Every fall monarch butterflies leave their summering locations in the United States and southern Canada and begin their migrations to two wintering spots located in southern California and Central Mexico. The largest group is funneled through Texas along two primary migration routes, one on the coast and the other through the western Edwards Plateau. I had happened upon a concentration of those Hill Country butterflies, feeding on nectar and resting in the vegetation along the Frio.

Monarchs are an eminently recognizable butterfly species, sporting orange wings patterned with black veins and white dots. The sexes are similar; males may be recognized by thinner black veins and the presence of a small black dot on the hind wings—a sac that secretes scents associated with mating. Monarchs' bright colors also reveal an unpleasant little secret. The warning coloration tells potential predators that monarch bodies contain toxic compounds, cardenolides that they pick up when their larvae feed on their specialty host plants—milkweeds. The warning is serious. In high concentrations, cardenolides can cause heart failure.

The Frio River monarchs were on their way to El Rosario Sanctuary and nearby fir forests in the mountains west of Mexico City, a trip totaling more than 2,500 miles for some of the butterflies. Scientists are not sure how monarchs know the way to El Rosario. Because adult butterflies only live a few months, the Frio River monarchs, like all wintering monarchs, had never been to Mexico. But their ancestors have gone there for thousands of years. Long known to local people, it was not until 1975 that two part-time naturalists living in Mexico City, Catalina Aguardo and Ken Brugger, managed to find the remote wintering locations and share the remarkable sightings with the rest of the world. It is estimated that up to a billion monarchs once sheltered in the mountains of Mexico each year, but, with numbers impossible to count, the monarchs are instead monitored by how many acres they occupy.

Figure 16.23. Female monarch (recognizable by thicker black wing veins) nectaring on a milkweed. Photo: USFWS (public domain).

In Mexico, the monarchs enter diapause, clustered as a continuous mass of butterflies coating trees in the sheltered environment of the forest. As spring approaches, the sleepy monarchs become more active. Mating takes place. The males then die, and the females begin flying northward. As they reach Texas, they seek out milkweed plants to lay their eggs. The females then die, just as the tiny, round eggs begin to hatch into tiny yellow, black, and white striped caterpillars with a set of tentacles on the front and the back. The

Figure 16.24. The life cycle of the monarch butterfly. Source: kazia89636, CC BY 2.0.

larvae feed voraciously on the milkweed leaves, going through several molts. Finally, they are ready to leave their host plant and find a nearby tree or bush where they secrete a beautiful jade green chrysalis adorned with gold dots that houses the pupa. After 9 to 15 days, the sleepy transformation occurs, and a fully-formed butterfly emerges.

The new butterfly born in Texas has never migrated before, but it continues its journey northward, stopping in new rangeland habitat, possibly in Kansas or Illinois or New York, to lay eggs. The egg-laying adults die, but a month later the next generation travels northward again, perhaps even reaching southern Canada. The final generation is born in late summer, and driven by instinct, they head south, passing through Texas, maybe along the Frio River, to Mexico. A few monarchs may reproduce in Texas in the fall and may remain during the winter, but the majority follow the ancient mysterious migration pattern.

Monarchs are much beloved, thanks in large part to aesthetic experiences such as my day on the Frio. Native peoples in Mexico welcome their return, which coincides with the Day of the Dead celebrations, believing the butterflies are the transformed souls of their ancestors. Citizen scientists are fascinated by their remarkable life history and participate in volunteer programs, such as the University of Kansas Monarch Watch, Journey North's online tracking program, and TPWD's Texas Milkweeds for Monarchs Project on iNaturalist. Thousands of homeowners choose to plant milkweed to provide monarch larval plants.

And yet, monarchs are in trouble. Habitat destruction and fragmentation in the wintering grounds have reduced roosting sites and made populations more vulnerable to adverse weather events. Changes in agricultural practices in the U.S., including the increasing use of herbicides during this century, have resulted in fewer weeds to serve as host and nectar plants for monarchs in the U.S. The prevalence of a debilitating parasite is also on the rise in monarchs. As a result, keeping monarchs in captivity is not recommended. Winter surveys in 2022 showed that monarchs only occupied about five acres of forest in Mexico, down from 10 acres when I saw the Frio River monarchs and from 45 acres when assessments first began.

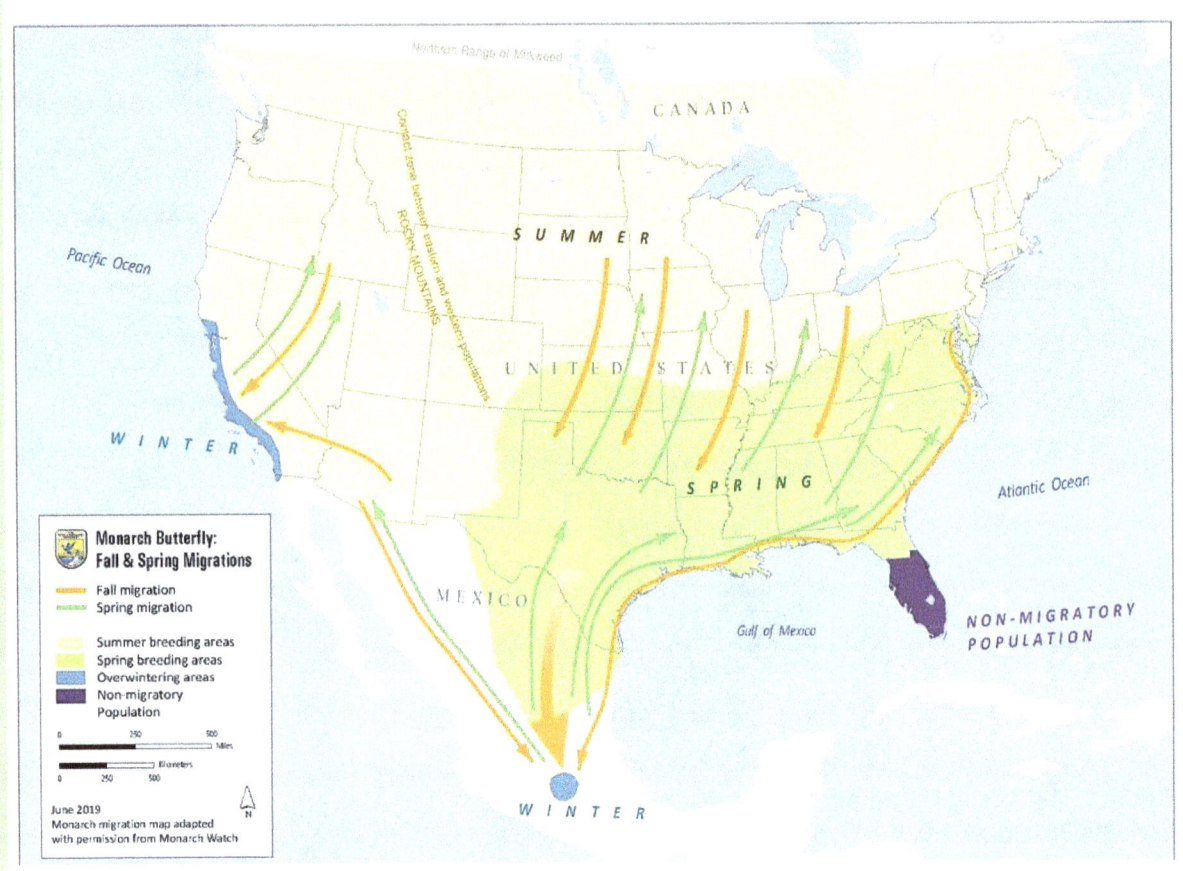

Figure 16.25. Monarchs follow two primary migration paths in Texas. Source: USFWS, CC BY 2.0.

Because of these trends, in 2014 the Xerxes Society, a nonprofit dedicated to invertebrate conservation, and several other partners petitioned USFWS to list the monarch as a threatened species. The actions encouraged the formation of a Monarch Joint Venture to bring together government agencies, universities, conservation organizations, and private landowners to collaborate on conservation, research, and education. There are many reasons to work together. Protecting monarchs provides habitat for a host of other pollinators, protects watersheds in Mexico, and ensures that our Texas state insect can transform the natural environment for future generations.

Figure 16.26. A cluster of resting monarchs transforms a tree into a flutter of orange and black during autumn in Texas. Photo: Jeff Forman, TPWD.

CHAPTER 17 – A ROLE FOR EVERYONE

Everyone has a "niche" when it comes to wildlife conservation. As noted throughout the book, private landowners in Texas play a key role in conserving wildlife habitat in the state, managing millions of acres that benefit game and nongame species. But there are many opportunities for all Texans to both experience wildlife and give back.

RESPONSIBLE RECREATION

Millions of people appreciate hunting, trapping, birding, photographing, or just watching wildlife, but it is possible to value and harm the state's natural diversity at the same time. For everyone, the goal should be responsible enjoyment of wildlife.

As noted in Chapter 2, hunters' expenditures have supported wildlife conservation programs for more than 80 years. One of the goals of state wildlife agencies is to both recruit hunters and develop them into informed, conscientious sportsmen and sportswomen. TPWD aims to build hunters who understand and appreciate the wildlife resource and its habitat while providing safe, ethical experiences in the field. To help achieve this goal, TPWD requires anyone born after 1971 to complete a hunter education program offered by certified Hunter Education instructors before they can hunt on their own.

The Hunter Education Program recognizes that people getting involved in the sport of hunting may proceed through several stages. The goal is to progress from just shooting wildlife to behaving as a sportsman that gives back to conservation.

Some hunters start off in what TPWD calls the "Shooting Stage." Novice hunters sometimes have an impulse to get off a shot quickly, usually at the first animal that appears. This eagerness can lead to bad decisions, such as shooting the wrong animal, taking a poor shot that wounds the animal, or even endangering nearby people. Target practice, good mentoring, studying wildlife identification, and more experience can lead most hunters out of this stage.

As hunters improve their shooting skills, they may move into the "Limiting Out Stage." These hunters are eager to bag the limit, sometimes causing them to take unsafe shots or misidentify targets in the zeal to limit out. More experience and hunting with mature hunters can lead hunters out of this stage.

At some point, a hunter may move into the "Trophy Stage." In this stage the hunter may seek quality over quantity. The hunter is selective and passes on many opportunities that do not match the desired trophy characteristics. Many trophy hunters focus on big game. The hunter's patience and commitment might be highly evolved, but wildlife managers need hunters that understand that harvest should be balanced in order to maintain the quality of the population.

Hunters who are looking for more of a challenge may embrace the "Method Stage." In this stage, the hunter's primary focus is the process and challenge of hunting. The hunter may choose a more challenging method, such as using a longbow, crossbow, or muzzleloader. The hunter may also decide to stalk or still hunt rather than sit in a stand next to a feeder.

As a hunter reaches the "Sportsman Stage," he or she values the total hunting experience. At this stage, the hunter has learned to appreciate immersion into nature; the diversity of wildlife species, behaviors, and habitats; and the experience shared with hunting companions. Hunting success may be less important than the richness of the experience.

Hopefully, all hunters eventually reach the "Give-back Stage." At this stage, the sportsman is motivated to pass on the proper

(overleaf) Figure 17.1. Visitors to the restored cienega at Balmorhea SP have the chance to draw closer to the world of wildlife.
Photo courtesy TPWD © 2025 (Chase A. Fountain, TPWD).

Figure 17.2. TPWD and other states offer youth hunting programs as a means of teaching hunting safety and ethics to new hunters. Source: USFWS, CC BY 2.0.

hunting values, safety skills, and responsible attitudes to others. They want to preserve the hunting heritage by introducing new hunters to the sport in the most rewarding manner. The sportsman may join a conservation organization that improves habitat through wildlife management. They may even donate the game they harvest to feed the hungry.

The traditions of many indigenous people have a strong ethic of respect for natural resources. Many teach the hunter to honor the species to be harvested, whether plant or animal, and to take only a portion, leaving the rest for other people, other creatures, or replenishment. Robin Wall Kemmerer, a member of the Citizen Potawatomi Nation, describes the Native American approach to hunting or plant collection as the "Honorable Harvest." She summarizes this collection of unwritten traditions with the following guidelines:

Know the ways of the ones who take care of you, so that you may take care of them.
Introduce yourself. Be accountable as the one who comes asking for life.
Ask permission before taking. Abide by the answer.
Never take the first. Never take the last.
Take only what you need.
Take only that which is given.
Never take more than half. Leave some for others.
Harvest in a way that minimizes harm.
Use it respectfully. Never waste what you have taken.
Share.
Give thanks for what you have been given.
Give a gift, in reciprocity for what you have taken.
Sustain the ones who sustain you and the earth will last forever.
~Robin Wall Kemmerer. *Braiding Sweetgrass: Indigenous Wisdom, Scientific Knowledge and the Teachings of Plants.* Milkweed Editions, publ. 2015.

Hunters are not the only ones who must keep responsible recreation in mind. Wildlife watchers are often as enthusiastic as hunters in pursuit of their quarry. Irresponsible birders and photographers can cause damage to habitat, interrupt wildlife feeding, cause animals to abandon habitat areas, or force parents to become separated from their young. The National Audubon Society offers recommendations for birders that can also apply to other wildlife watchers. Among their recommendations are:

- Use a telephoto lens or a wildlife blind to maintain enough distance to allow the subject to behave naturally.
- Never advance on wildlife to make them fly or run.
- An approach is too close if it causes an animal to flush (fly or run away) or change its behavior.
- Use flash sparingly as a supplement to natural light. Avoid using a flash on nocturnal birds such as owls and nighthawks.
- Remove GPS data from images and videos of rare or sensitive species. Be selective about sharing locations publicly.
- Do not use drones to photograph or record video footage of birds, especially at their nests.
- Avoid the use of sound or recordings to attract endangered or nesting birds.

Finally, several considerations apply to hunters and all wildlife watchers. Enthusiasts should be aware and respectful of their surroundings and avoid trampling sensitive vegetation. Everyone should respect private property rights, requesting permission to recreate on private property in Texas. When accessing public lands and leased lands, recreationists should learn the rules and laws that apply to the location.

GETTING INVOLVED

Just as sportsmen ultimately want to give back, many other Texans look for a way to contribute to wildlife conservation in Texas. Whether as a conservation volunteer, a citizen scientist, or a wildlife rehabilitator, there are many opportunities to make a difference.

Nearly all refuges, parks, wildlife agencies, and conservation organizations have volunteer opportunities. Often, these volunteer opportunities let the participants experience wildlife in a way that they could not have on their own. One opportunity in particular stands out. In 1996, an idea emerged in San Antonio to train a group of volunteers dedicated to helping with natural resource conservation in the state. TPWD and Texas A&M Agrilife soon joined together to become the second state in the nation to launch a Master Naturalist Program. The mission of the Texas Master Naturalist program is to "develop a corps of well-informed volunteers to provide education, outreach, and service dedicated to the beneficial management of natural resources and natural areas within their communities for the State of Texas."

Master Naturalists are dedicated to conservation efforts in Texas. New Master Naturalists attend 40 hours of training. Each year, they must obtain eight hours of advanced training and contribute at least 40 hours of volunteer service. The program has multiplied to include about 50 local chapters around the state. Master Naturalists have made significant contributions to the goals of wildlife agencies and organizations. When the program celebrated its 20th anniversary, it noted that it had amassed 4.4 million volunteer service hours valued at over $100 million. The Texas program has set a standard that inspires Master Naturalist programs all around the country.

One vital activity that Master Naturalists and other volunteers can undertake is collecting and reporting data on native plant and animal species. This activity is part of a broader movement to involve volunteers in "citizen science." The National Geographic Society defines citizen science as "the practice

Figure 17.3. The Texas Master Naturalist Program gives volunteers a chance to learn about Texas wildlife and to give back to its conservation. Volunteers here plant trees in a Hays County park. Photo: Tom Hausler.

of public participation and collaboration in scientific research to increase scientific knowledge." Citizen science is not really a new idea, as citizen scientists conducted the first Christmas bird count in 1900. However, the concept has expanded in recent years. A wide variety of organizations now implement citizen science programs. Data is recruited on invasive species, Texas horned lizards, monarch butterflies, weather, and many other natural occurrences. An easy way to get involved is to sign up for an account on the iNaturalist website. TPWD uses this interactive platform as the primary means for collecting data for its citizen science program called Texas Nature Trackers. Launched in 2008, as of 2025, iNaturalist had over 3.9 million observers who had accumulated nearly 283 million observations on more than 500,000 different species.

One other volunteer opportunity deserves special mention. When citizens find an injured wild animal, they often earnestly want to help. Patching up every individual animal is beyond the capabilities of agencies like TPWD or USFWS; however, an indispensable cadre of volunteers exists—wildlife rehabilitators. Wildlife rehabilitators are volunteers or nonprofits who have received a permit from TPWD to handle injured wildlife. Some rehabilitators are larger organizations with specialized services, such as Last Chance Forever, which rehabilitates birds of prey at its facility in San Antonio. Other rehabilitators are volunteers who operate out of their own homes using their own financial resources. When an injured animal is encountered, Texas citizens can look at the TPWD website to find a licensed rehabilitator near them who accepts that type of animal. The goal of rehabilitation is to return the animal to the wild. TPWD encourages citizens to observe carefully before capturing an animal for rehabilitation. Many species that seem helpless, such as fawns or young birds, can survive if left in place or returned to the nest.

WILDLIFE CONSERVATION AT HOME

Believe it or not, even suburban and urban homeowners can help provide wildlife habitat in Texas. Short green lawns and tall exotic trees do little for wildlife and use costly resources like water and pesticides. By switching to native species and considering landscaping structure, landowners can become "wildscapers" who conserve money and water and help native wildlife.

In the book *Nature's Best Hope*, University of Delaware entomology professor Doug Tallamy suggests that everyone can play a role in conserving biodiversity by restoring native plants. His idea? Start by replacing lawns with native plants. Non-native turf grass lawns cover 90% of the acreage of residential lots in the United States. In contrast to native environments, lawns fail miserably in providing ecosystem services such as producing oxygen, storing carbon dioxide, conserving water, supporting pollinators and predators of pests, rebuilding soils, and moderating climate. In contrast, Tallamy discovered that certain high-value groups of native plants, such as oaks, willows, wild plums and apples, pecans, goldenrods, sunflowers, yuccas, and native legumes support a high diversity of insects. Tallamy calls his campaign to replace lawns with native plants the "Homegrown National Park" initiative.

TPWD and the National Wildlife Federation offer guidance on landscaping to benefit wildlife species. The National Wildlife Federation provides recognition for wildscaped homes, schools, and businesses as "Certified Wildlife Habitats." To be recognized, properties must use native plants for food and cover. In addition, properties must provide water and places for wildlife to raise young, all while utilizing sustainable practices, such as collecting rainwater, composting, mulching, and avoiding chemical use.

Figure 17.4. This educational poster produced by TPWD illustrates some of the wildlife that can commonly be found in urban areas in many parts of the state.

TPWD's Wildscapes program has webpages that recommend native plants that flourish in different ecoregions and different amounts of light. There are also recommendations for plants that deer do not prefer to browse—an important consideration in many parts of Texas! The National Wildlife Federation provides further links to native plant providers and a list of species that have high pollinator value. Tallamy, TPWD, and the National Wildlife Federation offer the following design suggestions.

- Reduce turfgrass areas. Replace St. Augustine grass (in full sun locations) with grasses such as buffalograss.

- Remove other non-native species, especially invasive species. The website Texas Invasives can help identify species that are problems in Texas.

- Plant keystone genera that host a high number of invertebrate species (example: oaks) and plants that host specialist pollinators (example: sunflowers). The National Wildlife Federation offers suggestions for keystone plants for different ecoregions on their website.

- Plant generously. Clumps of trees provide better habitat and prevent tree falls better than individual trees.

- Plant to create a multilayered effect. Offering tall, medium, and short plants grouped in a tiered arrangement appeals to wildlife.

- Include evergreen plants in the design. Evergreens keep their leaves year-round, offering cover for wildlife throughout the year.

- Choose a selection of plants that bloom or fruit at various times of the year, so there is always food for wildlife. Supplement with clean bird feeders if natural food sources are lacking or not yet established.

- Snags (dead or dying trees) can be left standing to provide cavity nesting sites. Leave leaf litter on the ground and old stalks of annual plants to offer insect habitat and bird food.

- Install berms or mounds and use curved lines to add interest. Rock features and logs can be attractive while providing homes for butterflies, lizards, and other small wildlife.

- Install additional features to attract wildlife, such as water features with moving water, nest boxes, and small, scattered bee "hotels."

- Take action to protect invertebrates. Don't spray pesticides or fertilizers (native plants don't need fertilizer).

- Watch out for wildlife "traps," such as window wells and uncovered meter boxes, where wildlife can fall in and become trapped.

- Windows should also be made safe for birds. Large windows that reflect plants or sky can result in bird strikes. Homeowners should use screens, one-way film, hanging cords ("Zen curtains"), or decals/markings no more than two inches apart to help birds see windows and avoid deadly strikes.

- Work with neighbors. Use social media to invite others to wildscape. Provide data to homeowners associations to help them appreciate the value of native plants and balanced ecosystems. Homeowners can even put up a sign in their yard that indicates "Prairie in Progress," or some other goal.

Figure 17.5. Water features and native plants can provide multitudes of habitat niches for wildlife in the backyard.

Two more recommendations can make neighborhoods more wildlife-friendly. First, keep cats indoors. Cats are natural predators and can have a severe impact on local wildlife populations. These non-native predators kill between 1.3 and 4 billion birds and 6.3 to 22.3 billion mammals annually in the U.S. The Cornell Laboratory of Ornithology estimates that, after habitat loss, free-ranging cats are the greatest threat to wild bird populations. Globally, cats have been linked to the extinction of 63 species of birds, mammals, and reptiles. Keep Kitty inside!

Second, minimize outdoor lighting. Artificial light can shift reproductive patterns and alter the day-night patterns of animals (and humans), causing organisms to not get enough rest and recovery. Light can artificially attract some organisms, such as moths, frogs, and sea turtles, causing them to be more vulnerable to predators. Lights have also been shown to affect the migration of songbirds, resulting in more building strikes. Homeowners can protect themselves and wildlife by using lights linked to motion sensors and by using shields on lights to limit the scattering of illumination. Dark Sky International offers suggestions for effective, wildlife-friendly lighting.

Finally, everyday sustainable choices can help wildlife on a global scale. Reducing consumption of energy, water, plastic, and other goods places less demand on natural resources around the world. Reduce, re-use, and recycle are great recommendations for benefitting wildlife habitats.

CAREERS IN WILDLIFE SCIENCE

Some individuals may want to do more than volunteer. For the outdoor enthusiast drawn to the beauty of nature, the fascination of wild animals, the rich complexity of ecology, and a desire to leave the world a better place, there is no more rewarding career than one in wildlife science.

Preparation begins in high school, where interested students should pursue coursework that includes science and math. Skills in communication, including writing and speaking, are very beneficial as well.

Volunteer and work opportunities at camps and other outdoor locations provide experience, build skills, and strengthen a resume. Organizations, such as Texas Brigades and the Texas Chapter of the Wildlife Society, offer in-depth summer camps for students with an interest in wildlife management.

Most jobs require a degree in the field of wildlife biology. Although many universities offer degrees in biology, zoology, and ecology, some universities offer programs focusing on applied science concepts specifically associated with wildlife management. One way to identify a university that provides coursework in wildlife science is to look for a university with a student chapter of The Wildlife Society, the professional association for wildlife scientists. Some examples in Texas include:

- Abilene Christian University
- Stephen F. Austin State University
- Sul Ross State University
- Tarleton State University
- Texas A&M – College Station
- Texas A&M University – Commerce
- Texas A&M University – Kingsville
- Texas State University
- Texas Tech University
- West Texas A&M University

As students pursue a bachelor's, master's, or even doctoral degree, gaining experience through jobs, research, volunteer opportunities, and networking at meetings and conferences is essential. It's not all "book-learning."

A diversity of employers may seek out students with these degrees. Most often, agencies such as USFWS and TPWD come to mind. Other state and federal agencies—the U.S. Army Corps of Engineers, Texas Department of Agriculture, USDA Natural Resources Conservation Service, TFS, USFS, and many others—also hire wildlife scientists. Non-governmental organizations such as the Nature Conservancy, Texas Wildlife Association, and Ducks Unlimited also bring wildlife biologists onto their staff. Zoos seek graduates who understand the care of captive animals and how to implement field conservation. Local parks departments may also welcome skills in habitat management. Finally, in Texas, the private sector is a significant source of jobs in wildlife management, with many large ranches and several private foundations such as the Welder Wildlife Foundation having wildlife biologists on staff.

Figure 17.6. TPWD and other employers often offer internship opportunities for college students to gain experience. Students on Mason Mountain WMA may learn about radio telemetry or may help to build fences. Photo 1: Mark Mitchell, TPWD. Photo 2: Jeff Forman, TPWD.

These employers offer a variety of career options that exist in the wildlife conservation world, such as:

- Wildlife biologist – The typical job of a wildlife biologist involves conducting wildlife surveys, conducting wildlife research, and helping implement habitat management. Outreach to the public and to private landowners is essential.

- Wildlife refuge manager – Some wildlife scientists obtain jobs associated with one particular refuge or wildlife management area. These employees need to have a good understanding of managing habitat. They also must be able to supervise employees, negotiate paperwork for projects, oversee budgets, and build bridges to the community.

- Wildlife researcher – Scientists with an advanced degree may obtain a position at a university where they teach and seek grants for advanced research on wildlife species. Emerging fields include the science of genetics and modeling ecosystems. Some agencies and private foundations also look for scientists with a research focus.

- Wildlife technician – Wildlife technicians need practical skills to help maintain and run equipment used in research and management. They also need to be able to gather data and keep good records. Wildlife technician positions may not require a university degree, but formal education may open up more opportunities.

- Private lands biologist – TPWD and USFWS both have private lands programs where biologists focus on reaching out to private landowners, seeking conservation solutions that work in a "real-world" context. In addition, many private ranches in Texas hire wildlife biologists. Understanding how to balance competing needs and achieve conservation in a working landscape is crucial.

- Wildlife veterinarian – Zoos are valuable partners in conservation. Nearly all zoos have veterinarians with special training in handling and treating wildlife, and some nonprofit organizations, such as the International Crane Foundation, also have a veterinary staff.

- Game warden – Game wardens enforce wildlife laws, so they need law enforcement and wildlife identification skills. TPWD requires applicants to its Game Warden Academy to have a four-year degree in one of these areas. Many people turn to game wardens as their first source of information on wildlife, so they are valuable conservation liaisons. Game wardens are sometimes called upon to participate in other public safety and rescue situations, assisting other law enforcement officers.

- Outdoor educator – Conservation doesn't happen without public support. A wide variety of jobs exist in agencies, conservation organizations, schools, nature centers, and camps where a wildlife scientist can work to build public support, both now and in the future. Skills in sharing wildlife knowledge in a clear, fun, and engaging manner are key to these positions.

Figure 17.7. Wildlife biologists are partners with private landowners in wildlife conservation in Texas. The USFWS Partners in Wildlife and TPWD Private Lands Program work with these important collaborators. Photo 1: USFWS, CC BY 2.0. Photo 2: Lee Ann Linam.

A LAND ETHIC

Aldo Leopold has guided us through our tour of wildlife science—creating the discipline of wildlife management, recognizing species as parts of ecosystems, offering a toolbox for habitat management, and exploring complicated predator management issues—so it is appropriate that he provides our final words of inspiration.

In *The Sand County Almanac*, his famous book of essays inspired by a worn-out farm in Wisconsin that he worked to restore, Leopold called for developing a "land ethic." An ethic is a moral principle designed to enable people to live in community with one another. Leopold called on people to recognize the natural world as part of their community, observing, "When we see land as a community to which we belong, we may begin to use it with love and respect."

His definition of land included soils, waters, plants, and animals—an ecosystem-based view. He did not define his ethic according to a set of right or wrong actions. Instead, he believed that respect would be built by experience and by spending time in the natural world. He wrote, "We can only be ethical in relation to something we can see, understand, feel, love, or otherwise have faith in."

Leopold's land ethic does not require us to own land. It does not require us to become wildlife scientists. But it does invite each person to spend time experiencing the land, noting its diverse and essential parts, and then measure our choices in terms of their long-term impact on the land and the wildlife with whom we share it.

(overleaf) Figure 17.8. Seeing, understanding, and experiencing the natural world is the first step in developing a land ethic. Photo: Mark Alan Storey.

APPENDIX A – REFLECTION QUESTIONS

CHAPTER 1 – WHAT IS WILDLIFE SCIENCE?
1. Name three animal species that would not be classified as wildlife and tell why they do not meet the definition.
2. Look up the scientific name of one of your favorite Texas wildlife species. Find out what the Latin names for that species mean.
3. Referring to the whooping crane research problem example,
    a. What was the independent variable (the variable that the researchers changed)?
    b. What was the dependent variable (the variable that the researchers measured)?
    c. What was the control?
    d. What was the sample size?
    e. Name at least two variables that were kept constant.
4. How is wildlife management both a science and an art?
5. Thinking again of your favorite Texas wildlife species, what wildlife values can you associate with it?

CHAPTER 2 - A HISTORY OF WILDLIFE SCIENCE IN TEXAS
1. Define "naturalist."
2. Name two species that became extinct in North America after European colonization.
3. List three species that almost disappeared from North America, but have been recovered with wildlife management.
4. Several strategies have been important in the conservation of wildlife in North America. Give one example of each of the following:
    a. Land conservation
    b. Wildlife laws
    c. Wildlife research
    d. Better funding
5. Which program at TPWD do you think has been most important in wildlife conservation in the state? Why?

CHAPTER 3 – BASICS OF ECOLOGY
1. A scientist is studying how ferruginous hawk numbers change when prairie dog numbers increase. Is the biologist studying a species, population, community, ecosystem, or biome?
2. In 1979, an ecologist published a book entitled, *Why Big Fierce Animals Are Rare*. How does the trophic pyramid support the idea behind that book?
3. Biologists in Texas sometimes see tarantulas sharing a burrow with narrow-mouthed toads. There is a hypothesis that the toads eat ants that might bother the tarantula, while the tarantula might protect the small toad from predators. What type of symbiosis would that represent?
4. American alligators wallow holes in wetlands, providing water and habitat for other animals during dry periods. Do you think alligators are a keystone species? If so, what type?

5. Many ranchers are not fond of prairie dogs, worrying that they diminish grass for their cattle and create risks with their burrows. What reasons could you give a rancher that might help him or her decide not to exterminate their prairie dogs?

CHAPTER 4 – FOUNDATIONS OF ECOSYSTEMS

1. Rock quillwort is a tiny plant that grows only in granite outcrops in pools formed by spring rains. In what part of Texas would you expect to find rock quillwort?
2. Amphibians and reptiles are ectothermic—meaning they rely on their environment to help their bodies reach the correct temperature. Looking at Figure 4.2, would you expect the diversity and abundance of reptiles to be greater in the High Plains or the Gulf Coastal Plains? Why?
3. Amphibians also require a humid environment, as they use moisture to breathe through their skin. Looking at Figure 4.2, would you expect the diversity and abundance of amphibians to be greater in the High Plains or the Gulf Coastal Plains? Why?
4. Thinking about the major soil orders of Texas, where do you think Texas has the most crop agriculture?
5. Name some benefits when land management restores native plants.

CHAPTER 5 – ECOREGIONS OF TEXAS

1. In which Texas ecoregion do you live?
2. In addition to being large, what are the other reasons that Texas has so many different ecoregions?
3. Is only one type of vegetation found in each ecoregion? Give some examples to support your answer.
4. The Blackland Prairie and Oak Woods and Prairies ecoregions are found basically in the same part of the state and are even interspersed with each other. What are their characteristics that make them different from each other?
5. According to the Great Plains biome shown below, which Texas ecoregions are part of the larger Great Plains biome?

Source: William L. Farr, CC BY_SA 4.0 Intl

CHAPTER 6 – HABITAT MANAGEMENT

1. List five characteristics of plants that might be used in their identification.
2. List four categories into which plants can be placed based on their structure or growth form.
3. List four types of vegetation communities from driest to wettest.
4. List the four components of wildlife habitat.
5. List five tools of wildlife management suggested by Aldo Leopold.
6. If you wanted to improve habitat for a bird species on your property, what are some of the things you would need to know about the species?
7. Read through Appendix B. List four methods of sampling vegetation, along with the habitats where you might use them.

CHAPTER 7 – TEXAS FORESTS

1. If the Pineywoods are in the Eastern Deciduous Forest, why are they called the "Pineywoods?" (And how did they get in that condition?)
2. How can forests play a role in water quality?
3. Do you think there are more or less acres of woodland in Texas now than in 1800? Give reasons for your answer.
4. Which ecoregions do NOT have a significant amount of forest acreage? What are the reasons that forests probably do not occur in abundance there? (Perhaps look back at information in Chapter 5.)
5. If you are a landowner who hopes for a sustainable income from 200 acres of forest, what management strategies would you choose? Why?

CHAPTER 8 – TEXAS GRASSLANDS

1. Why are grasslands called "upside-down forests?"
2. You're looking to buy a little patch of prairie in Texas. Someone offers to sell you one. They don't really know much about Texas ecosystems, but they tell you that the soils are red and it rains about 25" per year there. It was heavily grazed in the past but hasn't been grazed or burned for about 25 years.

    a. In which ecoregion do you think the tract is located? What plants do you suspect would be growing at this location? Would you buy it for your prairie habitat?

    b. If you did buy the above prairie tract, what would be your management strategy if you wanted to see many meadowlarks there?

3. Attwater's prairie-chickens and bobwhite quail are two species that have been declining in the North American Breeding Bird Survey. Young quail and prairie-chickens eat insects; adult quail and prairie-chickens mostly eat forbs and seeds. Name some of the factors that could be causing the decline of these species.
4. Why do you think that Texas has so many introduced grass species?
5. What grazing strategy would you use if your grassland had a lot of woody vegetation?

CHAPTER 9 – TEXAS DESERTS AND SHRUBLANDS

1. Based on the average annual rainfall and elevation information listed below, do you think forest, grassland, or desert would occur at these West Texas locations?
    a. Marfa – elevation 4,724'; rainfall 16"
    b. Dell City – elevation 3,701'; rainfall 11"
    c. McDonald Observatory – elevation 6,800; rainfall 19"
2. Based on your reading, what other factors could affect the type of vegetation that you would actually find at these locations?
3. Pecos is famous for the cantaloupes that it produces. Why do you think someone would want to grow cantaloupes in such a dry area?
4. The night-blooming cereus is a spiny, elongated succulent with a huge underground root that only blooms once per year and at night. Describe its adaptations for living in the Chihuahuan Desert. How do you think it is pollinated?
5. How has hunting both helped and harmed desert bighorn sheep?

CHAPTER 10 – TEXAS WETLANDS

1. Even though wetlands only make up a small percentage of Texas land area, why would you say that they are one of our most important habitats?
2. What are some of the stereotypes associated with wetlands that might make it hard to convince people that they are important?
3. Which wetland types occur in the driest parts of Texas? Discuss their importance.
4. Constructed wetlands are often used for mitigation (when a project that destroys wetlands is required to try to compensate for the loss). Do you think that constructed wetlands do a good job of replacing destroyed wetlands? Why or why not? What factors could you consider to improve the mitigation?
5. Think about the story of the whooping crane and the story of wetlands in this country. What are some of the threats to wetlands that have impacted whooping cranes?

CHAPTER 11 – Population Management

1. Would you describe the population dispersion of red-cockaded woodpeckers as uniform, random, or clumped? What do you think would be an effective way to monitor RCW population numbers?
2. If alligators can lay an average of 35 eggs, why wouldn't they recover immediately after hunting was ceased? What factors would slow it down?
3. In the 1980s, waterfowl biologists were concerned about very high snow goose populations causing damage to their nesting habitat in Canada. As a management strategy, they increased the bag limits and the length of the hunting season for snow geese.
    a. Were they thinking that hunting was compensatory or additive mortality?
    b. After the "Light Goose Conservation Order" allowing greater harvest was in place, the overall survival of adult geese actually increased. Was the hunting actually compensatory or additive?
    c. As the number of geese increased, productivity went down and fewer young were produced. Were the limiting factors on geese density-dependent or density-independent?

4. Some individuals are not comfortable with the concept of hunting wild animals. Trying to think of the issue objectively, what are some of the reasons that hunting could be an acceptable tool?
5. If hunting were not an option, or if there were objections to hunting, what are some of the other ways you might deal with the overpopulation problems as depicted in Figure 11.11?

CHAPTER 12 – TEXAS BIRDS

1. Do you think penguins have hollow bones? Why or why not?
2. Birders love going out early in the morning when birds are active, but seeing colors in the low light is hard. What characteristics could you use to identify birds instead?
3. Many bird species are declining, but biologists have documented 70 species that have expanded their ranges in Texas, moving northward into new territory. Some examples include white-winged doves, green jays, cave swallows, and green parakeets. What kind of factors could explain that range expansion?
4. Joint Ventures have successfully worked together to conserve waterbird habitat, but funding is needed for such big plans. What sources of funding are available to conserve habitat for waterfowl?
5. Knowing that the Lower Rio Grande is very popular with birders, how could funding be raised to conserve habitat in the Lower Rio Grande Valley Joint Venture?

CHAPTER 13 – TEXAS MAMMALS

The following is a set of data collected on a Hahn line for white-tailed deer. Use the data to calculate the density of white-tailed deer on this route. The length of the route is 2 miles (3560 yards). There are 4,840 square yards in an acre.

Visibility Data

Stop	Visibility on right in yards	Visibility on left in yards
1 (0.2 mi)	30	70
2 (0.4 mi.)	250	100
3 (0.6 mi.)	50	10
4 (0.8 mi.)	40	50
5 (1.0 mi.)	80	200
6 (1.2 mi.)	100	250
7 (1.4 mi.)	30	10
8 (1.6 mi.)	50	40
9 (1.8 mi.)	20	80
10 (2.0 mi.)	10	30

Deer Data

Date	9/15/23	9/18/23	9/25/23
Bucks	4	2	3
Does	12	6	8
Fawns	5	3	3
Undetermined	2	3	2
Total	23	14	18

1. In the above data set, what is the fecundity? (Remember, fecundity is the number of offspring per female). What is the doe to buck ratio?
2. In what ecoregion do you think this sample was taken? Do you see any management issues for this population?
3. Name a mammal species that is a keystone species. Describe why.
4. There are a number of mammal management issues that create controversy. These include:
    a. Breeding white-tailed deer in captivity
    b. High fences around ranches
    c. Harvest of furbearers
    d. Removal of predators

Pick one or more of these issues and explore the pros and cons of each situation. Explain your own perspective on the issue.

CHAPTER 14 – TEXAS AMPHIBIANS

1. Why can amphibians be considered a good environmental barometer?
2. What do you think is the biggest threat to amphibians in Texas? Explain why you think so.
3. The typical amphibian life cycle is egg, aquatic larva, metamorphosis, and terrestrial adult. Give two examples of Texas amphibians that do not follow that pattern. Describe their life cycle.
4. What are some ways amphibians can avoid predators?
5. How are amphibian malformations an ecosystem problem?

CHAPTER 15 – TEXAS REPTILES

1. Reptiles prefer body temperatures between 90 and 95° F. Why would basking surveys be best done when temperatures are in the 70s or even the 60s?
2. Warning mimicry is a strategy in which non-dangerous animals copy the coloration of dangerous animals to warn predators away. Which genus of snakes in Texas seems to exhibit warning mimicry? Which snake are they copying?
3. Suppose you wanted to try to identify a snake by its shed skin. What clues could you draw from the skin to help you with the identification?
4. If alligators and turtles have many young, why is overharvest of the adults a threat?
5. Finding common ground regarding rattlesnake roundups has been difficult. Suppose TPWD invited you to participate in their Snake Harvest Working Group. Describe some compromise strategies that you think might be helpful.

CHAPTER 16 – TEXAS INVERTEBRATES

1. Think about E.O. Wilson's quote. What are some ways in which invertebrates "run the world?"
2. This chapter presented much data about non-native species. Why do you think there are so many non-native invertebrates?
3. When many people think of pollinators they think of European honey bees. How do you think they compare in importance to our native bees? What are some of the benefits of each?
4. Suppose you increased the number of native plants in your yard. What are some of the monitoring methods you could use to see if the invertebrate biodiversity is increasing?
5. What would you tell your neighbors if they were concerned about the increase in invertebrates in the neighborhood?

CHAPTER 17 - A ROLE FOR EVERYONE

Reflecting on this final chapter, how do you see yourself contributing to wildlife conservation now and in the future? Discuss some details of habits and lifestyle choices that Leopold and this book might inspire you to adopt.

APPENDIX B – VEGETATION SAMPLING METHODS

How will a habitat manager know whether their activities are effective? By monitoring changes. When habitat management activities are applied, it is desirable to measure the changes in the vegetation characteristics quantitatively. These sampling techniques should be used before and after management activities are implemented to measure changes in the habitat. Note that it is crucial to select a variety of sample points randomly. Computers can generate random latitude-longitude points within the sample area using websites such as Epitools (https://epitools.ausvet.com.au/rgcs).

POINT-CENTERED QUARTER METHOD

This technique can assess the dominant tree density in a woodland or savanna. Measurements are made from a randomly located center point.

1. Select random sample locations in the managed area.
2. At each location, use a compass to establish four quadrants based on north-south and east-west orientation.

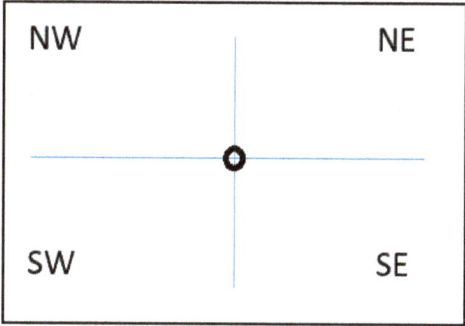

3. Use a measuring tape to measure the distance to the nearest tree in each quadrant. Note the distance and the tree species.
4. Calculate average tree density. Density = $1 / (\text{avg distance to tree})^2$
5. Count the number of hits on each species and divide by the total number of samples to calculate the percent dominance of each species.

LINE INTERCEPT METHOD

1. This technique can assess the dominant canopy coverage in a woodland or savanna.
2. Identify starting points random sample locations in the managed area.
3. At each location, use a compass and measuring tape to stretch out a transect 100 meters north from the starting point.
4. Go along the tape, stopping every meter.
5. At each meter, note the tallest plant hanging over that spot. Write down the name of the plant. Alternatively, classify the plant as tree, brush (shorter woody plants <25' tall), grass, grass-like plant, forb, litter (dead plants), or bare ground.
6. Repeat for each transect.
7. Add up the number of hits for each plant species or plant type. Include bare ground. The number of hits represents the dominance of that species or vegetation type in the canopy coverage. It provides insight into the type of vegetation community.

QUADRAT TRANSECT METHOD

This technique can assess the canopy coverage and plant diversity in a grassland.

1. Use ½" or ¾" PVC to build a 1 m. x 1 m. frame. This frame is the quadrat.
2. Identify random sample locations in the managed area.
3. At each location, use a compass and measuring tape to stretch out a transect 30 meters north from the starting point.
4. Go along the tape, stopping every 3 meters.
5. At each stopping point, place the quadrat alongside the measuring tape, alternating the side of the tape used.

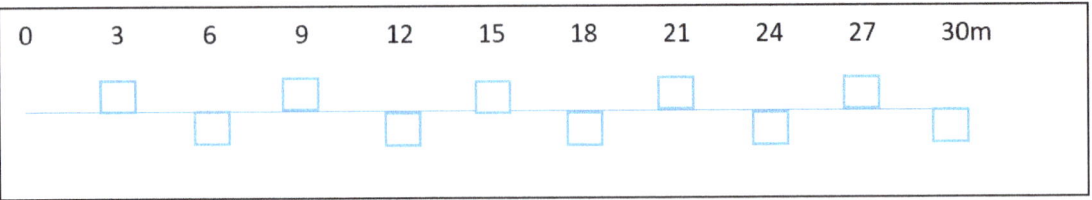

6. In each quadrat, estimate the percentage covered by each species. Alternatively, estimate the percentages covered by grass, grass-like plants, forbs, brush, litter, and bare ground. The percentages should add up to 100.
7. Count the total number of different plant species (dead or alive) in each quadrat. This count is the plant species richness, a measure of diversity.
8. Repeat for each transect.
9. Compile results from each quadrat.

SIMPLIFIED QUADRAT METHOD

1. Depending on the size of the sample area, pick a random number between one and 100. Walk that many paces into the grassland.
2. Face north.
3. Now, take a hula hoop and throw it over your head.
4. Go to the hula hoop.
5. Look in the hoop and estimate the percentage covered by grass, forbs, brush, litter, and bare ground. The percentages should add up to 100.
6. Count the number of different species in the hula hoop.
7. Repeat at least two more times, each time using a random number and a random starting point.
8. Calculate the average number of species and average percent cover for each plant type.

APPENDIX C – KEY REFERENCES

Chapman, B. R. and E. G. Bolen. 2018. The natural history of Texas. Texas A&M Press, College Station, Texas, USA.

Haggerty, M. M. and M. P. Meuth, editors. 2015. Texas Master Naturalist statewide curriculum. Texas A&M AgriLife Research and Extension Service Series, College Station, Texas, USA.

Hatch, R., editor. 2021. Texas almanac 2022-23. 71st edition. Texas State Historical Association, Austin, Texas, USA.

McMahan, C., Frye, R., and Brown, K. 1984. The vegetation types of Texas Including cropland. Texas Parks and Wildlife Department, PWD Bulletin 7000-120, Austin, Texas, USA.

Seek, M. 2011. Nature unbound: the impact of ecology on Missouri and the world. Missouri Department of Conservation, Jefferson City, Missouri, USA.

CHAPTER 1 – WHAT IS WILDLIFE SCIENCE?

Allen, T. and Southwick, R. 2007. The 2006 economic benefits of hunting, fishing and wildlife watching in Texas. Southwick Associates, Fernandina Beach, Florida, USA.

Brookshire, D. S., L. S. Eubanks and R. A. Randall. 1983. Estimating option prices and existence values in wildlife resources. Land Economics 69: 1-15.

National Wildlife Federation. 2006. Medicinal benefits of endangered species. NWF, Reston, Virginia, USA.

Phillips, M. 2012. Economics of wildlife recreation in Texas. Texas A&M AgriLife Extension Service, College Station, Texas, USA.

Southwick, T. A. and R. Soutwick. 2007. The 2006 economic benefit of hunting, fishing and wildlife watching in Texas. Southwick Associates prepared for Texas Parks and Wildlife Department, Austin, Texas, USA.

Southwick Associates. 2012. Hunting in America: an economic force for conservation. National Shooting Sports Foundation, Newtown, Connecticut, USA.

Stein, B. A. 2002. States of the union: ranking America's biodiversity. NatureServe, Arlington, Virginia, USA.

Texas A&M University Department of Rangeland, Wildlife and Fisheries Management and Texas A&M Natural Resources Institute. 2023. Economic values of white-tailed deer in Texas: 2022 Survey-Part II. College Station, Texas. USA.

U. S. Department of the Interior, U. S. Department of Commerce, and U. S. Census Bureau. 2018. 2016 National Survey of Fishing, Hunting, and Wildlife-Associated Recreation. U. S. Fish and Wildlife Service, Arlington, Virginia, USA.

CHAPTER 2 – HISTORY OF WILDLIFE SCIENCE IN TEXAS

Baker, Rollin H. 1995. Texas wildlife conservation - Historical notes. **East Texas Historical Journal** 33(1):12.

Borland, Hal. 1975. The history of wildlife in America. Arch Cape Press, Arch Cape, Oregon, USA.

Dubois, S. 2020. Wild Texas history website. https://wildtexashistory.com/. Accessed 4 December 2024.

Gray, G. G. 1985. Wildlife and people: the human dimensions of wildlife ecology. University of Illinois Press, Champaign, Illinois, USA.

Nelle, S. N.d. Historic wildlife of Texas: in numbers, numberless. Texas Wildlife. Texas Wildlife Association. https://www.texasbowhunter.com/discussions/forums/forum/topics/around-the-campfire/700164-wildlife-history-in-texas. Accessed 1 April 2020.

Organ, J. F., V. Geist, S. P. Mahoney, S. Williams, P. R. Krausman, G. R. Batcheller, T. A. Decker, R. Carmichael, P. Nanjappa, R. Regan, R. A. Medellin, R. Cantu, R. E. McCabe, S. Craven, G. M.

Vecellio, and D. J. Decker. 2012. The North American Model of Wildlife Conservation. The Wildlife Society Technical Review 12-04. The Wildlife Society, Bethesda, Maryland, USA.

CHAPTER 3 – BASICS OF ECOLOGY

Crosby, A.D., R. D. Elmore, D. M. Leslie, and R. E. Will. 2015. Looking beyond rare species as umbrella species: northern bobwhites (*Colinus virginianus*) and conservation of grassland and shrubland birds. Biological Conservation 186:233-240.

Forseth, I. 2010. Terrestrial biomes. Nature Education Knowledge 3(10):11.

Hassan, R., R. Scholes, and N. Ash, editors. 2005. Ecosystems and human well-being: current state and trends, Vol. 1. Island Press, Washington, D.C., USA.

Henke, S. E. 2002. Coyotes: friend or foe of northern bobwhite in southern Texas. Pages 57-60 in S. J. DeMaso, W. P. Kuvlesky, Jr., F. Hernandez, and M. E. Berger, editors. Quail V: proceedings of the fifth National Quail Symposium. Texas Parks and Wildlife Department, Austin, Texas, USA.

Henke, S. E. and F. C. Bryant. 1999. Effects of coyote removal on the faunal community in western Texas. *The Journal of Wildlife Management* 63(4): 1066-1081.

Sterling D. M. and J. F. Cully, Jr. 2001. Conservation of black-tailed prairie dogs (**Cynomys Ludovicianus**). **Journal of Mammalogy** 82(4):889–893.

CHAPTER 4 – FOUNDATIONS OF ECOSYSTEMS

Ferring, R. 2007. The geology of Texas. Cengage Learning, Boston, Massachusetts, USA.

Lindbo, D. 2008. Soil! Get the inside scoop. Soil Science Society of America, Madison, Wisconsin, USA.

Metzger, L., H. Post,, and K. Brady. 2023. Climate change and Texas: impacts, sources of pollution, and the path towards keeping our climate livable. Environment Texas Research & Policy Center, Austin, Texas, USA.

Spearing, D., and R. A. Sheldon. 1991. Roadside geology of Texas. Mountain Press Publishing Company, Missoula, Montana, USA.

CHAPTER 5 – TEXAS ECOREGIONS

Lyndon B. Johnson School of Public Policy. 1978. Preserving Texas's natural heritage. The University of Texas at Austin.

Texas A&M Forest Service. N.d. Trees of Texas: Texas ecoregions. http://texastreeid.tamu.edu/content/texasecoregions/. Accessed 1 December 2023.

Texas Parks and Wildlife Department. N.d. Hunter Education Online Course: Texas ecoregions. https://tpwd.texas.gov/education/hunter-education/online-course/wildlife-conservation/texas-ecoregions. Accessed 1 December 2023.

Texas State Historical Association. N.d. Texas almanac: Texas plant life. https://www.texasalmanac.com/articles/texas-plant-life. Accessed 1 December 2023.

CHAPTER 6 – HABITAT MANAGEMENT

Hegar, G. 2018. Guidelines for qualification of agricultural land in wildlife management use. Texas Comptroller of Public Accounts, Austin, Texas, USA.

Leopold, A. 1933. Game management. Charles Scribner's Sons. Reprinted in 1986 by University of Wisconsin Press, Madison, Wisconsin, USA.

Rowland, M. M. and C. D. Vojta, editors. 2013. A technical guide for monitoring wildlife habitat. U. S. Department of Agriculture Forest Service General Technical Report WO-89, Washington, D.C. USA.

Salafsky, N., R. Margoluis, and K. Redford. 2001. Adaptive management: a tool for conservation practitioners. BiodiversitySupport Program, Washington, D.C., USA.

CHAPTER 7 – TEXAS FORESTS

Food and Agricultural Organization of the United Nations. 2014. Youth guide to forests. FAO, Rome, Italy.

Simpson, H., E. Taylor, Y. Li, and B. Barber. 2013. Texas statewide assessment of forest ecosystem services. Texas A&M Forest Service, College Station, Texas, USA.

Texas A&M Forest Service. 2017. Texas forestry best management practices pocket guide. Texas A&M Forest Service, College Station, Texas, USA.

Texas A&M Forest Service. N.d. Texas A&M Forest Service Manage Forests and Lands. https://tfsweb.tamu.edu/content/landing.aspx?id=19856. Accessed 10 December 2023.

Texas Parks and Wildlife Department. N.d. Pineywoods Habitat Management. https://tpwd.texas.gov/landwater/land/habitats/pineywood/habitat_management/. Accessed 10 December 2023.

U.S. Department of Agriculture Forest Service. 2018. Benefits to people—National Forests and Grasslands in Texas. USDA Forest Service, Ecosystem Management and Coordination – Economics Group, Washington, D.C., USA.

Wildlife Alliance for Youth. 2016. Wildlife and recreation management study guide. Texas Soil and Water Conservation Board, Temple, Texas, USA.

CHAPTER 8 – TEXAS GRASSLANDS

Begosh, A., L. M. Smith, and S. T. McMurry. 2022. Major land use and vegetation influences on potential pollinator communities in the High Plains of Texas. Journal of Insect Conservation 26:231–241.

Diamond, D. 2019. Grasslands. Handbook of Texas online. Texas State Historical Association. https://www.tshaonline.org/handbook/entries/grasslands. Accessed 20 December 2023.

Hanberry, B. B., S. J. DeBano, T. N. Kaye, M. M. Rowland, C. R. Hartway, and D. Shorrock. 2021. Pollinators of the Great Plains: disturbances, stressors, management, and research needs. Rangeland Ecology and Management 78:220-234.

Hays, K. B, M. Wagner, F. Smeins, and R. N. Wilkins. 2005. Restoring native grasslands. Texas Agrilife Extension RWFM-PU-071, College Station, Texas, USA.

Horton, T. 1998. A prairie called Katy--land & people. Trust for Public Land website. https://www.tpl.org/resource/prairie-called-katy151landpeople. Accessed 1 October 2023.

Nelle, S. 2012. The great grassland myth of the Texas Hill Country. Texas Wildlife July 2012:46-51.

Peterjohn, B. and J. R. Sauer. 1999. Population status of North American grassland birds from the North American Breeding Bird Survey. Pp. 27-44 in P. D. Vickery and J. R. Herkert, editors. Ecology and conservation of grassland birds of the Western Hemisphere. Paxtuxent Wildlife Research Center, Laurel, Maryland, USA.

TPWD. N.d. Prescribed burning webpage. https://tpwd.texas.gov/landwater/land/habitats/post_oak/habitat_management/fire/. Accessed 20 December 2023.

U.S. Department of Agriculture. 2020. Texas agricultural statistics 2020. USDA National Agricultural Southern Plains Regional Field Office, Austin, Texas, USA.

Wilsey, C. B., J. Grand, J. Wu, N. Michel, J. Grogan-Brown, and B. Trusty. 2019. North American grasslands. National Audubon Society, New York, New York, USA.

CHAPTER 9 – TEXAS DESERTS

Hoyt, C. 2002. The Chihuahuan Desert: diversity at risk. Endangered Species Bulletin XXVII(2):16-17.

Schmidt, R. H. 2017. Trans-Pecos. Texas State Historical Association. Handbook of Texas Online. https://www.tshaonline.org/handbook/entries/trans-pecos. Accessed 30 December 2023.

Texas Parks and Wildlife Department. 2009. Big Bend country. PWD LF K0700-0139P. TPWD, Austin, Texas, USA.

CHAPTER 10 – TEXAS WETLANDS

Gray, M. J.; H. M. Hagy, J. A. Nyman, and J. D. Stafford. 2013. Chapter 4 Management of Wetlands for Wildlife. Pages 121-180 in J.T. Anderson and C.A. Davis, editors. Wetland techniques: volume 3: applications and management. USGS Staff -- Published Research. Springer Publishers, New York, New York, USA.

Lang, M. W., J. C. Ingebritsen, and R. K. Griffin. 2024. Status and trends of wetlands in the conterminous United States 2009 to 2019. U. S. Department of the Interior Fish and Wildlife Service, Washington, D.C., USA.

Playa Lakes Joint Venture. N.d. Playas webpage. https://pljv.org/playas/. Accessed 10 April 2024.

Rosen, R. A. 2015. Texas aquatic science. Texas A&M University Press, College Station, Texas, USA.

Texas Parks and Wildlife Department. 2012. Wetland management calendar for the Texas Central Coast. PWD BR W7000-1179A. TPWD, Austin, Texas, USA.

Wagner, M. 2004. Managing riparian habitats for wildlife. PWD BR W7000-306. Texas Parks and Wildlife Department, College Station, Texas, USA.

CHAPTER 11 – POPULATION MANAGEMENT

Bondavalli, C. and R. E. Ulanowicz. 1999. Unexpected effects of predators upon their prey: the case of the American alligator. Ecosystems 2:49-63.

Leopold, A. 1949. A Sand County almanac and sketches here and there. Oxford University Press, New York, New York, USA.

Outlaw, J. L., D. P. Anderson, M. L. Earle, and J. W. Richardson. 2017. Economic impact of the Texas deer breeding and hunting operations. Research Report 17-3. Agricultural and Food Policy Center, Department of Agricultural Economics, Texas A&M University, College Station, Texas, USA.

Bishop, R.C. 2004. The economic impacts of chronic wasting disease (CWD) in Wisconsin. Human Dimensions of Wildlife 9(3):181-192.

Texas Parks and Wildlife Department. N.d. Chronic wasting disease webpage. https://tpwd.texas.gov/huntwild/wild/diseases/cwd/positive-cases/. Accessed 10 January 2024.

Yarrow, G. 2018. Developing a wildlife management plan. Factsheet. https://hgic.clemson.edu/factsheet/developing-a-wildlife-management-plan/. Accessed 10 January 2024.

CHAPTER 12 – TEXAS BIRDS

Angelo, E., C. Elliot, and L. Brennan. 2005. Where have all the quail gone? PWD RP W7000-1025. The Texas Quail Conservation Initiative, Texas Parks and Wildlife Department, Austin, Texas, USA.

Peterson, R. T. 1957. How to know the birds. The New American Library, New York, New York, USA.

Roger Tory Peterson Institute. N.d. Roger Tory Peterson webpage. https://rtpi.org/about/about-roger-tory-peterson/. Accessed 20 January 2024.

Rosenberg K. V., A. M. Dokter, P. J. Blancher, J. R. Sauer, A. C. Smith, P. A. Smith, J. C. Stanton, A. Panjabi, L Helft, M. Parr, and P. P. Marra. 2019. Decline of the North American avifauna. Science 366(6461):120-124.

Shackelford, C. E., E. R. Rozenburg, W. C. Hunter, and M. W. Lockwood. 2005. Migration and the migratory birds of Texas: who they are and where they are going. PWD BK W7000-511 (11/05). TPWD, Austin, Texas, USA.

Southwick Associates. 2018. Hunting in America: An Economic Force for Conservation. National Shooting Sports Association, Newtown, Connecticut, USA.

Texas Birds Record Committee. N.d. Texas State List. https://www.texasbirdrecordscommittee.org/texas-state-list. Accessed 20 January 2024.

Welty, J. C. 1982. The life of birds, 3rd edition. Saunders College Publishing, New York, New York, USA.

CHAPTER 13 – TEXAS MAMMALS

Graves, R. A. 2019. The state of whitetails. Texas Parks and Wildlife Magazine, November 2019.

Schmidly, D. J. and R. D. Bradley, R. D. 2016. The Mammals of Texas, 7th edition. The University of Texas Press, Austin, Texas, USA.

Schmidly, D., R. Bradley, F. Yancey, and L. Bradley. 2024. Comprehensive Annotated Checklist of Recent Land and Marine Mammals of Texas, 2024, with Comments on Their Taxonomic and Conservation Status.

Shult, M. and B. Armstrong. 1999. Deer census techniques. PWD BK W7000-083 (10/99). Texas Parks and Wildlife Department, Austin, Texas, USA.

Smith-Rodgers, S. 2017. Exotic answers: landowners stock foreign breeds in quest to optimize family land. Texas Coop Power Magazine, September 2017.

Taylor, R. 2003. The feral hog in Texas. PWD BK W7000-195 (09/03). Texas Parks and Wildlife Department, Austin, Texas, USA.

Vaughan, T. A. 1978. Mammalogy. Saunders College Publishing, Philadelphia, Pennsylvania, USA.

CHAPTER 14 – TEXAS AMPHIBIANS

AmphibiaWeb. n.d. Amphibian Conservation webpage. https://amphibiaweb.org/declines/conservation.html. Accessed 1 February 2024.

Goin, C. J., O. B. Goin, and G. R. Zug. 1978. Introduction to herpetology, 3rd edition. W. H. Freeman and Company, San Francisco, California, USA.

Souder, W. 2000. A plague of frogs. University of Minnesota Press, Minneapolis, Minnesota, USA.

Texas Parks and Wildlife Department. 2001. Texas Amphibian Watch monitoring packet. PWD BK W7000-0493. TPWD, Austin, Texas, USA.

Tipton, B. L., T. L. Hibbitts, T. D. Hibbitts, T. J. Hibbitts, and T. J. Laduc, T. J. 2012. Texas amphibians: a field guide. University of Texas Press, Austin, Texas, USA.

Vandewege, M., T. Swannack, K. Greuter, D. Brown, and M. Forstner. 2013. Breeding site fidelity and terrestrial movement of an endangered amphibian, the Houston toad (*Bufo houstonensis*). Herpetological Conservation and Biology 8:435-446.

CHAPTER 15 – TEXAS REPTILES

Ceballos, C. P. and L. Fitzgerald. 2004. The trade in native and exotic turtles in Texas. Wildlife Society Bulletin 32(3):881–892.

Conant, R. and J. T. Collins. 1998. Reptiles and amphibians of Eastern/Central North America. Houghton Mifflin Company, New York, New York, USA.

Davis, J. M. 2016. Snake Harvest Working Group final report. Texas Parks and Wildlife Department, Austin, Texas, USA.

Klym, M. 2008. An introduction to Texas turtles. PWD BK W7000-1672. Texas Parks and Wildlife Department, Austin, Texas, USA.

McKinley, E. 2023. Inside the famous rattlesnake roundup: a case study for Texas's culture wars. San Antonio Express News, March 18, 2023.

CHAPTER 16 – TEXAS INVERTEBRATES

McGavin, G. C. 2002. Insects, spiders, and other terrestrial arthropods. DK Press, New York, New York, USA.

Neck, R. 2019. Snails. Handbook of Texas Online. https://www.tshaonline.org/handbook/entries/snails. Accessed 20 February 2024.

Quinn, M. 2019. Texas invertebrate species of concern webpage. https://www.texasento.net/TXsoc.htm. Accessed 20 February 2024.

Tallamy, D. W. 2007. Bringing nature home: how you can sustain wildlife with native plants. Timber Press, Portland, Oregon, USA.
Tallamy, D. W. 2019. Nature's best hope: a new approach to conservation that starts in your yard. Timber Press, Portland, Oregon, USA.
Van Dam, A. 2022. Wait, why are there so few dead bugs on my windshield these days? Washington Post, Oct. 21, 2022.

CHAPTER 17 – A ROLE FOR EVERYONE

National Audubon Society. n.d. Audubon's guide to ethical bird photography and videography. https://www.audubon.org/get-outside/audubons-guide-ethical-bird-photography. Accessed 20 January 2024.
National Wildlife Federation. n.d. Native plant finder. https://nativeplantfinder.nwf.org/. Accessed 20 January 2024.
Texas AgriLife Extension. n.d. Texas Master Naturalists. https://txmn.tamu.edu/. Accessed 20 January 2024.
Texas Parks and Wildife Department. n.d. Six stages of hunter development. https://tpwd.texas.gov/education/hunter-education/online-course/responsible-and-ethical-hunting/six-stages. Accessed on 20 January 2024.
Texas Parks and Wildife Department. n.d. Wildscapes: design tips. https://tpwd.texas.gov/wildlife/wildlife-diversity/wildscapes/design-tips/. Accessed 20 January 2024.

REFERENCES FOR SCIENTIFIC NAMES

Amphibians and Reptiles
Crother, B., J. Boundy, J. Campbell, K. de Queiroz, D. Frost, D. Green, R. Highton, J. Iverson, R. Mcdiarmid, P. Meylan, T. Reeder, M. Seidel, J. Sites, Jr., S. Tilley, and D. Wake. 2003. Scientific and standard English names of amphibians and reptiles of North America north of Mexico: update. Herpetological Review 34:196-203.

Birds
Chesser, R. T., S. M. Billerman, K. J. Burns, C. Cicero, J. L. Dunn, B. E. Hernández-Baños, R. A. Jiménez, A. W. Kratter, N. A. Mason, P. C. Rasmussen, J. V. Remsen, Jr., and K. Winker. 2023. Check-list of North American birds (online). American Ornithological Society. https://checklist.americanornithology.org/taxa/. Accessed 1 June 2024.

Invertebrates
iNaturalist. n.d. https://www.inaturalist.org. Accessed on 1 June 2024.

Mammals
Schmidly, D. J. and R. D. Bradley. 2016. The Mammals of Texas, 7th edition, online. https://www.depts.ttu.edu/nsrl/mammals-of-texas-online-edition/. Accessed 1 June 2024.
Schmidly, D.J., R.D. Bradley, F. Yancey, and L. Bradley. 2024. Comprehensive Annotated Checklist of Recent Land and Marine Mammals of Texas, 2024, with Comments on Their Taxonomic and Conservation Status, Museum of Texas Tech University Special Publication No. 80.

Plants
Ladybird Johnson Wildflower Center. 2023. Native plants of North America. https://www.wildflower.org/plants-main. Accessed 4 June 2024.
United States Department of Agriculture Natural Resources Conservation Service. 2024. The PLANTS database. http://plants.usda.gov. National Plant Data Team, Greensboro, North Carolina. Accessed 4 June 2024.

APPENDIX D - GLOSSARY

Abiotic – non-living

Adaptive management – a systematic process for continually improving management policies and practices by learning from the outcomes of previous practices

Aestivate - a dormant state for animals that happens in hot or dry conditions

Allen's Rule – an ecological principal stating that warm-blooded animals in warmer climates have longer appendages to maximize heat dissipation

Altricial - a bird or mammal that is born without full development that requires extensive parental care

Amplexus - literally "an embrace;" the posture when a male frog grabs a female frog for egg-laying and fertilization

Aquatic – of, or pertaining to, water

Apex predator - a predator at the top of a food chain, without natural predators of its own

Atmosphere - all the gases making up the Earth's air

Bag limit - the maximum harvest of fish or game allowed within a given time period

Bask - a "sunbathing" strategy used by reptiles and other animals to increase body temperature to ideal levels for activity

Batholith - a large igneous structure beneath the earth's surface; may be exposed by erosion

Bergmann's Rule – an ecological principal that warm-blooded animals in a given species tender to have larger body size farther from the equator

Binomial nomenclature - a standardized system of naming animals with Latin names consisting of a genus name and a species name

Bioaccumulation - the gradual accumulation of substances, such as pesticides or other chemicals, in an organism; can result in high concentrations in animals high in the trophic pyramid

Biodiversity- the variety of living things found in an area

Biogeochemical cycles - the movement and transformation of chemical elements and compounds between living organisms, the atmosphere, and the earth's crust

Biology - the study of living things

Biomass - the total quantity or weight of organisms in a given area

Biome - a distinct geographical region with specific climate, vegetation, and animal life

Biosphere – all the living organisms on Earth

Biotic – referring to something which is living

Bog - a wetland of soft, spongy ground consisting mainly of partially decayed plant matter called peat

Bolson - a large basin area, usually in arid regions, where water collects and evaporates

Bottomland hardwoods - forested riparian wetlands of East Texas where the streams or rivers occasionally flood beyond their channels into hardwood forested floodplains

Browse line - the boundary between normal plant growth at higher levels and lower portions of plants where they have been stripped back by wildlife or livestock

Browser – an animal that feeds on leaves, soft shoots, or fruits of woody plants

Brumate - a dormant state that reptiles and amphibians exhibit during cold weather

Brushlands - woodlands made up of woody species usually less than 25 feet tall

Buck - a male deer

Bunchgrass - a perennial grass that grows as an individual plant forming a clump of stems

Caldera - a volcanic crater, often resulting from the collapse of the volcano cone

Caliche - a shallow layer of soil or sediment that has been cemented together by the precipitation of calcium carbonate or other mineral material.

Carapace - the upper shell of a turtle consisting of numerous fused plates; the carapace is also fused to the turtle's backbone

Carnassial teeth - molar teeth in carnivores that are modified for shredding

Carnivore - an animal which eats only other animals; also, a member of the mammalian order Carnivora

Carrion - a decaying animal carcass

Carrying capacity - the maximum number of

individuals of a given species that a habitat can sustain indefinitely

Census - a complete count of animals in a wildlife population

Chrysalis - the pupa form of a butterfly in which metamorphosis from larva to adult takes place

Ciénega - shallow wetlands formed by spring seeps in desert regions

Clearcut - a timber harvest strategy in which all or nearly all trees are removed from a forest patch at the same time

Climate - the weather conditions prevailing in an area in general or over a long period

Climate change - a long-term change in average weather patterns

Climax community - a stable vegetation type that tends to develop over time and includes the animals and plants that are best adapted to the abiotic factors in an area

Coastal Plain - a band of flat topography that parallels the Texas Gulf Coast

Commensalism – a symbiotic relationship between two species in which one species benefits and the other is neither benefitted nor harmed

Community – an association of interacting populations of different organisms in a natural environment

Compensatory mortality - a situation in which hunting or predation reduces mortality rates from other factors; it applies when limiting factors are density-dependent

Competition – when two or more organisms have a demand for the same limited resources, such as nutrients, space or light. Competition between two members of the same species is intraspecific competition. Competition between two different species is interspecific competition.

Compound – a basic building block of matter composed of two or more elements bonded in a specific chemical formula

Conservation – wise use of natural resources

Consumer - an organism that must consume other organisms for their energy needs

Crepuscular – active in the early morning and/or late evening

Cryptic - hard to see as a result of coloration and/or behavior

Deciduous - a broad-leaved tree or shrub that sheds its leaves each winter

Decomposer - an organism that breaks down other dead organisms or parts of organisms in order to obtain energy

Decreaser grass - a species of grass that decreases under heavy grazing pressure; usually these species are desired for both livestock and wildlife

Density-dependent factors - a limiting factor that is increased as density of populations increase

Density-independent factors - a limiting factor that is not affected by population density

Diapause - a period of suspended development in an insect, other invertebrate, or mammal embryo, especially during unfavorable environmental conditions

Dioecious - a plant with imperfect flowers in which the male and female flowers are born on separate plants

Diurnal – active during the day

Doe - a female deer

Ecological niche - the function, position or role of a species within an ecosystem

Ecological pyramid - a graphical representation of the energy in an ecosystem that illustrates the abundance of organisms at each feeding level (also called trophic pyramid)

Ecological succession – include primary succession and secondary succession

Ecology - study of the relationships living organisms have with each other and with their environment, including both biotic and abiotic components

Economics - the study of how wealth is produced and distributed

Ecoregion - a major ecosystem defined by distinctive geography, climate, and biotic communities

Ecosystem – a biological community of a given area and the physical environment with which it interacts

Ecosystem engineers – species that create, modify, maintain, or destroy habitats through their physical activities

Ecosystem services - outputs from nature, such as clean air and clean water, that benefit human welfare

Ecotone - a transition zone between two different ecosystems or ecoregions where characteristics of both can be found

Ectotherm - an animal that uses its surroundings to adjust its internal body temperature

Edge species - an animal that tends to utilize ecotones, or the boundary between two different habitat types or conditions

Element – one of the basic building blocks of matter composed of atoms with the same chemical structure

Emergent vegetation - herbaceous wetland plants that protrude from the water

Emigration - when an animal moves out of a population area

Endemic - a species whose distribution is restricted to a certain area

Endothermic - an animal whose internal body temperature is controlled by its internal metabolism

Ethology – the study of animal behavior

Eutrophication - an overgrowth of algae and aquatic plants due to high levels of nutrients in the aquatic environment

Evergreen - a tree or shrub with leaves that remain green throughout the year

Exotic - a species that is not native to a given area

Exploitation – a symbiotic relationship that benefits one organism but harms another; may also refer to human use of a wildlife resource

Extinction - the dying out of a species or subspecies completely

Extirpation - the dying out of a species or subspecies in a particular area

Fauna - the animal life of a particular natural area

Fawn - a young deer under one year of age

Fecundity - a measure of the reproductive capacity of a population; ratio of young produced to adult females

Feral – an animal originating from domestication or captivity that has established in the wild

Flora – the plant life of a particular natural area

Focal species – a species of plant or animal that is monitored during wildlife management as a measure of the success of conservation strategies

Food chain - part of a food web, the series of linkages that shows how energy is transferred in sequence from a producer to several different consumers, including decomposers

Forb - herbaceous (non-woody), broad-leafed plants

Forest - a habitat with trees grouped in a way so their leaves, or foliage, shade the ground

Fragmentation - a situation in which natural habitats are disconnected or broken into small tracts

Frugivore – a fruit-eating animal

Furbearers- species which may be legally harvested for the sale of their hide or pelt

Game – animals for which hunting regulations, including bag limits, seasons, and methods of harvest, have been adopted

Generalist species - a species of wildlife that can eat many types of food and survive in many types of habitats

Genetic bottleneck – a loss of genetic diversity in a population or species because of a very low population size at some point in time

Genetic diversity - the variety of inheritable traits that occur within a species

Geologic fault - cracks in the earth's crust resulting from stress and differential movements between sections of crust, often resulting in visible changes in landforms on the earth's surface

Geosphere - the rocks and minerals making up the Earth's land

Gestation period - the amount of time that a mammal develops within the uterus while being nourished by the placenta

Gilgai - depressional areas that occur in blackland soils that can support moisture-loving species

Grasslands - areas dominated by grasses, with tree or shrub canopies covering less than 25 percent of the area

Greenhouse effect - the trapping of the warmth of the sun near the earth's surface due to the presence of atmospheric gases that prevent heat from escaping

Group selection – a forestry technique that harvests small groups of trees, creating light gaps in the forest, but maintaining forest diversity

Habitat – the natural environment for an organism that supplies all their life needs. Habitat must provide food, water, shelter, and adequate space.

Habitat corridor - a linear tract of habitat that is conserved or restored in order to allow wildlife movement between two other fragments of habitat

Halophyte - a plant that grows in environments with high salinity

Hardwood tree - flowering trees, such as oaks and elms, that are usually deciduous and slow-growing; they produce a dense, hard wood

Herbivore - an animal that eats only plants

Home range – a familiar tract of habitat that an individual animal passes through on a regular basis to obtain food, seek shelter, or find mates

Homeothermic - an animal that maintains a constant internal body temperature despite the external temperature variations

Humus - the organic layer of soil formed when soil microorganisms break down dead plants and animals

Hydric soil - soil that is permanently or seasonally saturated by water, resulting in anaerobic conditions, as found in wetlands

Hydrology - the distribution and movement of water in a system

Hydrophytic - water-loving

Hydrosphere - the water in the Earth's oceans, rivers and lakes

Igneous rock - a rock that is formed from cooled magma or lava; often with air bubbles or large crystals within the rock

Immigration - when animals move into a new area

Increaser grass - a species of grass that increases under disturbed conditions or heavy grazing pressure; usually they provide poorer grazing nutrition and wildlife structure

Indicator species - species whose presence or abundance in an ecosystem reflect a specific environmental condition

Induction—gathering observations and data from many sources and circumstances and then developing theories

Insolation - the amount of solar radiation reaching a given area

Intrusive igneous rock – rock formed by cooling of magma underground, producing various shapes, such as stocks, laccoliths, sills, and dikes, where magma pushed its way into overlying geologic formations

Invasive species - alien species whose introduction does or is likely to cause economic or environmental harm or harm to human health

Keystone species – species that significantly influence ecosystem structure, composition, and function through their activities

Lacustrine wetlands - wetlands that form along lakes or ponds

Lava dome - a mound of lava that is emerging from the vent of a volcano

Limiting factor – a feature of habitat, biology, or ecological communities that constrains a population's size and slows or stops it from growing.

Marsh - a wetland dominated by reeds and other grass-like plants

Mast - food sources produced by woody plants, such as acorns, nuts, and fruit

Mesopredator - a medium-sized animal that hunts other animals, but may not be at the top of the food chain; examples include foxes and raccoons

Metamorphic rock - rock that is formed by the transformation of other rock types under pressure and heat; often demonstrate flattened crystals or bent layers

Migration - a regular seasonal movement of animals back and forth between a breeding ground to a wintering ground, usually at different latitudes; some species also move between different elevations

Mima mounds - small mounds up to one meter tall that occur in tallgrass prairie and may support unique species

Minimum tillage – farming that that minimizes soil disturbance by reducing the intensity of plowing, leaving greater crop residue on the surface

Mitigation - creation or restoration of a habitat to make up for a habitat loss as the result of a project at another site

Model - a representation or simulation of a real-life phenomenon, organism, or process. Concept-process models can be used to predict. how changes in one part of a system may affect other parts of the system

Monoculture - a planting of crops or timber that is composed of only one type of crop or tree

Monoecious - a plant with imperfect flowers (separate male and female flowers) in which an individual plant has only male or female flowers

Mortality - the ratio of the number of deaths to the size of the population; death rate

Motte - a small grove or clump of trees in open prairie country

Mutualism - an association between two different species in which both species benefit

Myrmecophagous - an organism that specializes in feeding on ants

Natality - the ratio of the number of births to the size of the population; birth rate

Natural History - the scientific study of plants and animals, based primarily on observation of the natural world

Natural Selection - the process in nature whereby organisms that have more suitable characteristics tend to survive and reproduce

Naturalist - a person who observes and studies wild flora and fauna

Nature-based tourism – travel focusing on wildlife or other natural features; sometimes called ecotourism

Neotenic - retaining larval characteristics even as an adult

Nocturnal – active at night

Nongame - any of a variety of animals that do not have seasons and bag limits designed for harvest; most wildlife is nongame

Omnivore – an animal that eats both plants and animals

Organic - a substance containing carbon associated with a living organism

Ornithology – the study of birds

Oviparous - an organism that lays eggs; young develop within the egg through incubation

Ovoviviparous - an organism that produces eggs that are hatched in the body; young are born live; characteristic of some reptiles

Oxbow - a lake or wetland that is formed from the ancient meanders of a river

Perennial - a plant that lives for three years or more

Pheromone - a chemical substance produced by an animal that can affect the behavior of other members of its species

Photosynthesis - the process by which plants and other producers use chlorophyll, sunlight, water, and carbon dioxide to produce and store sugars

Physiography - the study of the physical abiotic features of the earth's surface

Plastron - the lower shell of a turtle consisting of several bony plates fused together

Plate tectonics - theory that the earth is made up of rigid plates at the surface that move on a fluid mantle beneath

Playa lake - round depressions in the Southern High Plains of the United States that fill with water when it rains forming shallow temporary lakes or wetlands

Poikilotherm - an animal whose internal body temperate can vary considerably

Population – a group of organisms belonging to the same species occupying a particular area at the same time

Prairie - a term, most often used in North America, to describe a wide expanse of grassland habitats

Prairie potholes - depressional wetlands scattered in a grassland habitat; may be formed by wind or glaciers

Precocial - a mammal or bird that is born well-developed with an ability to move about quickly

Predator – an animal that captures and eats other animals (their prey)

Prescribed fire - a controlled fire set intentionally for purposes of land management

Prey – an animal that is hunted by another animal (their predator)

Primary succession - type of ecological succession following a severe disturbance in which plants and animals colonize a barren habitat, often with no soil yet present

Proboscis - a flexible tubular elongated sucking mouthpart in insects

Producer - an organism that can produce its own food, usually by capturing energy from the sun to complete the process

Quadrat - a frame of some size in which organisms are counted in order to make estimates about the population in general

Rangeland - land on which the indigenous vegetation (climax or natural potential) is predominantly grasses, grasslike plants, forbs or shrubs and is managed as a natural ecosystem

Resaca - refers to an oxbow lake found in South Texas, often formed from the Rio Grande

Riparian - refers to an area along the bank of a river or stream

Savanna - patches of woodland scattered within a grassland, usually with 10%-50% woody canopy cover

Scat - fecal droppings from a wild animal

Science - a system for discovering general truths or laws about the natural world as obtained and tested through the scientific method

Scientific method – a method of learning about the natural world that involves observation, measurement, experimentation, and the formulation, testing, and modification of hypotheses

Secondary succession - a type of ecological succession that occurs after a disturbance disrupts a community but leaves some life forms present

Sedimentary rock - a rock formed when sediments deposited by wind or water consolidate; often show layers and/or sediment grains

Seed-tree cut - leaving a small number of mature trees during a timber harvest so that they can provide seeds for regeneration of the forest

Selective cutting – a forestry management technique that uses harvest of individual trees to maintain a diversity of tree ages and species in the forest

Sentinel species - a species whose population changes can indicate a change in ecosystem structure or function

Sere (or seral stage) - a series of different plant communities that change with time

Shelterwood cut - leaving a moderate number of mature trees during a timber harvest so that they can provide seeds for regeneration of the forest and some habitat for wildlife

Shrubland - habitat composed of woody plants generally less than nine feet tall scattered throughout arid or semi-arid regions (less than 30 percent woody canopy coverage)

Snag - dead trees that are left upright to decompose naturally

Sociology - the study of human society

Soil horizon - a layer within the soil with recognizable differences in chemistry, biology, and physical structure from the layers above and below it; often represents different ages of the soil

Soil profile – the vertical arrangement of the soil horizons from the soil surface down to the solid rock below

Songbird – a member of the avian order Passeriformes; usually small in size, also called perching birds or passerines

Specialist species – a wildlife species that has very specific habitat or food requirements; it may disappear if those resources are not available

Species - a group of living organisms consisting of similar individuals capable of interbreeding

Spring-seep - a term for wetland habitats formed when groundwater discharges to the surface

Subspecies - populations that have the potential to produce young, but live in different areas and vary in size, shape, or other physical characteristics

Sustainable harvest - a policy of harvesting or collecting animals or plants at a level that allows the population to regenerate and maintain itself

Swamp - forested wetland, usually with water present year-round

Symbiosis – a close relationship between two different species in one depends upon the other

Taxonomy - the science of naming and classifying organisms

Terrestrial – found living on land

Territory – a habitat area defended by an organism or a group of organisms from other members of the species or similar species

Thermoregulation - strategies that serve to adjust and maintain an internal body temperature; some species thermoregulate internally, whereas others use external means

Topography - the arrangement of the natural physical features of an area

Transect - a straight line that cuts through a natural landscape so that standardized observations and measurements can be made

Trophic level – refers to "feeding," it is a level in a food chain, food web, or ecological pyramid that depicts a stage in energy flow in the ecosystem

Umbrella species - species that have either large habitat needs or other requirements whose conservation results in many other species being conserved at the ecosystem or landscape level

Ungulate - a hooved animal

Viviparous - an organism that gives birth to live young that are nourished in the body by the mother

Water guzzler - self-filling, constructed watering facilities that collect, store, and make water available for wildlife

Waterfowl - ducks and geese; members of the order Anseriformes

Wet meadow - a type of shallow marsh that commonly occurs in poorly drained grassland areas

Wildlife – terrestrial or semi-terrestrial non-domesticated vertebrates existing in a natural or semi-natural wild environment

Wildlife Management - implementing changes to populations, habitat, and human use to produce some desired result

Wildlife science - the study of animals and their environments, with a focus on how to apply ecological knowledge to balance the needs of wildlife with the interests of humans

Zygodactylus - bird feet in which two toes point forward and two toes point backwards; found in woodpeckers and roadrunners

American snout butterfly on poverty weed.

Appendix D - Glossary | 329

APPENDIX E – LIST OF SCIENTIFIC NAMES

With an index to page numbers in the text

AMPHIBIANS

American Bullfrog, *Lithobates catesbeianus* 241, 250, 251
Barking Frog, *Craugastor augusti* 63, 245
Barred Tiger Salamander, *Ambystoma mavortium mavortium* 246
Black-spotted Newt, *Notophthalmus meridionalis* 60, 151, 247
Blanchard's Cricket Frog, *Acris blanchardi* 27, 238, 242
Burrowing Toad, *Rhinophrynus dorsalis* 244
Canyon Treefrog, *Hyla arenicolor* 242
Chihuahuan Green Toad, *Anaxyrus debilis* 252
Cliff Chirping Frog, *Eleutherodactylus marnockii* 63, 93, 245
Cope's Gray Treefrog, *Hyla chrysoscelis* 242
Couch's Spadefoot, *Scaphiopus couchii* 139, 242, 252
Crawfish Frog, *Lithobates areolatus* 241
Eastern Narrow-mouthed Toad, *Gastrophryne carolinensis* 244
Eastern Tiger Salamander, *Ambystoma tigrinum* 246, 247, 252
Gray treefrog, *Hyla versicolor* 242
Green Treefrog, *Hyla cinerea* 242
Gulf Coast Toad, *Incilius nebulifer* 243
Gulf Coast Waterdog, *Necturus beyeri* 248
Holbrook's Southern Dusky Salamander, *Desmognathus auriculatus* 248
Houston Toad, *Anaxyrus houstonensis* 30, 56, 79, 83, 92, 100-1, 243, 249, 250, 252
Hurter's Spadefoot, *Scaphiopus hurterii* 242
Lesser Siren, *Siren intermedia* 248
Marbled Salamander, *Ambystoma opacum* 246
Mexican Spadefoot, *Spea multiplicate* 242-3
Mexican White-Lipped Frog, *Leptodactylus fragilis* 95
Mole Salamander, *Ambystoma talpoideum* 246
Northern Leopard Frog, *Lithobates pipiens* 241, 251, 252
Pig frog, *Lithobates grylio* 152
Plains Leopard Frog, *Lithobates blairi* 241
Red-spotted Newt, *Notophthalmus viridescens viridescens* 247
Rio Grande Chirping Frog, *Eleutherodactylus cystignathoides* 245-6
Rio Grande Leopard Frog, *Lithobates berlandieri* 241
Rio Grande Lesser Siren, *Siren intermedia netting* 60, 248
San Marcos Salamander, *Eurycea nana* ix, 32
Sheep Frog, *Hypopachus variolosus* 244
Small-mouthed Salamander, *Ambystoma texanum* 246
South American Cane Toad, *Rhinella marina* 250
Southeastern Dwarf Salamander, *Eurycea quadridigitata* 248
Southern Leopard Frog. *Lithobates sphenocephalus* 241
Southern Mountain Yellow-legged Frog, *Rana muscosa* 250
Southern Red-backed Salamander, *Plethodon serratus* 248
Spotted Chirping Frog, *Eleutherodactylus guttilatus* 245
Spotted Salamander, *Ambystoma maculatum* 246
Spring Peeper, *Pseudacris crucifer* 239
Strecker's Chorus Frog, *Pseudacris streckeri* 242
Texas Blind Salamander, *Eurycea rathbuni* 248
Texas Toad, *Anaxyrus speciosus* 236, 243
Three-toed Amphiuma, *Amphiuma tridactylum* 248
Western Narrow-mouthed Toad, *Gastrophryne olivacea* 244
Western Slimy Salamander, *Plethodon albagula* 248

BIRDS

Acorn Woodpecker, *Melanerpes formicivorus* 69, 95, 206
Altamira Oriole, *Icterus gularis* 95
American Coot, *Fulica americana* 152, 195
American Robin, *Turdus migratorius* 196
American White Pelican, *Pelecanus erythrorhynchos* 189, 202
American Wigeon, *Anas americana* 152, 192
American Woodcock, *Scolopax minor* 190, 197
Aplomado Falcon, *Falco femoralis* 62, 182, 203
Arctic Tern, *Sterna paradisaea* 196
Attwater's Prairie-chicken, *Tympanuchus cupido attwateri*

Bailey's Mountain Chickadee, *Parus gambeli baileyae* 12
Bald Eagle, *Haliaeetus leucocephalus* 39, 59, 84-5, 96, 106, 168, 172, 190, 203-4
Black Rail, *Laterallus jamaicensis* 195
Black Skimmer, *Rynchops niger* 197
Black-capped Vireo, *Vireo atricapilla* 63, 93, 116, 205, 207
Black-chinned Hummingbird, *Archilochus alexandri* 201
Black-throated Sparrow, *Amphispiza bilineata* 137, 138
Blue-gray Gnatcatcher, *Polioptila caerulea* 189
Blue-winged Teal, *Spatula discors* 154, 192
Broad-winged Hawk, *Buteo platypterus* 196
Bronzed Cowbird, *Molothrus aeneus* 207
Brown Pelican, *Pelecanus occidentalis* 14, 59, 85, 202
Brown-head Cowbird, *Molothrus ater* 206-7
Buff-bellied hummingbird, *Amazilia yucatanensis* 201
Burrowing Owl, *Athene cunicularia* 32, 117
California Condor, *Gymnogyps californianus* 176
Canvasback, *Aythya valisineria* 159
Carolina Parakeet, *Conuropsis carolinensis* 14, 81, 96, 150
Cave Swallow, *Petrochelidon fulva* 313
Cinnamon Teal, *Spatula cyanoptera* 154
Colima Warbler, *Leiothlypis crissalis* 68, 94
Common Black Hawk, *Buteogallus anthracinus* 68, 94
Common Gallinule, *Gallinula galeata* 195
Common Snipe, *Gallinago Gallinago* 197
Cooper's Hawk, *Accipiter cooperii* 28, 203
Crested Caracara, *Caracara plancus* 203
Dickcissel, *Spiza americana* 114
Double-crested Cormorant, *Nannopterum auritum* 180, 202
Eastern Bluebird, *Sialia sialis* 106-7, 181
Eastern Meadowlark, *Sturnella magna* 114
Eastern Phoebe, *Sayornis phoebe* 189, 207
Emu, *Dromaius novaehollandiae* 23
Eurasian Collared-dove, *Streptopelia decaocto* 23, 193
European Starling, *Sternus vulgaris* 23
Ferruginous Hawk, *Buteo regalis* 32, 117, 181
Ferruginous Pygmy-owl, *Glaucidium brasilianum* 60
Forster's tern, *Sterna forsteri* 197
Gambel's Quail, *Callipepla gambelii* 198-9
Golden-cheeked Warbler, *Setophaga chrysoparia* 27, 63, 79, 82, 93, 99, 187, 205
Gray Hawk, *Buteo plagiatus* 203
Great Auk, *Pinguinus impennis* 13
Great Blue Heron, *Ardea herodius* 202, 203
Green Jay, *Cyanocorax yncas* 62, 95, 187
Green Parakeet, *Psittacara holochlorus* 313
Green-winged Teal, *Anas crecca* 192
Horned Lark, *Eremophila alpestris* 117
House sparrow, *Passer domesticus* 23
Interior Least Tern, *Sterna antillarum athalassos* 66
Ivory-billed Woodpecker, *Campephilus principalis* 81, 96, 150, 204
Killdeer, *Charadrius vociferus* 117, 197
Lark Sparrow, *Chondestes grammacus* 114
Laughing Gull, *Leucophaeus atricilla* 197
Lesser Prairie-chicken, *Tympanuchus pallidicinctus* 67, 83, 117, 132
Limpkin, *Aramus guarauna* 195
Loggerhead Shrike, *Lanius ludovicianus* 188
Louisiana Waterthrush, *Parkesia Motacilla* 82
Mallard, *Anas platyrhynchos* 38, 154, 192,
Mexican Jay, *Aphelocoma wollweberi* 94
Mississippi Kite, *Ictinia mississippiensis* 196
Montezuma Quail, *Cyrtonyx montezumae* 198-9
Mourning Dove, *Zenaida macroura* 28, 62, 193-4
Mottled duck, *Anas fulvigula* 39, 159, 160
Mountain Plover, *Charadrius montanus* 32, 117
Neotropic Cormorant, *Phalacrocorax brasilianus* 180
Northern Bobwhite, *Colinus virginianus* 28, 62, 79, 107, 115, 132, 185, 198-9, 291
Northern Cardinal, *Cardinalis cardinalis* 188, 205
Northern Flicker, *Colaptes auratus* 205
Northern Mockingbird, *Mimus polyglottos* 3, 27, 167, 188, 189, 205
Northern Pintail, *Anas acuta* 154, 159, 192
Osprey, *Pandion haliaetus* 85, 106-7, 152, 203
Painted Bunting, *Passerina ciris* 63, 116, 182, 206

Passenger Pigeon, *Ectopistes migratorius* 9, 11, 13, 14, 92
Pectoral Sandpiper, *Calidris melanotos* 196
Peregrine Falcon, *Falco peregrinus* 59, 78, 85, 152, 203, 204
Pileated Woodpecker, *Dryocopus pileatus* 31, 184, 205
Piping Plover, *Charadrius melodus* 59, 189
Plain Chachalaca, *Ortalis vetula* 32, 62, 95, 198
Prothonotary Warbler, *Protonotaria citrea* 106
Purple Gallinule, *Porphyrio martinicus* 195
Purple Martin, *Progne subis* 196
Red-bellied Woodpecker, *Melanerpes carolinus* 205
Red-cockaded Woodpecker, *Dryobates borealis* 20, 32, 55, 81, 91, 107, 168, 183-85
Reddish Egret, *Egretta rufescens* 152, 202
Redhead, *Aythya americana* 192
Red-headed Woodpecker, *Melanerpes erythrocephalus* 205
Red-shouldered Hawk, *Buteo lineatus* 96
Red-winged Blackbird, *Agelaius phoeniceus* 196
Ring-necked Duck, *Aythya collaris* 192
Ring-necked Pheasant, *Phasianus colchicus* 198
Rock Pigeon (Rock Dove), *Columba livia* 193
Roseate Spoonbill, *Platalea ajaja* 202
Royal Tern, *Thalasseus maximus* 197
Ruby-throated Hummingbird, *Archilochus colubris* 196, 201
Ruddy Duck, *Oxyura jamaicensis* 154
Rufous Hummingbird, *Selasphorus rufus* 201
Sandhill Crane, *Antigone canadensis* 195
Sandwich Tern, *Thalasseus sandvicensis* 197
Scaled Quail, *Callipepla squamata* 198-9
Scissor-tailed Flycatcher, *Tyrannus forficatus* 337
Sharp-shinned Hawk, *Accipiter striatus* 203
Shiny Cowbird, *Molothrus bonariensis* 207
Snowy Egret, *Egretta thula* 202
Southwest Willow Flycatcher, *Empidonax traillii extimus* 207
Spotted Sandpiper, *Actitis macularius* 189
Turkey Vulture, *Cathartes aura* 28, 138
Western Cattle Egret, *Bubulcus ibis* 202
White-tipped Dove, *Leptotila verreauxi* 193
White-winged Dove, *Zenaida asiatica* 11, 193
Whooping Crane, *Grus americana* 5-6, 8, 32, 59, 78-9, 152, 157, 163-5, 168, 170, 172, 195
Wild Turkey, *Meleagris gallopavo* 11, 14, 19, 77, 82, 106, 198, 200-1, 209
Wood Duck, *Aix sponsa* 106, 159
Yellow Rail, *Coturnicops noveboracensis* 195
Yellow Warbler, *Setophaga petechia* 207
Yellow-bellied Sapsucker, *Sphyrapicus varius* 205

FISH

Comanche Springs Pupfish, *Cyprinodon elegans* 135
Fountain Darter, *Etheostoma fonticola* 31, 32
Guadalupe Bass, *Micropterus treculii* 263cient
Pecos Gambusia, *Gambusia nobilis* 135

INVERTEBRATES

American Burying Beetle, *Nicrophorus americanus* 92, 285
American Dog Tick, *Dermacentor variabilis* 283
American Snout Butterfly, *Libytheana carinenta,* 331
Asian Tramp Snail, *Bradybaena similaris* 278
Atlantic Blue Crab, *Callinectes sapidus* 5, 164
Baldfaced Hornet, *Dolichovespula maculata* 287
Black Velvet Leatherleaf, *Belocaulus angustipes* 278
Blacklegged (Deer) Tick, *Ixodes scapularis* 283
Brown Garden Snail, *Cornu aspersum* 278
Brown Recluse, *Loxosceles reclusa* 281, 282
Chocolate-band Snail, *Eobania vermiculata* 278, 279
Comal Springs Riffle Beetle, *Heterelmis comalensis* 285
Common Earthworm (Nightcrawlers), *Lumbricus terrestris* 278
Common Pill Woodlouse, *Armadillidium vulgare* 280
Decollate Snail, *Rumina decollate* 278
Desert Cicada, *Diceroprocta apache* 139
Desert Termite, *Gnathamitermes tubiformans* 139
Emerald Ash Borer, *Agrilus planipennis* 99
Fiery Skipper, *Hylephila phyleus* 286
Giant Red Velvet Mite, *Dinothrombium pandorae* 139
Giant Vinegaroon, *Mastigoproctus giganteus* 283

House Centipede, *Scutigera coleoptrata* 280
Island Apple Snail, *Pomacea maculata* 157, 195, 279
Leopard Slug, *Limax maximus* 278
Mitchell's Wentletrap, *Amaea mitchelli* 12
Monarch Butterfly, *Danaus plexippus* 71, 77, 128, 275, 286, 294-6
New World Screw-worm Fly, *Cochliomyia hominivorax* 219
Parkhill Prairie Crayfish, *Procambarus steigmani* 58, 114
Red Imported Fire Ant, *Solenopsis invicta* 122-2, 264, 272, 282, 290-1, 292
Red Wiggler, *Eisenia fetida* 278
Southern Pine Beetle, *Dendroctonus frontalis* 98-9
Striped Bark Scorpion, *Centruroides vittatus* 283
Texas Redheaded Centipede, *Scolopendra heros* 280
Western Honey Bee, *Apis mellifera* 287, 290-1
White-lined Sphinx Moth, *Hyles lineata* 286
Widow Skimmer, *Libellula luctuosa* 289
Yellow Cellar Slug, *Limacus flavus* 278

MAMMALS

American Badger, *Taxidea taxis* 30, 32, 59, 178, 226
American Beaver, *Castor canadensis* 8, 11, 13, 14, 19, 31, 140, 148, 214, 216, 217, 219, 224-5, 292
American Bison, *Bos bison* 8, 10, 11, 13, 14, 29, 31, 32, 65, 67, 81, 109, 115, 117, 119, 127, 214, 216
American Black Bear, *Ursus americanus* 31, 55, 68, 94, 227-8, 234
American Mink, *Vison vison* 39, 226

Aoudad, *Ammotragus lervia* 140, 143-4, 232
Axis Deer, *Axis axis* 232
Bailey's Pocket Mouse, *Chaetodipus baileyi* 12
Banner-tailed Kangaroo Rat, *Dipodomys spectabilis* 138
Blackbuck, *Antilope cervicapra* 232
Black-footed Ferret, *Mustela nigripes* 32, 77, 234
Black-tailed Jackrabbit, *Lepus californicus* 11, 138, 228
Black-tailed Prairie Dog, *Cynomys ludovicianus* 8, 11, 14, 31-2, 78, 117, 224, 292
Bobcat, *Lynx rufus* 3-4, 170, 213, 214-5, 227
Brazilian Free-tailed Bat, *Tadarida brasiliensis* 31, 63, 167, 181, 212, 231
Caribbean Monk Seal, *Monachus tropicalis* 234
Collared Peccary (Javelina), *Dicotyles tajacu* 77, 215, 216, 223,
Common Gray Fox, *Urocyon cinereoargenteus* 216, 226
Common Muskrat, *Ondatra zibethicus* 217, 224-5,
Coyote, *Canis latrans* 27, 30, 31, 178, 214, 215, 227, 231, 264
Davis Mountains Cottontail, *Sylvilagus robustus* 94, 228
Desert Bighorn Sheep, *Ovis canadensis mexicana* 29, 137, 143-4, 176, 213, 216, 218, 222
Desert Cottontail, *Sylvilagus audubonii* 228
Eastern Cottontail, *Sylvilagus floridanus* 30, 78, 228
Eastern Elk, *Cervus canadensis canadensis* 13
Eastern Fox Squirrel, *Sciurus niger* 215, 219, 224
Eastern Gray Squirrel, *Sciurus carolinensis* 178-9, 219, 224
Eastern Red Bat, *Lasiurus borealis* 82
Fallow Deer, *Dama dama* 23, 232
Feral Hog, *Sus scrofa* 180-1, 216, 230, 233-4
Gray Wolf, *Canis lupus* 234

Gray-footed Chipmunk, *Tamias canipes* 94
Grizzly Bear, *Ursus arctos* 11, 234
Hairy-legged Vampire Bat, *Diphylla ecaudata* 231
Hispid Cotton Rat, *Sigmodon hispidus* 11, 78
House Mouse, *Mus musculus* 225
Jaguar, *Panthera onca* 11, 14, 18, 55, 234
Jaguarundi, *Puma yagouaroundi* 62, 234
Japanese Macaque, *Macaca fuscata* 232
Kit Fox, *Vulpes macrotis* 138, 181, 226
Lynx, *Lynx canadensis* 177
Margay, *Leopardus wiedii* 234
Merriam's Elk, *Cervus canadensis merriami* 11, 13, 87, 190, 219, 234
Mexican Long-nosed Armadillo (Nine-banded Armadillo), *Dasypus mexicanus* 8, 19, 211, 216, 229-30
Mexican Long-nosed Bat, *Leptonycteris nivalis* 8, 31, 71, 137-8, 231, 234
Mexican Long-tongued Bat, *Choeronycteris mexicana* 231
Mountain Lion, *Puma concolor* 1, 11, 14, 18, 68, 78, 87, 119, 142, 143, 217-8, 226-7, 234
Mule Deer, *Odocoileus hemionus* 220-1
Nilgai, *Boselaphus tragocamelus* 23, 232
Nine-banded Armadillo (see Mexican Long-nosed Armadillo)
North American Porcupine, *Erethizon dorsatum* 216, 225
Northern Raccoon, *Procyon lotor* 8, 27, 28, 31, 39, 77, 209, 214-5, 226, 231
Northern River Otter, *Lontra canadensis* 148, 226
Norway Rat, *Rattus norvegicus* 225
Nutria, *Myocastor coypus* 157, 232
Ocelot, *Leopardus pardalis* 62, 79-80, 95, 234

Palo Duro Mouse, *Peromyscus truei Comanche* 66, 96, 234
Pronghorn, *Antilocapra americana* 26, 32, 67, 117, 127, 169, 213-4, 216, 217, 222
Rafinesque's Big-eared Bat, *Corynorhinus rafinesquii* 96
Red Fox, *Vulpes vulpes* 232
Red Wolf, *Canis rufus* 55, 227, 234
Ringtail, *Bassariscus astutus* 214, 226
Rocky Mountain Elk, *Cervus canadensis nelsoni* 31, 221
Roof Rat (Black Rat), *Rattus rattus* 225
Sika Deer, *Cervus nippon* 23, 232
Snowshoe Hare, *Lepus americanus* 177
Southern Flying Squirrel, *Glaucomys volans* 224, 230
Southern Yellow Bat, *Dasypterus ega* 95
Striped Skunk, *Mephitis mephitis* 209, 226
Swamp Rabbit, *Sylvilagus aquaticus* 228
Swift Fox, *Vulpes velox* 32, 181, 226
Texas Kangaroo Rat, *Dipodomys elator* 66, 115, 234
Virginia Opossum, *Didelphis virginiana* 212, 213, 214-5, 216, 229
White-nosed Coati, *Nasua narica* 62, 234
White-tailed Deer, *Odocoileus virginianus* 7, 8, 13, 14, 29, 62, 63, 79, 82, 95, 99, 127, 169, 177, 180, 219-20, 221, 235
Yellow-nosed Cotton Rat, *Sigmodon ochrognathus* 117

PLANTS

Agarita (Algerita), *Mahonia trifoliolata* 73
Alkali Sacaton, *Sporobolus airoides* 117, 137
Alligator Juniper, *Juniperus deppeana* 94
Alligatorweed, *Alternanthera philoxeroides* 157
American Beautyberry, *Callicarpa americana* 74
American Sycamore, *Platanus occidentalis* 96, 151
Anacua, *Ehretia anacua* 61, 95
Arrowhead, *Sagittaria sp.* 147
Ashe Juniper, *Juniperus ashei* 27, 63, 74, 81, 93, 119
Bahiagrass, *Paspalum notatum* 121
Bailey's Ball-moss, *Tillandsia baileyi* 60
Bald Cypress, *Taxodium distichum* 55, 96, 149-50, 151
Basin Bellflower, *Campanula reverchonii* 64
Bermudagrass, *Cynodon dactylon* 57, 121, 122
Big Bluestem, *Andropogon gerardii* 57, 108, 109, 113, 126
Bigtooth Maple, *Acer grandidentatum* 94, 96
Black Hickory, *Carya texana* 56, 64
Black Lace Cactus, *Echinocereus reichenbachii var. albertii* 59
Black Willow, *Salix nigra* 96
Blackbrush Acacia, *Acacia rigidula* 61, 95
Blackgum, *Nyssa sylvatica* 96, 151, 154
Blackjack Oak, *Quercus marilandica* 56, 64, 92
Blue Grama, *Bouteloua gracilis* 65, 67, 109, 117
Bluewood Condalia, *Condalia hookeri* 95
Bracted Twistflower, *Streptanthus bracteatus* 178
Brownseed Paspalum, *Paspalum plicatulum* 59
Buffalograss, *Bouteloua dactyloides* 67, 109, 117, 303
Buffelgrass, *Pennisetum ciliare* 76, 116, 293
Bushy Bluestem, *Andropogon glomeratus* 30, 74, 147
Camphor Weed, *Heterotheca subaxillaris* 60
Candelilla, *Euphorbia antisyphilitica* 138
Cane Bluestem, *Bothriochloa barbinodis* 116
Canyon Mock Orange, *Philadelphus ernestii* 93
Carolina Wolfberry, *Lycium carolinianum* 164
Catclaw acacia, *Senegalia sp.* 137
Cattail, *Typha sp.* 147, 152, 160, 225
Cedar Elm, *Ulmus crassifolia* 59, 63, 64
Cedar Sage, *Salvia roemeriana* 12
Cedar Sedge, *Carex planostachys* 74
Cenizo, *Leucophyllum frutescens* 95
Cherrybark Oak, *Quercus pagoda* 96
Chinaberry Tree, *Melia azedarach* 76
Chinese Tallow, *Triadica sebifera* 59, 76, 114
Common Buttonbush, *Cephalanthus occidentalis* 149
Common Sunflower, *Helianthus annuus* 128
Common Water Hyacinth, *Eichhornia crassipes* 157
Cottonwood, *Populus deltoides* 96, 140, 151
Creosote Bush, *Larrea tridentata* 68, 133, 137, 138
Dayflower, *Commelina sp.* 128
Douglas-fir, *Pseudotsuga menziesii* 94
Dropseed, *Sporobolus sp.* 57, 113, 114, 126
Dwarf Palmetto, *Sabal minor* 56, 86
Dwarf Saltwort, *Salicornia bigelovii* 151
Eastern Gamagrass, *Tripsacum dactyloides* 57, 113, 126

Eastern Red Cedar, *Juniperus virginia* 52, 96, 99, 119
Eastern Swamp Privet, *Forestiera acuminata* 149
Elbow Bush, *Forestiera angustifolia* 73
Elephant Ear, *Colocasia esculenta* 155
Eurasian watermilfoil, *Myriophyllum spicatum* 157
Floating Water Primrose, *Ludwigia peploides* 155
Fourwing Saltbush, *Atriplex canescens* 137
Giant Dagger Yucca, *Yucca faxoniana* 130
Giant Reed, *Arundo donax* 76
Giant Salvinia, *Salvinia molesta* 157
Glen Rose Yucca, *Yucca necopina* 58, 114
Golden Gladecress, *Leavenworthia aurea* 55
Granjeno, *Celtis ehrenbergiana* 61
Guajillo, *Senegalia berlandieri* 61, 95
Guayacan, *Guaiacum angustifolium* 95
Guineagrass, *Urochloa maxima* 76
Gulf Cordgrass, *Spartina spartinae* 59, 114
Gulfdune Paspalum, *Paspalum monostachyum* 60
Hardstem bulrush, *Schoenoplectus acutus* 152
Honey Mesquite, *Prosopis glandulosa* 11, 32, 59, 61, 64, 65, 81, 95, 96, 114, 115, 116, 119, 127, 133
Huisache (Sweet Acacia), *Vachellia farnesiana* 95, 127, 151
Hydrilla (Waterthyme), *Hydrilla verticillata* 157
Illinois bundleflower, *Desmanthus illinoensis* 128
Indiangrass, *Sorghastrum nutans* 57, 109, 113, 126
Iodinebush, *Allenrolfea occidentalis* 137

Japanese Honeysuckle, *Lonicera japonica* 76
Japanese Privet, *Ligustrum japonicum* 76
Kidneywood, *Eysenhardtia texana* 95
King Ranch (Yellow) Bluestem, *Bothriochloa ischaemum* 76, 116, 124
Kudzu, *Pueraria montana* 76
Large-fruit Sand-verbena, *Abronia macrocarpa* 30, 56, 92
Leatherplant, *Jatropha cuneata* 138
Lechuguilla, *Agave lechuguilla* 68, 131, 139
Lehmann Lovegrass, *Eragrostis lehmanniana* 293
Lime Prickly Ash, *Zanthoxylum fagara* 95
Lindheimer's Silktassel, *Garrya ovata* ssp. *lindheimeri* 73
Little Bluestem, *Schizachyrium scoparium* 57, 60, 64, 65, 109, 114-6, 126
Live Oak, *Quercus virginiana* 59, 96, 114, 119, 147
Loblolly Pine, *Pinus taeda* 55, 91, 100-2, 107
Longleaf Pine, *Pinus palustris* 55, 81, 88, 91, 98, 99, 105, 183, 272
Macartney Rose, *Rosa bracteate* 59, 76, 1114
Maidencane, *Panicum hemitomon* 152
Marsh Pennywort, *Hydrocotyle umbellate* 155
Mexican Buckeye, *Ungnadia speciosa* 93
Mexican Pinyon Pine, *Pinus cembroides* 68, 94
Mormon Tea, *Ephedra viridis* 138
Mustang Grape, *Vitis mustangensis* 96
Navasota ladies'-tresses, *Spiranthes parksii* 56
Netleaf hackberry, *Celtis laevigata* 96, 151
Ocotillo, *Fouquieria splendens* 68, 137, 222

Olney Three-square Bulrush, *Schoenoplectus americanus* 152
Oneseed Juniper, *Juniperus monosperma* 96
Overcup Oak, *Quercus lyrata* 55, 96
Pale Yucca, *Yucca pallida* 58, 114
Partridge Pea, *Chamaecrista fasciculata* 128
Pecan, *Carya illinoinensis* 11, 55, 91, 96, 140, 147, 151, 263, 302
Poison Ivy, *Toxicodendron radicans* 155
Ponderosa Pine, *Pinus ponderosa* 94
Pondweed, *Potamogeton* sp. 152
Post Oak, *Quercus stellata* 56, 92
Poverty Weed, *Baccharis neglecta* 331
Prairie Dawn, *Hymenoxys texana* 59
Purple Loosestrife, *Lythrum salicaria* 76
Purpletop Tridens, *Tridens flavus* 126
Ragweed, *Artemisia* sp. 128
Red Maple, *Acer rubrum* 154
Redbay, *Persea borbonia* 96
Redberry Juniper, *Juniperus pinchottii* 96, 115
Retama (Paloverde), *Parkinsonia aculeata* 151
Rock Quillwort, *Isoetes lithophila* 64
Rocky Mountain Juniper, *Juniperus scopulorum* 96
Sabal Palm, *Sabal palmetto* 62, 95
Saltcedar, *Tamarix ramosissima* 76, 140
Saltgrass, *Distichlis spicata* 151
Saltmeadow Cordgrass, *Spartina patens* 152
Sand Lovegrass, *Eragrostis trichodes* 126
Sand Sagebrush, *Artemisia filifolia* 67, 117, 132
Saw Greenbrier, *Smilax bonanox* 170

Sea Ox-eye Daisy, *Borrichia frutescens* 151
Seacoast Bluestem, *Schizachyrium littorale* 60
Seaoats, *Unioloa paniculata* 114
Seashore Paspalum, *Paspalum vaginatum* 152
Shinnery Oak (Havard Shin-oak), *Quercus havardii* 67, 68, 132, 136 272
Shortleaf Pine, *Pinus echinata* 55, 91, 98
Sideoats Grama, *Bouteloua curtipendula* 58, 65, 67, 113, 115, 116, 126
Slash Pine, *Pinus elliottii* 55
Slender Rushpea, *Hoffmannseggia tenella* 59
Smooth Cordgrass, *Spartina alterniflora* 151
Sotol, *Dasylirion sp.* 137, 222
Southern Lady's-slipper, *Cypripedium kentuckiense* 55
Southern Naiad, *Najas guadalupensis* 152
Spanish Dagger, *Yucca aloifolia* 95
Spiny Hackberry, *Celtis ehrenbergiana* 95
Star Cactus, *Astrophytum asterias* 62
Sudan Grass, *Sorghum bicolor* 121
Sugar Hackberry, *Celtis laevigata* 59
Swamp Smartweed, *Polygonum hydropiperoides* 152
Sweetgum, *Liquidambar styraciflua* 55
Switchgrass, *Panicum virgatum* 57, 109, 113, 114, 126
Tepeguaje (Great Leadtree), *Leucaena pulverulenta* 95
Texas Ash, *Fraxinus albicans* 73
Texas Bluegrass, *Poa arachnifera* 126
Texas Ebony, *Ebenopsis ebano* 61, 95
Texas (Escarpment, Plateau) Live Oak, *Quercus fusiformis* 63, 73, 93, 99
Texas Madrone, *Arbutus xalapensis* 93, 94
Texas Mountain Laurel, Sophora secundiflora 73, 93, 95
Texas Persimmon, *Diospyros texana* 61, 95
Texas Poppymallow, *Callirhoe scabriuscula* 66, 115
Texas Prickly Pear Cactus, *Opuntia lindheimeri* 12
Texas Red Oak, *Quercus buckleyi* 93
Texas Snowbell, *Styrax platanifolius texanus* 93
Texas Trailing Phlox, *Phlox nivalis texensis* 32, 55, 91
Texas Wildrice, *Zizania texana* 31, 32
Texas Wintergrass, *Nassella leucotricha* 115, 126
Tobosagrass, *Pleuraphis mutica* 117
Turk's Cap, *Malvaviscus arboreus drummondii* 96
Virginia wildrye, *Elymus virginicus* 126
Walter's barnyard grass, *Echinochloa walteri* 152, 159
Water Elm (Planertree), *Planera aquatica* 149
Water Hickory, *Carya aquatica* 96
Water Lettuce, *Pistia stratiotes* 157
Water Oak, *Quercus nigra* 96
Water Tupelo, *Nyssa aquatica* 55, 96, 149
Wax Myrtle, *Morella cerifera* 154
Western (Cuman) Ragweed, *Ambrosia psilostachya* 28
Western Soapberry, *Sapindus saponaria drummondii* 96
Western Wheatgrass, *Pascopyrum smithii* 126
White Bladderpod, *Lesquerella pallida* 55
Widgeongrass, *Ruppia maritima* 152
Willow Oak, *Quercus phellos* 96
Windmillgrass, *Chloris verticillata* 116
Winged Elm, *Ulmus alata* 56, 92
Yaupon, *Ilex vomitoria* 56, 92, 96, 99
Zapata Bladderpod, *Lesquerella thamnophila* 62

REPTILES

Alligator Snapping Turtle, *Macrochelys temminckii* 269, 270, 272
American Alligator, *Alligator mississipiensis* 8, 23, 148, 152, 169, 172-4, 175, 178, 179, 255, 256, 257-9
Berlandier's (Texas) Tortoise, *Gopherus berlandieri* 12, 32, 272
Big Bend Tree Lizard, *Urosaurus ornatus schmidti* 138
Brazos River Watersnake, *Nerodia harteri* 66, 272
Broad-banded Copperhead, *Agkistrodon laticinctus* 265-6
Brown Anole, *Anolis sagrei* 260, 272
Brown Tree Snake, *Boiga irregularis* 272
Bullsnake, *Pituophis catenifer sayi viii*, 266
Cagle's Map Turtle, *Graptemys caglei* 32, 71, 270
Chihuahuan Nightsnake, *Hypsiglena jani* 266
Coachwhip, *Coluber flagellum* 267
Common Side-blotched Lizard, *Uta stansburiana* 137, 273
Common Spotted Whiptail, *Aspidoscelis gularis* 261-2
Concho Watersnake, *Nerodia paucimaculata* 66, 272
DeKay's Brownsnake, *Storeria dekayi* 266
Diamond-backed watersnake, *Nerodia rhombifer* 268
Diamond-backed terrapin, *Malaclemys terrapin* 59, 151, 270
Dunes Sagebrush Lizard, *Sceloporus laticinctus* 67, 136, 262, 272

Eastern Collared Lizard, *Crotaphytus collaris* 115, 261, 273
Eastern Copperhead, *Agkistrodon contortrix* 265
Eastern Hog-nosed Snake, *Heterodon platirhinos* 267
Four-lined Skink, *Plestiodon tetragrammus* 261
Gray-banded Kingsnake, *Lampropeltis alterna* 268
Great Plains Ratsnake, *Pantherophis emoryi* 267
Greater Short-horned Lizard, *Phrynosoma hernandesi* 262, 264, 272
Green Anole, *Anolis carolinensis* 260, 272
Kemp's Ridley Sea Turtle, *Lepidochelys kempii* 59
Little Striped Whiptail, *Aspidoscelis inornata* 137
Long-nosed Leopard Lizard, *Gambelia wislizenii* 137, 261
Louisiana Pinesnake, *Pituophis ruthveni* 32, 91, 185, 272
Mediterranean Gecko, *Hemidactylus turcicus* 261, 272
North American Snapping Turtle, *Chelydra serpentina* 269

Northern Cat-eyed Snake, *Leptodeira septentrionalis* 95
Northern Cottonmouth, *Agkistrodon piscivorus* 265, 266
Ornate Box Turtle, *Terrapene ornata* 271, 273
Red-eared Slider, *Trachemys scripta elegans* 271, 273
Reticulate Banded Gecko, *Coleonyx reticulatus* 261
Round-tailed Horned Lizard, *Phrynosoma modestum* 262
Slender Glass Lizard, *Ophisaurus attenuatus* 260
Speckled Racer, *Drymobius margaritiferus* 95
Spectacled Caiman, *Caiman crocodilus* 259
Spiny Softshell, *Apolone spinifera* 270, 273
Spot-tailed Earless Lizard, *Holbrookia lacerate* 262
Tamaulipan Black-striped Snake, *Coniophanes imperialis imperialis* 266
Texas Alligator Lizard, *Gerrhonotus infernalis* 260
Texas Banded Gecko, *Coleonyx brevis* 261
Texas Coralsnake, *Micrurus tener* 266

Texas Horned Lizard, *Phyrnosoma cornutum* 27, 65, 77, 115, 171, 254, 262-4, 272, 291, 301
Texas Indigo Snake, *Drymarchon melanurus erebennus* 32, 266
Texas Lyresnake, *Trimorphodon vilkinsonii* 266
Texas Map Turtle, *Graptemys versa* 270
Texas Spiny Lizard, *Sceloporus olivaceus* 262
Texas Tortoise (see Berlandier's Tortoise)
Three-toed Box Turtle, *Terrapene triunguis* 271
Trans-pecos Ratsnake, *Bogertophis subocularis* 140
Tuatara, *Sphenodon punctatus* 255
Western Diamond-backed Rattlesnake, *Crotalus atrox* 115, 266
Western Milksnake, *Lampropeltis gentilis* 268
Western Ratsnake, *Pantherophis obsoletus* 267
Western Ribbonsnake, *Thamnophis proximus* 268
Yellow Mud Turtle, *Kinosternon flavescens* 273

Scissor-tailed flycatcher. Photo: Mark Alan Storey.

INDEX

See Appendix E for an index to individual species in the text

Adaptive management 6, 182
Additive mortality 179-80
Aerial surveys 169-70, 172, 204, 216
Africanized honey bees 291
Agriculture and wildlife habitat
 in grasslands 120, 128
 in wetlands 161
Allen's Rule 138
Ambystomatidae 246
Amniotic eggs 188, 256-7
Amphibians
 conservation and management 252
 definition and characteristics 237
 monitoring 238
 Texas frogs and toads 240
 Texas salamanders 246
 threats 249
Amphiumidae 248
Angelina National Forest 91, 155
Anguidae 260
Anseriformes 192-3
Antlers 223
 cycle in whitetails 219-20
Apex predators 31
Arachnida 281
Aransas NWR 96, 163
Arthropoda 279
Artificial lighting 292, 304
Artificial nest structures 101-2, 181, 184-85
Artiodactyla 220
Ashe juniper-oak forests 93
Attwater's Prairie Chicken NWR 114, 123
Audubon, John James 11
Bailey, Florence Merriam 12, 15, 190-1
Bailey, Vernon Orlando 12
Baker, Rollin 19
Balcones Canyonlands NWR 15, 93
Balcones Fault 41, 44, 63
Balmorhea SP 135-6, 155, 297
Basking 255, 271
Basking surveys 169, 257-8
Bastrop County Complex Fire 100-3
Bastrop SP 101-2, 103
Bats 231-2
Beaufort Wind Scale 48
Bergmann's Rule 235

Berlandier, Jean Louis 12
Best Management Practices, forestry 101, 163
Big Bend National Park 45, 94, 130, 139, 227
Big Bend Ranch SP 45, 67, 139
Big Thicket 15, 87, 91, 154, 155
Biodiversity 2-3
Biogeochemical cycles 33-37
Bioindicators, see Indicator species
Bird banding 193-4, 208-9
Bird City Program 210
Bird flu 193, 209
Birds
 conservation and management 209-10
 definition and characteristics 187-9
 identification 189-94
 monitoring 208-9
 Texas bird diversity 192
 threats 209
Blackland Prairie ecoregion 57-8, 113-4
Black Gap WMA 19, 139, 143
Blind snakes 266
Bogs 154-5
Bottomland hardwoods 96-7, 151
 threats 103, 156
 management strategies 160, 162
Brood parasitism 206-7
Bufonidae 243
Bureau of Land Management 16
Cabeza de Vaca, Álvar Núñez 11
Caddo Lake NWR, SP, & WMA 96
Caddo National Grasslands 114
Camera traps 166, 169, 216, 228
Candy Cain Abshier WMA 96
Caprock Canyons SP 81, 115, 117, 219
Captive breeding 176-7
 Attwater's prairie-chicken 123
 deer 177, 221
 desert bighorns 143
 falcons 203-4
 Houston toad 101
 whooping cranes 163-4

Carbon cycle 34-5
 carbon storage 88, 111, 148, 151, 302
 management activities 37
Careers in wildlife science ix, 2, 304-5
Carrying capacity 76, 174-5
Centipedes 280
Certified Wildlife Habitats 302
Chaparral WMA 96
Chelydridae 269
Chenier Plain NWR 38
Chihuahuan Desert 131-2, 137-9
Chiroptera 231-2
Choke Canyon SP 96
Chronic wasting disease 177, 221
Chytrid fungus 250, 252
Citizen science 301
Clean Water Act 17, 158
Clearcutting 98, 105, 107
Climate change 37, 47, 100, 157, 249
Climate in Texas 46
Coastal oak woodlands 60, 96
Coastal Sand Plain ecoregion 60
Coleoptera 285
Colony Collapse Disorder 291
Coluber 267
Comanche Springs 135
Commensalism 29
Compensatory mortality 175
Competition 30, 173
Consumptive use value 7-8
Cool-season grasses 126
Cooper WMA 114
Cranes 195
Cross Timbers 56, 92
Crotaphytidae 261
Crustacea 279
Dactyloidae 260
Davis Mountains SP 177
Davy Crockett National Forest 87
DDT 84-5, 180, 190, 203, 204
Dental formula 212
Deserts 74-5, 131
 animal adaptations 138-9
Diamond Y Preserve 134, 155
Dichotomous key 73
Diplopoda 280
Diptera 286
Direct development in frogs 245

338 | Index

Distance sampling 170, 193, 216
Domestic cat predation 234, 304
Doves 194
Ducks Unlimited 158, 163, 192
Ecological classification levels 25
Ecological models 29
Ecological niche 27
Ecological pyramid 28
Ecological succession 75-6
   in forests 103-4
Ecological value of wildlife 8
Ecology definition 25
Economic impact of birding in Texas 187
Ecoregions of Texas
   EPA map 54
   Gould ecoregions map 54
   natural regions map 53
   natural subregions map 69
   vegetation map 52
Ecosystem engineers 31, 225
Ecosystem definition 25
Ecosystem services 88
   failures in urban settings 302
   in forests 88
Ectothermy 238, 255
Edge species 79
Edwards Plateau 62-3, 93, 104, 115
El Nino-La Nina cycle 47
Elapidae 266
Eleutherodactylidae 245
Emergency Wetlands Resources Act 157
Emerson, Ralph Waldo 15
Emydidae 270
Endangered Species Act 17, 23
Energy flow in ecosystems 27
Eublepharidae 261
Existence value 8
Exotic mammals 232-4
Exploitation (relationships) 30
Exploitation of wildlife (historic) 13-4
Exponential growth 174
Exploration period 11
Falconry 204
Farm Bill Conservation Reserve Program 17, 121, 220
Feather trade 13, 15, 180, 190
Feathers 187-8
Federal Aid in Wildlife Restoration Act 16-17, 19, 20
Feeding wildlife 175

Fire
   in forest management 105
   in grassland management 118-9, 124
   in habitat management 80-1
   wildfire 100-3
Food chains and food webs 28-9
Forested wetlands 149-51
Forests
   definition 87
   importance 88-89
   management 103
   threats to 98
   types and extent in Texas 89-97
Fort Worth Nature Center 92, 114
Franklin Mountains SP 144
Frogs and toads 240
Game animals 23
   game mammals 219
   migratory game birds 192-5
   upland game birds 198-9
Game warden 306
Gastropoda 278-9
Geckos/Gekkonidae 261
Gene Howe WMA 117
Generalist species 27
Genetic diversity 2, 27
Geology of Texas 42
Goodnight, Molly & Charles 117
Goodrum, Phil 19
Goose Island SP 96
Granger WMA 114
Grant, Ulysses S. 15
Graptemys 270
Grasslands
   definition 109-110
   importance 110-11
   management 124
   threats to 118
   types and extent in Texas 112-7
Grazing
   historical overgrazing 119, 140
   in habitat management 81
   rotational grazing 126-7
   stocking rates 125
   types of grazing animals 127
Gruiformes 195
Guadalupe Mountains National Park 94, 96, 100
Gulf Coast Prairies and Marshes ecoregion 58-9
Gus Engeling WMA 92, 155

Habitat
   definition 71
   components 76
   Food 77
   Water 77-8
   Shelter 78
   Adequate space 78-9
Habitat corridors 79
Habitat management tools 80
Hagerman NWR 114
Halophytes 147
Harvestmen 283
Hemiptera 288
Heterodon 267
Highly Pathogenic Avian Influenza (HPAI or bird flu) 193, 209
High Plains 66-7, 117
Historic milestones 21
Homeothermic 187, 213
Home range 78-79
Honey Creek State Natural Area 155
Honorable Harvest 299
Hornaday, William 15
Horned lizards 262
Horns (in wildlife) 223
Hummingbirds 201-2
Hunter education program 298
Hunting
   as a management tool 82, 178-79
   by Native Americans 10
   contribution to wildlife funding 16-7
   economic value 7
   ethical hunting 298-9
   historic overharvest 13-4
   potential disturbance factor 4-5
Huntsville SP 91
Hydrosphere 33
Hylidae 242
Hymenoptera 287
Igneous geology 42, 45, 64
iNaturalist 277, 301
Increaser grass 65, 115, 119
Indicator species 30-1
   environmental quality bioindicators 8, 249, 250
   for wetlands 147
Indigenous people
   relationship to wildlife 8, 10, 22, 300
   grasslands and fire 111, 118
Inductive reasoning 6
Insecta 284

Insolation 68, 132
Introduced grass species in Texas 312
Introduced species 251
Invasive species
   Ashe juniper 93
   feral hogs 180, 233-4
   in grasslands 114
   in plant communities 76
   in wetlands 157
   management 105, 124, 125, 127, 160
   red imported fire ants 291
   Texas Invasives Project 121, 278
Invertebrates
   conservation and management 292
   ecological importance 276
   invertebrate diversity 276
   monitoring 289
   threats 290
Isopoda 280
J.D. Murpree WMA 38
Joint Ventures 158, 209-10
Kerr WMA 116
Keystone species 31-2
   prairie dogs 32
   role in invertebrate communities 292
Kickapoo Caverns SP 116
Kinosternidae 270
Lacey Act 16
Lacustrine wetlands 153-4
Laguna Atascosa NWR 79
Lake Tawakoni SP 114
Lake Whitney SP 114
Lampropeltis 267
Land and Water Conservation Act 17
Land conservation actions 15
Land ethic 307
Landscape-scale conservation 83
Law of Conservation of Mass 33
Lay, Dan 18, 19, 20, 21
Lead poisoning 38-9
Legal categories of wildlife in Texas 23
Legal protection of wildlife 16
Lehmann, Val 19
Leopard frogs 241

Leopold, Aldo ix,
   definition of game management 6
   habitat management tools 80
   land ethic 307
   start of wildlife management profession 17
   views on predator removal ("Green Fire" essay) 178
Lepidoptera 286
Leptodactylidae 245
Leptotyphlopidae 266
Life table 168, 172-3
Light traps for insects 290
Limiting factor 321
Limiting factors 76, 173
   density-dependence 174
Lincoln-Petersen estimate 171
Lindheimer, Ferdinand Jacob 12
Lizards 260-4
Llano Uplift 64
   geology 44-5
Logistic growth 174
Lower Rio Grande Valley NWR 95
Malformations in amphibians 251
Mammals
   characteristics 212
   conservation and management 234
   identification 213
   monitoring 216
   regulatory categories 219
   Texas diversity 218
   threats 234
Map turtles 270
Marking wildlife 170-71
   amphibians 240
   birds 209
   mammals 218
   reptiles 258
Marsh, George Perkins 15
Marshes 151-2
Marsupials 229
   reproduction 212
Mason Mountain WMA 305
Mast 72
Matador WMA 115
Mechanical brush control 82, 127
Megafauna 10
Mesopredators 31, 178
Metamorphic rock 43, 45, 64

Metamorphosis
   amphibians 237
   atypical amphibian 245
   insects 284-5
Microhylidae 244
Mid-coast NWR 96, 151
Migration
   birds 196-7
   monarch butterfly 294-6
Migratory Bird Hunting Stamp Act 16
Migratory Bird Treaty Act 16, 23
Migratory game birds 23, 192
Millipedes 280-1
Mima mounds 114
Mitchell, Joseph Daniel 12
Mitigation 161
Mixed-grass prairie 109, 115-6
Mollusks 278
Monahans Sandhills SP 136
Monocultures
   forest 98, 107
   grassland 121-2
Mortality 168-72
   compensatory versus additive 179-80
Mountain forests 94
Muir, John 15, 16, 25
Multiple use philosophy 103
Mutualism 29, 31, 244
Myriapoda 280-1
Natality 172
National Audubon Society 13, 208, 300
National parks 15
National Environmental Policy Act 17
National Wildlife Refuge System 15
Naturalists in Texas 10-2
Natural Resources Conservation Service 121, 163
Natural selection 27
Nature-based tourism 8
Neches River NWR 96
Neotenic salamanders 248
Nerodia 268
Nitrogen cycle 35
Non-consumptive use value 8
Nongame 20, 23
Non-native species (see also Invasive species)
   earthworms 278
   grasses 121
   insects 291
   invertebrates 315
   plants 293

North American Model of Wildlife Conservation 22
North American Waterfowl Management Plan 158, 192
Nuisance wildlife/Wildlife Services program 181
Oak forests 92
Oak Woods and Prairies ecoregion 56-7
Odonata 288
Old Tunnel WMA 231
Orthoptera 288
Outdoor educator jobs 306
Overcollection
   amphibians 250
   cacti 68, 140
   reptiles 273
Overharvest and wildlife decline 12
Oviparous vs. ovoviviparous 260
Oxbow lakes 151
Palmetto SP 87
Palo Duro Canyon SP 66
Pantherophis 267
Pat Murphy WMA 117
Perching birds 205
Pelecaniformes 202-3
Peterson, Roger Tory 191
Phosphorus cycle 36
Photosynthesis 27, 35
Phrynosomatidae 262
Piciformes 204-5
Pinchot, Gifford 15, 103
Pine-hardwood forests 91
Pineywoods ecoregion 55-6
   forest types 91, 96
Pitfall traps 240, 258, 289
Pittman-Robertson Act 16-17, 19, 20
Plant identification 71-72
Plate tectonics 44
Playa lake 153
Playa Lakes WMA 154
Plethodontidae 248
Powderhorn SP & WMA 152
Plow
   grasslands 120, 128
   impact on wildlife habitats 81
Pocket prairies 129
Poikilothermy 238, 255

Pollinators
   bats as 231
   declines in 122
   desert pollinators 137
   importance of grasslands 111
   invertebrate importance 276
   keystone role 31
   link to native plants 293, 303
   plant reproduction 71-2
   western honeybees 291
Pollution 36
   DDT 84-5
   light pollution 141
   water and wetlands 251-2
Post oak savanna 56, 92
   Lost Pines 100-1
Prairie (see Grasslands)
Prairie potholes 154
Predator avoidance 30, 238, 259, 262, 264
Predator management 177
Predator-prey dynamics 178
Prescribed fire (see Fire)
Private Lands Assistance and Technical Guidance Program 20
Private lands biologist jobs 306
Proteidae 247
Quadrat sampling 169, 316
Quail species 198
Rabbits 228
Rabies 231
Radio-telemetry 170
Ranavirus 251
Ranidae 241
Rangeland (see Grasslands)
Raptors 203
Rattlesnake roundups 273
Reservoirs, wetland impact 156
Rhinophrynidae 244
Richland Creek WMA 159, 162
Riparian forests 96, 151
Rock cycle 42-3
Rodents 224
Roemer, Ferdinand 12
Rolling Plains ecoregion 65-6, 96, 115
Roosevelt, Theodore 15, 16
Rotational grazing 126
Safe Harbor Habitat Conservation Plans 185
Salamandridae 247
Salt Basins subregion 137
San Bernard NWR 96
Sand Hills subregion 136
Sauria 260-4
Savanna definition 87
Scat 170, 213-4, 258

Scientific method 4-6
Scientific names 3
Scientific value of wildlife 8
Scincidae 261
Scutes 257
   alligators 259
   turtle shells 269
Scorpions 283
Sedimentary rock 42-44
   in phosphorous cycle 36
Seed-tree cut 105
Selective cutting 105
Sentinel species (see Indicator species)
Seres 103-4
Serpentes 265
Shelterwood and seed-tree cuts 101
Shin-oak communities 136
Shorebirds 197
Shortgrass prairie 116-7
Sierra Diablo WMA 19
Silent Spring 85
Sirenidae 247
Skinks 261
Snakes 265
Soil
   chemistry 49
   horizons 48
   orders 49
   texture 49
South Texas Brush Country 61-2
   woodland habitats 95
   grassland habitats 116
Spadefoot toads 242
Spanish exploration of Texas 11
Specialist species 27, 77, 234, 289, 303
Species of Greatest Conservation Concern list 290
Spiders 281
Spring-seeps 155
Squirrels 224
   compensatory mortality 178-9
Succession, primary and secondary (see Ecological succession)
Stages of hunting development 298
State and Tribal Wildlife Grants 17
Stocking wildlife 176
Stockton Plateau subregion 134
Strecker, John Kern 12
Structural plant groups 73

Subsidence and saltwater intrusion 156
Subspecies 199
Subtropical forests 95
Sustainability 302, 304
    sustainable harvest 7, 234, 252, 273, 299
Swampbuster bill, 156, 158
Swamps 55, 96, 149-50
Symbiosis 29
Tallgrass prairie 109, 113-4
Tallamy, Doug 292-3, 302-3
Taxonomy 3
Teiidae 261
Testudinidae 269
Texas Master Naturalist Program 300
Texas Nature Trackers 278
Texas Parks and Wildlife Department – history 19-20
Texas Wetlands Conservation Plan 158
Thamnophis 268
The Aransas Project 165
Thoreau, Henry David 15
Threatened and endangered species 23
Ticks and mites 283
Timber harvest strategies
    even-aged management 102
    uneven-aged management 101
Topographic maps 42
Tracks of mammals 214
Transect surveys 216, 315-6
Trans-Pecos
    desert and shrub communities 132-139
    ecoregion description 67-8
    geology 44-5
    grasslands 117
    management strategies 142
    threats 140-1
    woodlands 94

Treefrogs 242
Trinity River NWR 96, 151
True frogs 241
True toads
Turtles 269
Typhlopidae 266
Umbrella species 32
Ungulates (see Artiodactyla)
United States Forest Service 15
Upland game birds 98
UV radiation and amphibians 250
Vegetation sampling methods
    line intercept method 315
    point-centered quarter method 315
    quadrat transect method 316
    simplified quadrat method 316
Venomous spiders 282
Village Creek SP 96, 151
Viperidae 265
W.G. Jones State Forest 91, 107
Warm-season grasses 126
Warning mimicry in snakes 314
Wastewater treatment wetlands 162
Waterbirds 202-3
Water control structures 158
Water cycle 33-4
Waterfowl 192-3
Water guzzler 77
Water rights and environmental flows 156
Wetlands
    characteristics 146-8
    conservation efforts 158
    management 158
    Texas wetland types 149
    threats 156
    values 148
Wet meadow 154

Whip scorpions and pseudoscorpions 283
Wildlife biologist jobs 306
Wildlife conservation
    conservation at home 302
    distinction from preservation 7
    funding 20
    individual roles and lifestyle choices 315
    milestones 21
    organizations 20
    role of private landowners in Texas 298
Wildlife Management Areas 19
Wildlife management plans 182
Wildlife populations 25
    censuses and surveys 168-9
    density 168
    dispersion 167-8
    dynamics 172
    management tools 175
Wildlife profession (beginning) 17
Wildlife refuge manager jobs 306
Wildlife rehabilitators 301
Wildlife researcher jobs 306
Wildlife science
    definition of 6
    origin and development 17
    scope and interdisciplinary nature 7
Wildlife science degree programs 305
Wildlife signs 158, 213-6
Wildlife tax valuation 183
Wildlife technician jobs 306
Wildlife veterinarian jobs 306
Wildlife watching 8
Wildscaping 304
Wilson, Andrew 11
Woodpeckers 188, 189, 204-5
Worm phyla 277-8

www.ingramcontent.com/pod-product-compliance
Lightning Source LLC
Chambersburg PA
CBHW080516030426
42337CB00023B/4544